Commutative algebra and combinatorics

Ramanujan Mathematical Society Lectures Notes Series

Vol. 1: Number theory

Vol. 2: The Riemann zeta function and related themes

Vol. 3: Formal language aspects of natural computing

Vol. 4: Commutative algebra and combinatorics

Vol. 5: Convexity in discrete structures

Vol. 6: Number theory and discrete geometry

Vol. 7: Discrete mathematics

Vol. 8: Lectures on operator theory

Vol. 9: Essays on geometric group theory

Vol. 10: Techmüller theory and moduli problems

Ramanujan Mathematical Society

Lecture Notes Series

Volume 4

Commutative algebra and combinatorics

Proceedings of the international conference
on computational algebraic geometry,
Harish-Chandra Research Institute,
Allahabad, December 2003

Volume editors

R. V. Gurjar
S. A. Katre
Ravi A. Rao
J. K. Verma

International Press
www.intlpress.com

Ramanujan Mathematical Society
Lecture Notes Series, Volume 4
Commutative algebra and combinatorics

Volume Editors:
R. V. Gurjar
S. A. Katre
Ravi A. Rao
J. K. Verma

Published in 2010 by International Press, Somerville, Massachusetts, U.S.A. under license from the Ramanujan Mathematical Society.

This work is also published in the Republic of India, exclusively by the Ramanujan Mathematical Society.

ISBN 978-1-57146-189-6

Typeset using the LaTeX system.
Printed in the United States of America.

Foreword

One of the most useful functions of the International Joint Meetings run by the American Mathematical Society with various countries in recent years is the opportunity for specialized "shoulder" conferences mixing participants from many different countries. Despite the wonders of modern email and the web, anyone who has gone to such an event knows that the level of collaboration and interaction developed in such face-to-face meetings greatly exceeds what is possible by any other means.

Not only do senior researchers from the different countries have opportunities for discovery of mutual interests and the initiation of cooperation, but students can interact with teachers who bring a truly international perspective on the field, one not accessible in any one place. Introductory and survey lectures on a variety of topics of intense current interest bring a broad perspective that can dramatically influence the course of a young person's work.

The current volume makes available a record of the expository and survey talks at just such a meeting, the International Conference on Commutative Algebra and Combinatorics, organized by the Bhaskaracharya Pratishthana, Pune, and the Harish-Chandra Research Institute, in Allahabad just before the joint meeting in Bangalore, in December 2003. The 5 survey articles and the extended tutorial cover a remarkable range of current computational commutative algebra and algebraic geometry.

The central objects of study could be described with only the slightest oversimplification as ideals of monomials (Stanley-Reisner Theory) and rings generated by monomials (Toric Geometry). There is a clear description of the basic problems and constructions of current interest, as well as some of the major applications of this theory, from the enumeration of "magic squares" to algebraic statistics. Throughout there are plenty of open problems to orient newcomers to the area to the exciting work that is–and will be!–going on. I think that these expositions and surveys will be useful to any student (and lots of mature mathematicians) interested in aspects of these fields.

DAVID EISENBUD
Berkeley, California
September 25, 2007

International Workshop on Computational Algebraic Geometry
and
International Conference on Commutative Algebra & Combinatorics
(December 8–13, 2003)

International Organizing Committee

Chairman	*W. Bruns*, University of Osnabrück, Germany.
Members	*A. Conca*, University of Genoa, Italy.
	I. Swanson, New Mexico State University, Las Cruces, U.S.A.
	T. Hibi, Osaka University, Japan.
	R. V. Gurjar, T.I.F.R., India.
	S. D. Adhikari, H.R.I., Allahabad.
	R. Thomas, University of Washington at Seattle, U.S.A.
	N. V. Trung, Hanoi Mathematical Institute, Vietnam.
	S. A. Katre, University of Pune, India.
	R. A. Rao, T.I.F.R., India.

Local Organizing Committee

Patrons	*C. S. Inamdar*, Custodian, Bhaskaracharya Pratishthana, Pune,
	Ravi Kulkarni, Director, H.R.I., Allahabad.
Chairman	*R. V. Gurjar*, T.I.F.R., Mumbai, & Trustee, Bhaskaracharya Pratishthana.
Secretary & Treasurer	*S. A. Katre*, University of Pune, Pune. & Trustee, Bhaskaracharya Pratishthana.
Local Secretary	*S. D. Adhikari*, H.R.I., Allahabad.
Coordinators	*Ravi A. Rao*, T.I.F.R., Mumbai,
	J. K. Verma, I.I.T. Bombay, Powai, Mumbai.

Preface

On the occasion of the first International Meeting of the American Mathematical Society in India organised by the Indian Academy of Sciences and Indian National Science Academy at Indian Institute of Science, Bangalore from December 17–21, 2003, the Bhaskaracharya Pratishthana felt it would be an ideal opportunity to further its interest in the spread of computational and combinatorial methods in commutative algebra.

The Bhaskaracharya Pratishthana appropriately has taken the lead in developing this frontier area of research in accordance with its traditions begun by its founder, and continued and extended by its present Managing Committee.

The Bhaskaracharya Pratishthana, Pune and Harish-Chandra Research Institute, Allahabad (H.R.I.) jointly organized an international workshop on Computational Algebraic Geometry from December 8–13, 2003 at the Harish-Chandra Research Institute, on the banks of the river Ganga. During the same period an International Conference on Commutative Algebra and Combinatorics was also organised by BP and HRI at the same venue.

An international organising committee was constituted to oversee the academic aspects of the school type workshop planned in the morning sessions accessible to graduate students, and in choosing the speakers for the international conference being planned for the afternoon sessions. In the morning session the participants were exposed to several front-line areas of commutative algebra, and some of the recent developments and open problems in those areas. A computational laboratory was set up at H.R.I. with 15 computers and special free computational software such as *Macaulay 2, CoCoa 4.1, 4ti2, CaTS, Gfan, Polymake*. The computational workshop was funded by a generous grant from the Department of Science and Technology (D.S.T.), Government of India. The international conference was partially supported by the National Board for Higher Mathematics (N.B.H.M.), International Mathematical Union, National Science Foundation, U.S.A., and the organizing institutions.

The international organising committee decided to focus on expository lectures in a few areas of commutative algebra which were prominent today, and to lead to open problems which could be tackled by students. It was also decided to separate the areas of research being covered in this school from those being covered in the session on Commutative Algebra being organised at the A.M.S. meet in Bangalore. There were tutorial sessions in the evenings in order to emphasize the computational aspects and to include some practical training.

The topics and speakers for the morning and afternoon sessions were:

Speakers (Expository talks)

1. Irena Swanson, New Mexico State Univ., USA, Primary Decompositions.
2. N. V. Trung, Institute of Mathematics, Hanoi, Vietnam, Regularity, Castelnuovo-Mumford Regularity and Related Invariants.

3. Serkan Hoşten, San Francisco State University, A Survey of Toric Initial Ideals.
4. Winfried Bruns, University of Osnabrück, Linear Diophantine Equations and the Anand-Gupta-Dumir Conjectures.

Speakers (Computational Sessions)
1. Diane Maclagan, Rutgers University, U.S.A.
2. Rekha Thomas, University of Washington, Seattle, U.S.A.

Associates (Practical Sessions on Macaulay 2)
1. Amit Khetan (UMass, Amherst, U.S.A.)
2. Leah Gold (Texas A&M Univ., U.S.A.)
3. A. V. Jayanthan (I.I.T. Bombay)
4. Tony Puthenpurackal (I.I.T. Bombay)
5. Sara Faridi (Dalhousie University, Canada)

Part I of the present volume is a compilation of the material disseminated in the computational part, under the leadership of Diane Maclagan and Rekha Thomas. The school was based on the book Gröbner Bases and Convex Polytopes by Bernd Sturmfels (A.M.S. 1995). The topics in this book provide an ideal combination of combinatorics, algebra and computation. There were six main lectures in the school, one each day, followed by three to four hours of hands on computational exercises using *Macaulay 2* and *CoCoA*. The main topics were as follows: Gröbner basics, The Gröbner fan, toric ideals, triangulations, and minimal free resolutions.

Part II of this volume consists of articles based on the expository talks. This part is edited by W. Bruns and it also contains the notes supplied by T. Hibi, who was unable to attend the conference.

In view of the fact that 2003 was the 60-th year of two of India's major contributors in Commutative Algebra and Combinatorics related to it, viz. R. C. Cowsik and V. C. Dumir, it was felt appropriate to have the international conference on Commutative Algebra and Combinatorics at this juncture, in their honour. This conference was held in parallel to the workshop. Professor V. C. Dumir was felicitated in this conference. Unfortunately, Prof. Dumir passed away all of a sudden on February 3, 2006. This volume is dedicated to his memory.

R. V. GURJAR S. A. KATRE RAVI A. RAO J. K. VERMA

This volume is dedicated to the memory of

Vishwa Chander Dumir

(1943 – 2006)

Vishwa Chander Dumir
1943–2006

Professor Vishwa Chander Dumir, an eminent mathematician and a distinguished scholar, left for his heavenly abode suddenly, on February 3, 2006.

Professor Dumir was born on June 7, 1943 at Mighiana, district Jhang now in Pakistan. He was the second son of Sh. Bhagwan Dass Dumir and Smt. Parkash Vati. He got his early education at Roorkee (U.P.). In 1953, his father was transferred to Chandigarh and Dr. Dumir had the opportunity to see this beautiful new city grow. He used to speak nostalgically about it. He had a brilliant academic career. He graduated from Government College Chandigarh and stood first in B.A. Honours in Mathematics. He did his M.A. (Mathematics) from Department of Mathematics, Panjab University in 1962 and stood first in the University. He joined the Department of Mathematics, Panjab University, Chandigarh for research and later started working under the guidance of Professor R.P. Bambah and went to Ohio State University, Columbus, Ohio, in 1964 with him, where he was awarded Ph.D. in 1965. The title of his thesis was *Diophantine Inequalities for Quadratic and other Forms*. He served as Assistant Professor at Ohio State University (1966) and at University of Illinois at Urbana-Champaign from 1966 to 1968. He joined the Department of Mathematics, Panjab University as a reader in 1968 in July 1968 and was promoted as a professor in 1979.

Professor Dumir made significant contributions in number theory especially in the fields of geometry of numbers and discrete geometry. Some of his major achievements are: proof of a conjecture of G.L. Watson on non-homogeneous minima of indefinite quadratic forms, proof of a conjecture of Bambah, Dumir and Hans-Gill on positive values of non-homogeneous indefinite quadratic forms, proof of a conjecture of L. Fejes Toth on saturated system of circles and convex domains, proofs of conjectures of Mahler and Jackson. He proved results on lattice double packings and lattice double coverings in the plane, an analogue of a problem of Mordell and general covering density of star domains. He worked extensively on view-obstruction problems, developed a general theory and generalized it to subspaces and flats, obtained several special results on these and on the Billiard Ball motion problem of Schoenberg. He also contributed to the determination of idempotent generators of minimal cyclic codes. He collaborated with Professors R.P. Bambah, H. Gupta, R.J. Hans-Gill, J.B. Wilker, A.C. Woods, D.S. Khassa, V.K. Grover, Madhu Raka and Dr. Ranjeet Sehmi, Dr. Harsh Anand, Dr. Gurmeet Kaur and Ms. Anuradha Sharma.

He was elected Fellow of Indian National Science Academy in 1983 and of the National Academy of Sciences in 1985. He was awarded Srinivasa Ramanujan Birth Centenary Award by Indian Science Congress Association in 2004. He visited Universities of Cambridge, Oxford and London under the British Council Exchange Programme in 1972. He visited University of Illinois in 1977–78 under the Indo-American Fellowship of UGC. He was visiting professor to Ohio State University (1986–88), University of Colorado (1998), University of Illinois (2000–2001). He delivered invited talks in seminars and conferences at many places in India and abroad. He was honoured on the occasion of his 60th birthday felicitations for his significant contributions to geometry of numbers and discrete mathematics during the International Conference on Commutative Algebra and Combinatorics held at Harish-Chandra Research Institute, Allahabad, December 8–13, 2003. He delivered 16th Srinivasa Ramanujan Memorial award lecture at the annual conference of the Indian Mathematical society held at Roorkee in December 2005.

Professor Dumir has published more than 45 research papers in a variety of areas of his interest. He was an excellent teacher and eminent researcher. As a person, Professor Dumir was very cordial, hospitable and friendly. The Mathematics community has suffered irreparable loss due to his sudden demise. He is survived by his wife, Saroj, and two sons, Sanjeev and Neeraj, both of them are accomplished engineers.

Rajinder J. Hans-Gill

Part I

COMPUTATIONAL ALGEBRA AND COMBINATORICS OF TORIC IDEALS

Diane Maclagan Rekha R. Thomas

With the co-operation of

Sara Faridi Leah Gold A. V. Jayanthan
Amit Khetan Tony Puthenpurakal

List of Participants at the International Workshop on Computational Algebraic Geometry, 8–13 Dec. 03

(This list consists of participants who attended computational part as well as the expository part)

Sr. No. Name	Institute
Foreign Participants:	
1. Sabzrou Hossein	Amir Kabur Univ., Iran
Indian Participants:	
2. Rabeya Basu	TIFR, Mumbai
3. Chanchal Kumar	Jammu Univ.
4. S. Sivakumar	Ramanujan Inst. Chennai
5. Tamil Selvi	
6. Pavinder Singh	Jammu Univ.
7. A. Joseph Kennedy	Ramanujan Inst. Chennai
8. Rajendra Deore	North Maharashtra Univ.
9. Patil Arunkumar	IIT, Bombay
10. Sartaj Ul Hasan	IIT, Bombay
11. Habeeb Basha Syed	IIT, Bombay
12. Pooja Singla	IIT, Bombay
13. Amitava Bhattacharya	TIFR, Mumbai
14. Sivarama Krishnan Sivasubramanian	TIFR, Mumbai
15. Vinay V. Wagh	Bhaskaracharya Pratishthana
16. Shiva Datta	Govt. P.G. College, Nainital
17. Archana Morye	IIT, Mumbai

Preface

This part of the volume is based on six lectures and tutorials that were prepared for a Workshop in Computational Commutative Algebra held at the Harish Chandra Research Institute (HRI), Allahabad, India in December 2003. The workshop was aimed at graduate students and was conducted in parallel to the Conference on Commutative Algebra and Combinatorics held at HRI from December 8–13, 2003. The material in the early chapters is based heavily on the research monograph *Gröbner Bases and Convex Polytopes* [Stu96] by Bernd Sturmfels. We have attempted to explain the key concepts in this monograph to students who are not familiar with either Gröbner bases or convex polytopes by building up the basics of these theories from scratch. The tutorials and examples are meant to help this development. There is a special emphasis on actual computations via various software packages.

Lectures 1, 3, and 5 were written and delivered by Rekha R. Thomas (University of Washington), and Lectures 2, 4, and 6 were written and delivered by Diane Maclagan (Rutgers University). The tutorials for the lectures were prepared and conducted by Tony Puthenpurakal of IIT Bombay and A.V. Jayanthan of HRI (Tutorial 1); Amit Khetan of the University of Massachusetts, Amherst (Tutorial 2); Leah Gold of Cleveland State University (Tutorials 3 and 5); and Sara Faridi of Dalhousie University (Tutorial 4). Amit Khetan also contributed to Lecture 6.

We thank the host institutions Bhaskaracharya Pratishthana and Harish-Chandra Research Institute, India for the invitation to conduct this workshop and their very warm hospitality. We also thank the students at the workhop for helpful comments and corrections. Special thanks to Amitava Bhattacharya, Tristram Bogart, Anders Jensen and Edwin O'Shea who helped to proofread the final version. Finally, we acknowledge Sara Faridi for editing and creating an index for the book, and Leah Gold for stellar proofreading.

DIANE MACLAGAN (Rutgers University)
REKHA R. THOMAS (University of Washington).

Contents

1

Gröbner Basics

1.1 Introduction

This first chapter aims to give a brief introduction to the basics of Gröbner basis theory. There are many excellent books on Gröbner bases and their applications such as [AL94], [CLO97], [CLO98], [GP02] and [KR00]. Our account here will be brief, biased and focused.

Throughout this book, we let $S := \mathbf{k}[x_1, \ldots, x_n] =: \mathbf{k}[\mathbf{x}]$ be a polynomial ring over a field $\mathbf{k}$. For the most part we may take $\mathbf{k}$ to be the field of complex numbers $\mathbb{C}$, the field of rational numbers $\mathbb{Q}$, or the field of real numbers $\mathbb{R}$. A nonempty subset $I \subset S$ is called an *ideal* of S if I is closed under

1. addition — i.e., for all $f, g \in I$, $f + g \in I$, and
2. multiplication by elements of S — i.e., $h \in S$, $f \in I$ implies $hf \in I$.

An ideal I is *finitely generated* if there exists a subset $\{f_1, \ldots, f_t\} \subset I$, called a *basis* of I, such that $I = \left\{\sum_{i=1}^{t} h_i f_i \,:\, h_i \in S\right\}$. We write $I = \langle f_1, \ldots, f_t\rangle$. In the 1890s David Hilbert proved that every polynomial ideal in S has a finite basis. This fact is known as the Hilbert basis theorem. In this lecture we define and construct special bases of I called Gröbner bases.

The simplest polynomials in S are *monomials*, which are the polynomials $x_1^{a_1} x_2^{a_2} \cdots x_n^{a_n}$ where $\mathbf{a} := (a_1, \ldots, a_n) \in \mathbb{N}^n$. Here $\mathbb{N}$ stands for the set of nonnegative integers. We denote this monomial by $\mathbf{x}^{\mathbf{a}}$. If $c \in \mathbf{k}$, then $c\mathbf{x}^{\mathbf{a}}$ is called a *term* in S. Every polynomial is a finite sum of terms. The *support* of a polynomial $f \in S$ is the set $\mathrm{supp}(f) := \{\mathbf{a} \in \mathbb{N}^n \,:\, f = \sum c_{\mathbf{a}} \mathbf{x}^{\mathbf{a}},\ c_{\mathbf{a}} \neq 0\}$.

Example 1.1.1. If $f = \frac{3}{2}x_1^6 x_5^8 - \sqrt{2}x_1 x_3^{16} - 12 \in \mathbb{R}[x_1, \ldots, x_6]$, then $\mathrm{supp}(f) = \{(6, 0, 0, 0, 8, 0), (1, 0, 16, 0, 0, 0), (0, 0, 0, 0, 0, 0)\} \subset \mathbb{N}^6$.

The *variety* of $f_1, \ldots, f_t \in S$ is the set

$$\mathcal{V}(f_1, \ldots, f_t) := \{\mathbf{p} \in \mathbf{k}^n \,:\, f_1(\mathbf{p}) = f_2(\mathbf{p}) = \cdots = f_t(\mathbf{p}) = 0\}.$$

The variety of the ideal $I = \langle f_1, \ldots, f_t\rangle$ is

$$\mathcal{V}(I) = \{\mathbf{p} \in \mathbf{k}^n \,:\, f(\mathbf{p}) = 0,\ \forall f \in I\} = \mathcal{V}(f_1, \ldots, f_t).$$

1.2 Motivation

Given an ideal $I \subset S$ and a polynomial $f \in S$, a fundamental problem is to decide whether f belongs to I. This is known as the *ideal membership problem.* We will see shortly that Gröbner bases can be used to solve this. We begin by examining algorithms for ideal membership in the two well-known families of univariate and linear ideals. For ease of exposition we choose the field $\mathbf{k} = \mathbb{C}$ for the first family and $\mathbf{k} = \mathbb{R}$ for the second.

(i) **Univariate Ideals:** ([CLO97, §1.5]) Since $\mathbb{C}[x]$ is a principal ideal domain, every ideal in $\mathbb{C}[x]$ is of the form $I = \langle g \rangle$ where $g \in \mathbb{C}[x]$. The polynomial g is of the form

$$g = c_d x^d + c_{d-1}x^{d-1} + \cdots + c_1 x + c_0$$

where $c_i \in \mathbb{C}$. Without loss of generality, we may assume that the leading coefficient c_d is one. When this is the case, we say that g is *monic*. The degree of g, denoted as $\deg(g)$, is then d.

- **Finding the basis**: It is not hard to show theoretically that I is generated by any polynomial g in I of least degree. If I is given as $I = \langle f_1, \ldots, f_t \rangle \subset \mathbb{C}[x]$, then $g = \gcd(f_1, \ldots, f_t)$, the greatest common divisor of $f_1, \ldots, f_t$. The gcd can be computed by the Euclidean Algorithm.
- **Ideal membership**: If $f \in \mathbb{C}[x]$ then, by the usual division algorithm for univariate polynomials, there exist unique polynomials h and $r \in \mathbb{C}[x]$ such that $f = hg + r$ where r is the remainder and $\deg(r) < \deg(g)$. The polynomial f lies in $I = \langle g \rangle$ if and only if $r = 0$. Thus ideal membership can be determined by the division algorithm.
- **Solving** $\{f_1 = f_2 = \cdots = f_t = 0\}$: Let $g = \gcd(f_1, \ldots, f_t)$. Then the variety $\mathcal{V}(f_1, \ldots, f_t)$ equals $\mathcal{V}(g)$. The roots of a univariate polynomial can be found via radicals when its degree is small and by numerical methods otherwise. See [Stu02] for a recent survey of methods for solving polynomial equations.

(ii) **Linear ideals:** (See Example 1.5 in [Stu96]). Let $A \in \mathbb{Z}^{d \times n}$ be a matrix with ith row $A_i = (a_{i1}, \ldots, a_{in})$ and $\operatorname{rank}(A) = d$. Consider the *linear ideal*

$$I = \left\langle \sum_{j=1}^{n} a_{ij}x_j \; : \; i = 1, \ldots, d \right\rangle \subseteq \mathbb{R}[x_1, \ldots, x_n].$$

The variety $\mathcal{V}(I)$ is the $(n-d)$-dimensional vector space $\ker_{\mathbb{R}}(A) = \{\mathbf{p} \in \mathbb{R}^n \; : \; A\mathbf{p} = \mathbf{0}\}$.

A nonzero linear form f in I is a *circuit* of I if f has minimal support (with respect to inclusion) among all linear forms in I. The coefficient vector of a circuit of I is a vector in the row span of A of minimal support. For $J \subseteq [n] := \{1, \ldots, n\}$ with $|J| = d$, let $D[J] := \det(A_J)$ be the determinant of A_J, where A_J is the submatrix of A with column indices J. The following algorithm computes the circuits of I.

Algorithm 1.2.1 ([Stu02, Chapter 8.3]). Let B be an integer $(n-d) \times n$ matrix whose rows form a basis of $\ker_{\mathbb{R}}(A)$. Then every vector in the row span of A (which includes the rows of A) is a linear dependency of the columns of B since $BA^t = 0$. Thus the coefficient vectors of the circuits of I are the minimal dependencies of the columns of B which are also called circuits of B.

For any $(n-d+1)$-subset $\tau = \{\tau_1, \ldots, \tau_{n-d+1}\} \subseteq [n]$ form the vector

$$C_\tau := \sum_{i=1}^{n-d+1} (-1)^i \, \det(B_{\tau \setminus \{\tau_i\}}) \, \mathbf{e}_{\tau_i}$$

where $\mathbf{e}_j$ is the jth unit vector of $\mathbb{R}^n$. If C_τ is nonzero, then compute the primitive vector obtained by dividing through with the gcd of its components. The resulting vector is a circuit of B and all circuits of B are obtained this way. Can you prove this?

Example 1.2.2. Let $A = \begin{pmatrix} 1 & 2 & 3 & 4 & 5 \\ 6 & 7 & 8 & 9 & 10 \end{pmatrix}$. Then

$$I = \langle x_1 + 2x_2 + 3x_3 + 4x_4 + 5x_5,\ 6x_1 + 7x_2 + 8x_3 + 9x_4 + 10x_5 \rangle.$$

The rows of

$$B = \begin{pmatrix} 3 & -4 & 0 & 0 & 1 \\ 2 & -3 & 0 & 1 & 0 \\ 1 & -2 & 1 & 0 & 0 \end{pmatrix}$$

form a basis for $\ker_{\mathbb{R}}(A)$. Check that $BA^t = 0$. Let us compute one of the circuits of B. For $\tau = \{1, 2, 3, 4\}$,

$$C_\tau = -\det\begin{pmatrix} -4 & 0 & 0 \\ -3 & 0 & 1 \\ -2 & 1 & 0 \end{pmatrix} \mathbf{e}_1 + \det\begin{pmatrix} 3 & 0 & 0 \\ 2 & 0 & 1 \\ 1 & 1 & 0 \end{pmatrix} \mathbf{e}_2$$
$$- \det\begin{pmatrix} 3 & -4 & 0 \\ 2 & -3 & 1 \\ 1 & -2 & 0 \end{pmatrix} \mathbf{e}_3 + \det\begin{pmatrix} 3 & -4 & 0 \\ 2 & -3 & 0 \\ 1 & -2 & 1 \end{pmatrix} \mathbf{e}_4$$

which equals $(-4, -3, -2, -1, 0)$. Hence $-4x_1 - 3x_2 - 2x_3 - x_4$ or equivalently, $4x_1 + 3x_2 + 2x_3 + x_4$ is a circuit of I.

The above algorithm provides a linear algebraic method for computing the circuits of a linear ideal. The software package `4ti2` [Hem] computes circuits via a different method. We now prove that Gaussian elimination computes circuits of I. Let I_d be the $d \times d$ identity matrix and $C = (c_{ij})$ be the Gauss-Jordan form of A. Since the rank of A is d, we can assume, after possibly permuting columns, that C has the form $[I_d | *]$.

Proposition 1.2.3. *Let $C = (c_{ij}) = [I_d | *] \in \mathbb{R}^{d \times n}$ be the Gauss-Jordan form (reduced row-echelon form) of A and $g_i = x_i + \sum_{j=d+1}^{n} c_{ij} x_j$, $i = 1, \ldots, d$ be the linear forms corresponding to the rows of C. Then*

1. *$\{g_1, \ldots, g_d\}$ is a minimal generating set for I, and*
2. *the linear forms $g_1, \ldots, g_d$ are circuits of I.*

Proof.

1. Since every row of C is a linear combination of rows of A and vice-versa, every g_i is a linear combination of the f_i's and every f_i a linear combination of the g_i's. Thus $I = \langle f_1, \ldots, f_d \rangle = \langle g_1, \ldots, g_d \rangle$.
2. Suppose g_1 is not a circuit. Then there exists a linear polynomial $g \in I$ such that $\text{supp}(g) \subsetneq \text{supp}(g_1)$. However, $g = t_1 g_1 + \cdots + t_d g_d$ for scalars $t_1, \ldots, t_d \in \mathbb{R}$. Since $\text{supp}(g) \subset \text{supp}(g_1)$, $t_2 = t_3 = \cdots = t_d = 0$. This implies that $g = t_1 g$, $t_1 \neq 0$ and hence $\text{supp}(g) = \text{supp}(g_1)$, a contradiction. The same argument can be repeated for $g_2, \ldots, g_d$. □

Proposition 1.2.4. *Assume the same setup as in Proposition 1.2.3.*

1. *A polynomial $f \in S$ is an element of I if and only if successively replacing every occurrence of x_i , $i = 1, \ldots, d$ in f with $-\sum_{j=d+1}^{n} c_{ij} x_j$ results in the zero polynomial.*
2. *The linear system $A\mathbf{x} = \mathbf{0}$ can be solved by back-solving the "triangularized" system $g_1 = g_2 = \cdots = g_d = 0$.*

Proof.

1. Let f' be obtained from f by successively replacing every occurrence of x_i, $i = 1, \ldots, d$ in f with $-\sum_{j=d+1}^{n} c_{ij} x_j$. Then $f' \in \mathbb{R}[x_{d+1}, \ldots, x_n]$. This implies that $f = \sum_{i=1}^{d} h_i g_i + f'$ where $h_i \in \mathbb{R}[\mathbf{x}]$. If $f' = 0$ then clearly $f \in I$. Conversely, if $f \in I$, then $f' = f - \sum_{i=1}^{d} h_i g_i \in I \cap \mathbb{R}[x_{d+1}, \ldots, x_n] = \{0\}$. The last equality follows from Proposition 1.2.3(1) which implies that no polynomial combination of $g_1, \ldots, g_d$ can lie in $\mathbb{R}[x_{d+1}, \ldots, x_n]$.
2. This is the familiar method of solving linear systems by Gaussian elimination from linear algebra. □

The proof of Proposition 1.2.4(1) is employing a division algorithm for linear polynomials in many variables that succeeds in determining ideal membership for linear ideals. Note that when we perform Gauss-Jordan elimination on A to obtain $C = [I_d \mid *]$, we are implicitly ordering the variables in $\mathbb{R}[\mathbf{x}]$ such that $x_1 > x_2 > \cdots > x_n$. The division algorithm used in the proof of Proposition 1.2.4(1) replaces every occurrence of the "leading term" x_i in g_i with $x_i - g_i$, which is the sum of the "trailing terms" in g_i.

Thus the questions we started with have well-known algorithms and answers when I is either a univariate principal ideal or a multivariate linear ideal. We now seek a common generalization of these methods to multivariate polynomials of arbitrary degrees, and their ideals. This leads us to Gröbner bases of polynomial ideals.

1.3 Gröbner bases

In order to mimic the procedures from the last section, we first need to impose an ordering of the monomials in S so that the terms in a polynomial are always ordered. This is important if the generalized division algorithm is to replace the leading term of a divisor by the sum of its trailing terms.

Definition 1.3.1. A *term order* $\succ$ on S is a total order on the monomials of S such that

1. $\mathbf{x}^{\mathbf{a}} \succ \mathbf{x}^{\mathbf{b}}$ implies that $\mathbf{x}^{\mathbf{a}}\mathbf{x}^{\mathbf{c}} \succ \mathbf{x}^{\mathbf{b}}\mathbf{x}^{\mathbf{c}}$ for all $\mathbf{c} \in \mathbb{N}^n$, and
2. $\mathbf{x}^{\mathbf{a}} \succ \mathbf{x}^{\mathbf{0}} = 1$ for all $\mathbf{a} \in \mathbb{N}^n \backslash \{\mathbf{0}\}$.

Part 2 of Definition 1.3.1 and the Gordan-Dickson Lemma [CLO97, §4.2] imply that every term order is a *well-ordering* on the monomials of S which will be important for the finite termination of the algorithms described in this chapter.

Example 1.3.2. The most common examples of term orders are the *lexicographic (lex) order* and the *graded reverse lexicographic order* on S with respect to a fixed ordering of the variables, such as $x_1 \succ x_2 \succ \cdots \succ x_n$.

In the lex ordering, $\mathbf{x}^{\mathbf{a}} \succ \mathbf{x}^{\mathbf{b}}$ if and only if the left-most nonzero term in $\mathbf{a} - \mathbf{b}$ is positive. For example, if $x \succ y \succ z$, then

$$x^3 \succ x^2y \succ x^2z \succ xy^2 \succ xyz \succ xz^2 \succ y^3 \succ y^2z \succ yz^2 \succ z^3.$$

In the (graded) reverse lexicographic order, $\mathbf{x}^{\mathbf{a}} \succ \mathbf{x}^{\mathbf{b}}$ if and only if either $\deg(\mathbf{x}^{\mathbf{a}}) = |\mathbf{a}| > \deg(\mathbf{x}^{\mathbf{b}}) = |\mathbf{b}|$, or $\deg(\mathbf{x}^{\mathbf{a}}) = \deg(\mathbf{x}^{\mathbf{b}})$ and the right-most nonzero term in $\mathbf{a} - \mathbf{b}$ is negative. Here $|\mathbf{a}|$ denotes the 1-norm of $\mathbf{a}$ which is the degree of $\mathbf{x}^{\mathbf{a}}$ under the total-degree grading of S. Again if $x \succ y \succ z$, then

$$x^3 \succ x^2y \succ xy^2 \succ y^3 \succ x^2z \succ xyz \succ y^2z \succ xz^2 \succ yz^2 \succ z^3.$$

The reverse lexicographic order defined here is *degree-compatible* or *graded* which means that it first compares two monomials by degree and then breaks ties using the rule described. We call it *grevlex*. Note that there are $n!$ lex and grevlex orders on S.

We will see in the next chapter that vectors in $\mathbb{R}^n_{\geq 0}$ can be used to define term orders. We now fix a term order $\succ$ on S. Given a polynomial $f = \sum c_{\mathbf{a}}\mathbf{x}^{\mathbf{a}} \in S$, the *initial term* or *leading term* of f with respect to $\succ$ is the term $c_{\mathbf{a}}\mathbf{x}^{\mathbf{a}}$ in f such that $\mathbf{x}^{\mathbf{a}} \succ \mathbf{x}^{\mathbf{a}'}$ for all $\mathbf{a}' \in \mathrm{supp}(f)$ different from $\mathbf{a}$. It is denoted by $\mathrm{in}_{\succ}(f)$. The *initial monomial* of f is the monic term $\mathbf{x}^{\mathbf{a}}$. We can write $f = \mathrm{in}_{\succ}(f) + f'$.

Example 1.3.3. Let $f = 3x_1x_3^2 + \sqrt{2}x_3^2 - x_1x_2^2 \in \mathbb{C}[x_1, x_2, x_3]$ and $\succ$ be the reverse lexicographic order with $x_1 \succ x_2 \succ x_3$. Then $x_1x_2^2$ is the initial monomial of f and $-x_1x_2^2 = \mathrm{in}_{\succ}(f)$ is the initial term of f.

We can now attempt to write down a division algorithm for multivariate polynomials. The following algorithm is from [CLO97, Theorem 3].

Algorithm 1.3.4. Division algorithm for multivariate polynomials [CLO97, Theorem 3]

INPUT: A dividend $f \in S$, an ordered set of divisors $\{f_1, \ldots, f_s\}$, $f_i \in S$, and a term order $\succ$.

OUTPUT: Polynomials $a_1, \ldots, a_s, r \in S$ such that $f = \sum_{i=1}^{s} a_i f_i + r$ where either $r = 0$ or no term in r is divisible by $\mathrm{in}_{\succ}(f_1), \ldots, \mathrm{in}_{\succ}(f_s)$.

INITIALIZE: $a_1 := 0, \ldots, a_s := 0, r := 0$; $p := f$

WHILE $p \neq 0$ DO

```
i := 1
divisionoccurred := false
WHILE i ≤ s AND divisionoccurred = false DO
  IF in_≻(f_i) divides in_≻(p) THEN
    a_i := a_i + in_≻(p)/ in_≻(f_i)
    p := p − (in_≻(p)/ in_≻(f_i)) f_i
    divisionoccurred := true
  ELSE
  i := i + 1
IF divisionoccurred = false THEN
  r := r + in_≻(p)
  p := p − in_≻(p)
```

Example 1.3.5 (Taken from [CLO97, Chapter 3, §3]). Dividing $f = x^2y+xy^2+y^2$ by the ordered list of polynomials $\{f_1 = xy - 1, f_2 = y^2 - 1\}$, we get $f = (x + y)f_1 + f_2 + x + y + 1$. Switching the order of the divisors and redoing the division gives $f = (x + 1)f_2 + x(f_1) + 2x + 1$. We list all polynomials in the two division algorithms according to lex order with $x \succ y$. Note that the remainders are different. This example shows that the above division algorithm for multivariate polynomials has several drawbacks one of which is that the remainder depends on the choice of ordering of the polynomials $\{f_1, \ldots, f_s\}$. This makes it impossible to check ideal membership of f in $\langle f_1, f_2\rangle$ by dividing f by the generators f_1, f_2.

The above example shows that arbitrary generating sets of ideals and a naive extension of the usual division algorithm cannot be used for ideal membership. We will see that this difficulty disappears when the basis of the ideal is a Gröbner basis.

A *monomial ideal* is an ideal generated by monomials. The *initial ideal* of an ideal $I \subset S$ is the monomial ideal

$$\mathrm{in}_\succ(I) := \langle \mathrm{in}_\succ(f) \; : \; f \in I \rangle \subseteq S.$$

A monomial ideal in S can be depicted by its *staircase diagram* in $\mathbb{N}^n$ which is the collection of all exponent vectors of monomials in the ideal. Clearly, this set of "dots" is closed under the addition of $\mathbb{N}^n$ to any of the dots. Equivalently, the complement in $\mathbb{N}^n$ is a *down-set* or *order ideal* in $\mathbb{N}^n$.

Example 1.3.6. Let $I = \langle x^2 - y, x^3 - x\rangle \subset \mathbf{k}[x, y]$ and $\succ$ be the lexicographic order with $x \succ y$. The polynomial $x(x^2 - y) - (x^3 - x) = -xy + x \in I$ which shows that $\mathrm{in}_\succ(I) \supset \{x^2, xy\}$. We will see later that in fact, $\mathrm{in}_\succ(I) = \langle x^2, xy, y^2\rangle$. The monomial ideal $\langle x^2, xy\rangle$ has the following staircase diagram.

By the Gordan-Dickson Lemma [CLO97, §4.2], all monomial ideals of S have a unique minimal finite generating set consisting of monomials. Hence there exists $g_1, \ldots, g_s \in I$ such that $\mathrm{in}_\succ(I) = \langle \mathrm{in}_\succ(g_1), \ldots, \mathrm{in}_\succ(g_s)\rangle$.

Definition 1.3.7.

1. A finite set of polynomials $\mathcal{G}_\succ(I) = \{g_1, \ldots, g_s\} \subset I$ is a *Gröbner basis* of I with respect to $\succ$ if $\mathrm{in}_\succ(I) = \langle \mathrm{in}_\succ(g_1), \ldots, \mathrm{in}_\succ(g_s)\rangle$. We may assume that each $\mathrm{in}_\succ(g_i)$ is a monomial.

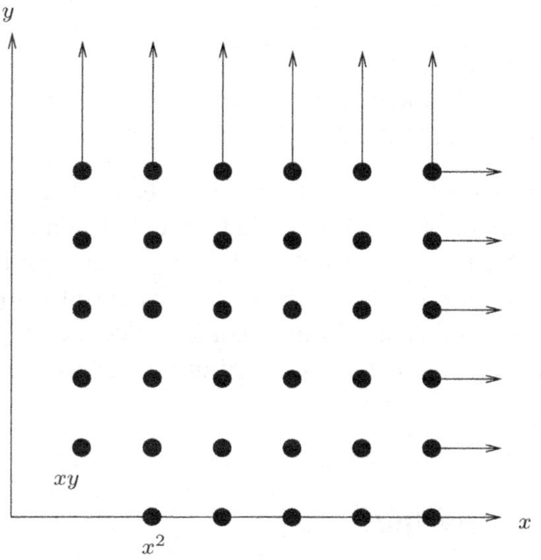

2. If $\{\text{in}_\succ(g_1), \ldots, \text{in}_\succ(g_s)\}$ is the unique minimal generating set of $\text{in}_\succ(I)$, we say that $\mathcal{G}_\succ(I)$ is a *minimal* Gröbner basis of I with respect to $\succ$.
3. A minimal Gröbner basis $\mathcal{G}_\succ(I)$ is *reduced* if no non-initial term of any g_i is divisible by any of $\text{in}_\succ(g_1), \ldots, \text{in}_\succ(g_s)$.
4. The monomials of S that do not lie in the initial ideal $\text{in}_\succ(I)$ are called the *standard monomials* of $\text{in}_\succ(I)$.

Every term order produces a unique reduced Gröbner basis of I. Unless otherwise specified, when we refer to a Gröbner basis with respect to a term-order, we will mean the reduced Gröbner basis.

Lemma 1.3.8. *If $\mathcal{G}_\succ(I)$ is a Gröbner basis of I with respect to the term order $\succ$, then the remainder of any polynomial after division by $\mathcal{G}_\succ(I)$ is unique.*

Proof. Suppose $\mathcal{G}_\succ(I) = \{g_1, \ldots, g_s\}$ and we divide a polynomial $f \in S$ by $\mathcal{G}_\succ(I)$ obtaining two remainders $r_1, r_2 \in S$. We get two expressions

$$f = \sum a_i g_i + r_1 = \sum a_i' g_i + r_2$$

which implies that $r_1 - r_2 \in I$ and that no term of $r_1 - r_2$ is divisible by $\text{in}_\succ(g_i)$ for any $g_i \in \mathcal{G}_\succ(I)$. However this implies that $r_1 - r_2 = 0$ since otherwise the nonzero term $\text{in}_\succ(r_1 - r_2) \in \text{in}_\succ(I)$ and some $\text{in}_\succ(g_i)$ would divide it. □

Corollary 1.3.9. If $\mathcal{G}_\succ(I) = \{g_1, \ldots, g_s\}$ is a Gröbner basis of I with respect to the term order $\succ$, then $\mathcal{G}_\succ(I)$ generates I.

Proof. Fix $f \in I$, and let r be the remainder on dividing f by $\mathcal{G}_\succ(I)$, so $f = \sum_{i=1}^{s} a_i g_i + r$. Then $r \in I$, but $\text{in}_\succ(r) \notin \text{in}_\succ(I)$, as otherwise we could further divide by $\mathcal{G}_\succ(I)$. This is only possible if $r = 0$, so $f = \sum_{i=1}^{s} a_i g_i$, and thus $\mathcal{G}_\succ(I)$ generates I. □

Corollary 1.3.10. Gröbner bases solve the ideal membership problem.

Definition 1.3.11.

1. The unique remainder of a polynomial $f \in S$ obtained after dividing f with the Gröbner basis $\mathcal{G}_\succ(I)$ is called the *normal form* of f with respect to $\mathcal{G}_\succ(I)$.
2. The division of f by $\mathcal{G}_\succ(I)$ is called *reduction* by $\mathcal{G}_\succ(I)$.

The passage from an ideal to one of its initial ideals is a "flat" deformation that preserves several invariants such as Hilbert function, dimension and degree. This property has allowed Gröbner bases and initial ideals to become an important theoretical tool in algebra as it allows one to study a complicated ideal by passing to monomial ideals. The details of this deformation point of view can be found in Chapter 15 of [Eis94]. The connection was first worked out by Dave Bayer in his Ph.D. thesis [Bay82].

1.4 Buchberger's algorithm

In [Buc65] Buchberger developed an algorithm to compute the reduced Gröbner basis of an ideal $I = \langle f_1, \ldots, f_t \rangle$ with respect to any prescribed term order $\succ$ on S. The algorithm needs as a subroutine the calculation of the S-pair of two polynomials f and g, denoted by S-pair(f, g). Let $\mathrm{lcm}(\mathrm{in}_\succ(f), \mathrm{in}_\succ(g))$ be the least common multiple of $\mathrm{in}_\succ(f)$ and $\mathrm{in}_\succ(g)$. This lcm is the product of the coefficients in $\mathrm{in}_\succ(f)$ and $\mathrm{in}_\succ(g)$ and the lcm of the initial monomials in $\mathrm{in}_\succ(f)$ and $\mathrm{in}_\succ(g)$. Then

$$\text{S-pair}(f, g) = \frac{\mathrm{lcm}(\mathrm{in}_\succ(f), \mathrm{in}_\succ(g))}{\mathrm{in}_\succ(f)} f - \frac{\mathrm{lcm}(\mathrm{in}_\succ(f), \mathrm{in}_\succ(g))}{\mathrm{in}_\succ(g)} g.$$

We also let $\mathrm{rem}_G(h)$ denote the remainder obtained by dividing the polynomial h by an ordered list of polynomials G.

We now describe Buchberger's algorithm. The algorithm hinges on the important fact that a set of polynomials G form a Gröbner basis with respect to $\succ$ if and only if for each pair $f, f' \in G$, $\mathrm{rem}_G(\text{S-pair}(f, f')) = 0$. The proof can be found in any of the books mentioned in the introduction. We reproduce the algorithm from [CLO97, Theorem 2].

Algorithm 1.4.1. Buchberger's algorithm [CLO97, Theorem 2]
INPUT: $F = \{f_1, \ldots, f_t\}$ a basis of the ideal $I \subset S$ and a term order $\succ$.
OUTPUT: The reduced Gröbner basis $\mathcal{G}_\succ(I)$ of I with respect to $\succ$.

$G := F$
REPEAT
  $G' := G$
  For each pair $\{p, q\}$, $p \neq q$ in G' do
    $S := \mathrm{rem}_{G'}(\text{S-pair}(p, q))$
    If $S \neq 0$ then $G := G \cup \{S\}$
UNTIL $G = G'$.
G is a Gröbner basis for I at this point.

Producing a minimal Gröbner basis with respect to $\succ$.
Let G be a Gröbner basis of I. Make all elements of G monic by dividing each by its leading coefficient. For each $g \in G$, remove it from G if its leading term is divisible by the leading term of another element $g' \in G$.

Producing the reduced Gröbner basis with respect to $\succ$.
$G' := G$ where G is a minimal Gröbner basis of I, $\mathcal{G}_\succ(I) := \emptyset$
For each $g \in G$ do
$g' = \text{rem}_{G' \setminus g}(g)$; $\mathcal{G}_\succ(I) = \mathcal{G}_\succ(I) \cup \{g'\}$; $G' = G' \setminus \{g\} \cup \{g'\}$.

Example 1.4.2. For the ideal $I = \langle f_1 := x^2 - y, f_2 := x^3 - x\rangle$ with the lex order $x \succ y$, we begin by setting $G = \{f_1, f_2\}$. The first step of the Buchberger algorithm computes S-pair$(f_1, f_2) = f_2 - x(f_1) = xy - x$. Note that $\text{rem}_G(xy - x) = xy - x$. Thus we define $f_3 := xy - x$ and update G to $G = \{f_1, f_2, f_3\}$. Next check that S-pair$(f_1, f_3) = y^2 - x^2$ which reduces modulo f_1 to $f_4 := y^2 - y$. Therefore, $G = \{f_1, f_2, f_3, f_4\}$. All other S-pairs reduce to zero modulo G. The reduced Gröbner basis of I with respect to $\succ$ is $\{f_1 = x^2 - y, f_3 = xy - x, f_4 = y^2 - y\}$.

Definition 1.4.3. A finite subset of I is a *universal* Gröbner basis of I if it is a Gröbner basis of I with respect to every term order.

In Chapter 2 we will see that every ideal has a universal Gröbner basis. We construct a universal Gröbner basis for linear ideals.

Proposition 1.4.4. (Linear ideals revisited)

1. *The set of linear forms* $\{g_1, \ldots, g_d\}$ *computed from the Gauss-Jordan form* C *of the matrix* A *in Proposition 1.2.3 is the reduced Gröbner basis of the linear ideal* $I = \langle g_1, \ldots, g_d\rangle$ *with respect to any term order such that* $x_1 \succ \cdots \succ x_n$.
2. *The set of all circuits of* I *is a minimal universal Gröbner basis of* I. *([Stu96, Proposition 1.6]).*

Proof.

1. Note that the terms of each $g_i = x_i + \sum_{j=d+1}^{n} c_{ij} x_j$ are already ordered in decreasing order with respect to the above lex order and that $G = \{g_1, \ldots, g_d\}$ is reduced in the sense that no term of any g_i other than x_i lies in $\text{in}_\succ(I) = \langle x_1, \ldots, x_d\rangle$. To show that G is a Gröbner basis it suffices to show that the remainder obtained by dividing S-pair(g_i, g_j) with respect to G is zero for all $i \neq j \in [d]$. This follows from the following general fact: If $f = \text{in}_\succ(f) + f'$ and $g = \text{in}_\succ(g) + g'$ are two monic polynomials such that $\text{in}_\succ(f)$ and $\text{in}_\succ(g)$ are relatively prime, then
$$\text{S-pair}(f, g) = \text{in}_\succ(g)\, f - \text{in}_\succ(f)\, g$$
which reduces to zero modulo $\{f, g\}$. This is an important criterion for avoiding S-pairs that will eventually reduce to zero, known as *Buchberger's first criterion.*
2. (Proof from [Stu96]) The argument in part 1 shows that every reduced Gröbner basis of I arises from a Gauss-Jordan form. We argued earlier that all the elements of these Gröbner bases are circuits of I. Thus the circuits of I form a universal Gröbner basis of I.

To prove minimality, we need to argue that each circuit l appears in some reduced Gröbner basis of I. Let $\succ$ be a term order such that $\{x_i : i \notin \text{supp}(l)\} \succ \{x_i : i \in \text{supp}(l)\}$ and $\mathcal{G} := \mathcal{G}_{\succ}(I)$. The term order makes any monomial not containing a variable in the first group of variables cheaper, regardless of its degree, than any monomial containing a variable in the first group. Such term orders are known as *elimination orders*. The lex order is an example of an elimination order with $\{x_1\} \succ \{x_2\} \succ \cdots \succ \{x_n\}$. Suppose l does not appear in $\mathcal{G}$. Then there exists $l' \in \mathcal{G}$ such that $\text{in}_{\succ}(l) = \text{in}_{\succ}(l')$. By the elimination property of $\succ$, $\text{supp}(l') \subseteq \text{supp}(l)$ and hence $\text{supp}(l - l') \subsetneq \text{supp}(l)$. However this contradicts that l is a circuit of I as $l - l'$ is a nonzero linear form with strictly smaller support. □

Here is another example of a universal Gröbner basis.

Proposition 1.4.5 ([Stu96, Example 1.4]). *Consider a polynomial ring in* $2m$ *indeterminates:*

$$\begin{pmatrix} x_{11} & x_{12} & \cdots & x_{1m} \\ x_{21} & x_{22} & \cdots & x_{2m} \end{pmatrix}$$

and the ideal I *generated by the* 2×2 *minors* $D_{ij} = x_{1i}x_{2j} - x_{1j}x_{2i}$, $1 \le i < j \le m$ *of the above matrix. Then* $\{D_{ij}\}$ *is a universal Gröbner basis of* I.

Finding universal Gröbner bases is a hard task in general. Chapter 2 gives a general algorithm for finding a universal Gröbner basis of an ideal. In special cases, a universal Gröbner basis can be described using intrinsic features of the ideal. In Chapter 3 we will construct a universal Gröbner basis for the special class of ideals called toric ideals.

1.5 Tutorial 1

Macaulay 2 [GS] is a mathematical software package, freely available from

`http://www.math.uiuc.edu/Macaulay2/`

Start Macaulay 2 with the command "M2" (without quotes), and you will be provided with an input prompt. We begin by illustrating some of the basic commands that we need in this tutorial. Most often the commands are self explanatory. We have provided an appropriate description whenever they are not.

```
i1 : 2+2
o1 = 4
i2 : 3/5 + 7/11
     68
o2 = --
     55
o2 : QQ
i3 : 2*3*4
o3 = 24
i4 : 2^8
o4 = 256
```

```
i5 : 6!
o5 = 720
```

Note that an input to Macaulay 2 is on a line that starts with an `i` and the outputs are on lines that start with `o`. For instance, `o2` is the answer to the input $3/5 + 7/11$ in `i2`. The second output line `o2` tells you what sort of object the output is. In this case, it is an element of the ring $\mathbb{Q}$ which in Macaulay 2 is denoted as `QQ`.

Rings and Fields in Macaulay 2.
`ZZ` - The ring of integers.
`QQ` - The field of rationals.
`RR` - The field of reals.
`CC` - The field of complex numbers.
`ZZ/p` - Finite field of order `p`, where `p` is a prime number.

Warning. Macaulay 2 does not accept rings of order a composite integer. Polynomial rings are defined over `ZZ`, `ZZ/p` and `QQ`.

```
i6 : k = ZZ/32003
o6 = k
o6 : QuotientRing
i7 : R = k[x,y,z]
o7 = R
o7 : PolynomialRing
i8 : R=QQ[a_0..a_6]
o8 = R
o8 : PolynomialRing
```

Different term orders in Macaulay 2. The default order in Macaulay 2 is the graded reverse lexicographic order. Other orders one can use are:

GRevLex	Graded Reverse Lexicographic order (default)
GLex	Graded Lexicographic order
Lex	Lexicographic order
Eliminate n	Elimination order, eliminating first n variables
ProductOrder {n1, ..., nv}	Product order
Weight Order	Represent the term order using a vector (See Chapter 2 for details)

```
i9 : R = QQ[x,y,z, MonomialOrder => GLex]
o9 = R
o9 : PolynomialRing
i10 : R = QQ[a..e, Weights => {2,89,100,23,1},
          MonomialSize => 16]
o10 = R
o10 : PolynomialRing
```

In line `i10`, the numbers in the curly brackets represent the weight for each variable. Setting `MonomialSize => n` specifies that monomial exponents may be as large as $2^{(n-1)} - 1$. The default value is 8, allowing for exponents up to 127. Currently the maximum value is 16, allowing for exponents up to 32767.

Ideals of a polynomial ring.

```
i11 : R = QQ[x,y,z]
o11 = R
o11 : PolynomialRing
i12 : I = ideal (x^2+y^2, x^3*y+y^3)
                 2    2   3     3
o12 = ideal (x  + y , x y + y )
o12 : Ideal of R
```

The command `leadTerm(I)` gives the generators of the initial ideal of I with respect to the term order specified in the ring `R`.

```
i13 : leadTerm(I)
o13 = | x2 xy3 y5 |
              1         3
o13 : Matrix R  <--- R

i14 : gb I
o14 = | x2+y2 xy3-y3 y5+y3 |
o14 : GroebnerBasis
i15 : J = ideal(x^3*y+x*y^2,x^4,x^3*y^2)
              3          2   4   3 2
o15 = ideal (x y + x*y , x , x y )
o15 : Ideal of R
i16 : I + J
              2    2   3     3   3          2   4   3 2
o16 = ideal (x  + y , x y + y , x y + x*y , x , x y )
o16 : Ideal of R
i17 : I*J
              5     3 3     3 2       4    6      4 2
o17 = ideal (x y + x y  + x y  + x*y , x  + x y ,
              5 2     3 4     6 2     4 3     3 4       5
             x y  + x y ,  x y  + x y  + x y  + x*y ,
              7     4 3    6 3    3 5
             x y + x y , x y  + x y )
o17 : Ideal of R
```

To suppress the output, put a semicolon at the end of the command.

```
i18 : intersect(I, J);
o18 : Ideal of R

i19 : I:J
```

```
                 2                       2
o19 = ideal (y   + x, x*y - y, x   - x)
o19 : Ideal of R
```

1. Define the polynomial ring $\mathbb{Q}[x, y, z, w]$, ideals $I = \langle x^2 + 2xy^3, z^2 - w^3, xz - 3yw\rangle$ and $J = \langle y^3 - 2zw^2, z^2 - 3yw, x^2y - z^2w\rangle$. Compute initial ideal and Gröbner bases of I, J, $I + J$, IJ, $I : J$ and $I \cap J$ with respect to the monomial orderings lex and grevlex.

Do the following exercises without the help of Macaulay 2. They have been taken from [CLO97].

2. Use Buchberger's algorithm to compute Gröbner basis of the ideal $I = \langle y - z^2, z - x^3\rangle \in \mathbb{Q}[x, y, z]$ with grevlex and lex orders.
3. Let $f, g \in S$ be such that $\text{in}_\succ(f)$ and $\text{in}_\succ(g)$ are relatively prime and the leading coefficients of f and g are 1. Show that $\text{S-pair}(f, g) = -(g - \text{in}_\succ(g))f + (f - \text{in}_\succ(f))g$ and hence reduce to zero modulo $\{f, g\}$. Deduce that the leading monomial of $\text{S-pair}(f, g)$ is a multiple of the leading monomial of either f or g in this case.
4. Show that the polynomials $f_1 = x - y^2w$, $f_2 = y - zw$, $f_3 = z - w^3$, $f_4 = w^3 - w \in \mathbb{Q}[x, y, z, w]$ with lex ordering where $x \succ y \succ z \succ w$ form a Gröbner basis for the ideal they generate. Show that they do not form a lex Gröbner basis if $w \succ x \succ y \succ z$.
5. (Division algorithm)
   Divide $f = xy^2 + 1$ by $f_1 = xy + 1$ and $f_2 = y + 1$ using lex order with $x \succ y$.
6. Solve the linear equations

$$3x + 4y - z + w = 0$$
$$x - 3y + 3z - 4w = 0$$
$$x - y + z - w = 0$$

   by computing a Gröbner basis of the ideal generated by the polynomials

$$f_1 = 3x + 4y - z + w, \quad f_2 = x - 3y + 3z - 4w, \quad \text{and} \quad f_3 = x - y + z - w.$$

   Note that the matrix corresponding to the Gröbner basis is a row reduced Gauss-Jordan form of the matrix of the original equations.

The next 3 exercises are from [Stu96, p. 6].

7. Compute all circuits in the following ideal of linear forms:

$$I = \langle 2x_1 + x_2 + x_3, \; x_2 + 2x_4 + x_5, \; x_3 + x_5 + 2x_6\rangle.$$

8. Let $\mathcal{U}$ be a universal Gröbner basis for an ideal I in $\mathbf{k}[x_1, \ldots, x_n]$. Show that for every subset $Y \subseteq \{x_1, \ldots, x_n\}$ the elimination ideal $I \cap \mathbf{k}[Y]$ is generated by $\mathcal{U} \cap \mathbf{k}[Y]$.

9. Note that for a term order $\succ$, $\mathrm{in}_{\succ}(f)\,\mathrm{in}_{\succ}(g) = \mathrm{in}_{\succ}(fg)$ for $f, g \in S$. Show that $\mathrm{in}_{\succ}(I)\,\mathrm{in}_{\succ}(J) \subseteq \mathrm{in}_{\succ}(IJ)$ for any two ideals I and J. Find I and J such that this containment is proper.
10. **Saturation of ideals:** [CLO97, Ex. 8, p. 196], The following exercise illustrates an algorithm to compute the saturation of ideals. We need this exercise in Lecture 3. Let $I \subset S$ be an ideal, and fix $f \in S$. Then the *saturation* of I with respect to f is the set

$$(I : f^\infty) = \{g \in S \mid f^m g \in I \text{ for some } m > 0\}.$$

(a) Prove that $(I : f^\infty)$ is an ideal.
(b) Prove that we have the ascending chain of ideals

$$(I : f) \subseteq (I : f^2) \subseteq (I : f^3) \subseteq \cdots$$

(c) By part (b) and the Ascending Chain Condition we have $(I : f^N) = (I : f^{N+1}) = \cdots$ for some integer N. Prove that $(I : f^\infty) = (I : f^N)$.
(d) Prove that $(I : f^\infty) = (I : f^m)$ if and only if $(I : f^m) = (I : f^{m+1})$.

When the ideal I is homogeneous and f is one of the variables x_n then one can use the following strategy to compute the saturation [Stu96, Lemma 12.1].

Fix the grevlex order induced by $x_1 \succ x_2 \succ \ldots \succ x_n$ and let G be a reduced Gröbner basis of a homogeneous ideal $I \subset S$.

(a) Show that the set $G' =$

$$\{f \in G \,:\, x_n \text{ does not divide } f\} \cup \{f/x_n \,:\, f \in G \text{ and } x_n \text{ divides } f\}$$

is a Gröbner basis of $(I : x_n)$.
(b) Show that a Gröbner basis of $(I : x_n^\infty)$ is obtained by dividing each element $f \in G$ by the highest power of x_n that divides f.

1.6 Solutions to Tutorial 1

1. We provide Macaulay 2 code for the solution of Exercise 1. The calculation shown here is only for the lex order and only for some of the ideals. The commands can be repeated on the other ideals. The grevlex order can be done similarly. We show how to change the monomial order in the ring at the end.

```
i1 : S = QQ[x,y,z,w, MonomialOrder=>Lex,
                     MonomialSize=>16]

o1 = S
o1 : PolynomialRing
i2 : I = ideal(x^2+2*x*y^3, z^2-w^3, x*z-3*y*w)
                   2        3   2     3
o2 = ideal (x  + 2x*y , z  - w , x*z - 3y*w)
o2 : Ideal of S
```

```
i3 : J = ideal(y^3-2*z*w^2, z^2-3*y*w, x^2*y-z^2*w)
               3         2              2   2      2
o3 = ideal (y  - 2z*w , - 3y*w + z , x y - z w)
o3 : Ideal of S
i4 : IplusJ = I + J;
i5 : toString IplusJ
o5 = ideal(x^2+2*x*y^3,z^2-w^3,x*z-3*y*w,y^3-2*z*w^2,
-3*y*w+z^2,x^2*y-z^2*w)
i6 : ItimesJ = I*J;
o6 : Ideal of S
i7 : IcolonJ = I:J;
o7 : Ideal of S
i8 : IintersectJ = intersect(I,J);
Abort (y/n)? y
returning to top level
```

The command `intersect(I,J)` took more than several seconds and so it was aborted by the user by typing `CTRL C`. The reader can try it. Sometimes Macaulay 2 will bring you back to the session, but other times it will quit which is what happened here. So we restart the program and redo our calculations to continue. We next calculate the reduced Gröbner basis and initial ideal with respect to lex order for the ideal I. The commands can be repeated on the other ideals.

```
i7 : GI = gens gb I
o7 = | z2-w3 xz-3yw x2+2xy3 xw3-3yzw xyw+2y4w
       y4zw+3/2y2w2 y4w3+3/2y2zw |
             1      7
o7 : Matrix S  <--- S
i8 : leadTerm GI
o8 = | z2 xz x2 xw3 xyw y4zw y4w3 |
             1      7
o8 : Matrix S  <--- S
i9 : leadTerm I
o9 = | z2 xz x2 xw3 xyw y4zw y4w3 |
             1      7
o9 : Matrix S  <--- S
```

Note that the commands in `i8` and `i9` produced the same answer. We now change the monomial order to grevlex and import ideals into this new ring `R`. Note that since grevlex is the default order, no monomial order needs to be specified.

```
i14 : R = QQ[gens S, MonomialSize => 16]
o14 = R
o14 : PolynomialRing
i15 : I = substitute(I,R)
```

```
                   3       2      3     2
o15 = ideal (2x*y    + x , - w   + z , x*z - 3y*w)
o15 : Ideal of R
i16 : GI = gens gb I
o16 = | xz-3yw w3-z2 xy3+1/2x2 y4w+1/2xyw y4z2+3/2y2zw|
                   1       5
o16 : Matrix R   <--- R
i17 : leadTerm I
o17 = | xz w3 xy3 y4w y4z2 |
                   1       5
o17 : Matrix R   <--- R
```

We now retry the computation of $I \cap J$ in the ring R with grevlex order and see that it does compute the intersection. We show just the initial ideal of $I \cap J$ with respect to grevlex.

```
i18 : J = substitute(J,R);
i19 : IintersectJ = intersect(I,J);
i20 : numgens IintersectJ
o20 = 23
i21 : leadTerm IintersectJ
o21 = | xz3 z2w3 xz2w2 x2zw2 xy2z2 x3z2 xy3z x2y2z
        x3yz y3w3 x2yw3 y4w2 x2yz2w y4zw y5w xy4w
        y4z2 xy5 x2y4 |
          1        19
o21 : Matrix R   <--- R
```

Now that we have the ideal $I \cap J$ in Macaulay 2, the reader can try to change the term order in the ring back to lex and compute the required initial ideal and Gröbner basis.

2. With respect to the lex ordering $x \succ y \succ z$, the leading terms of $f = y - z^2$ and $g = z - x^3$ are y and $-x^3$ respectively. By the division algorithm, we obtain that S-pair$(f, g) = x^3z^2 - yz$ and $x^3z^2 - yz = -z^2(z - x^3) - z(y - z^2)$. Hence $\{f, g\}$ is a Gröbner basis for I.

   Similarly, one can see that $\{f, g\}$ is a Gröbner basis with respect to the grevlex order as well.
3. Since $\text{in}_\succ(f)$ and $\text{in}_\succ(g)$ are relatively prime, S-pair$(f, g) = \text{in}_\succ(g)f - \text{in}_\succ(f)g$. Thus, every monomial of S-pair(f, g) is divisible by either $\text{in}_\succ(f)$ or $\text{in}_\succ(g)$. In particular, this is true for the leading monomial of S-pair(f, g).
4. The leading terms of f_i for $1 \le i \le 4$ are x, y,z and w^3 respectively. We tabulate the S-pairs below.

$$\text{S-pair}(f_1, f_2) = -y^3w + xzw = -y^2w(f_2) + zw(f_1).$$

$$\text{S-pair}(f_1, f_3) = -y^2wz + xw^3 = w^3(f_1) - y^2w(f_3).$$

$$\text{S-pair}(f_1, f_4) = -y^2w^4 + xw = w(f_1) - y^2w(f_4).$$

$$\text{S-pair}(f_2, f_3) = -z^2w + yw^3 = w^3(f_2) - zw(f_3).$$

$$\text{S-pair}(f_2, f_4) = -zw^4 + yw = w(f_2) - zw(f_4).$$

$$\text{S-pair}(f_3, f_4) = -w^6 + zw = w(f_3) - w^3(f_4).$$

The second equalities above are obtained by applying the division algorithm. Hence, the f_i's form a Gröbner basis for I.

If the term order is $w \succ x \succ y \succ z$, then the leading terms of the f_i's are $-wy^2, -zw, -w^3$ and w^3 respectively. Note that the S-pair$(f_1, f_2) = -zx + y^3$ is not divisible by the leading term of any of the f_i's.

5. Check that $f = y(f_1) - f_2 + 2$.
6. Fix any term order with $x \succ y \succ z \succ w$. Recall that $f_1 = 3x + 4y - z + w$, $f_2 = x - 3y + 3z - 4w$ and $f_3 = x - y + z - w$. It can be checked that S-pair$(f_1, f_2) = m_1 = 13y - 10z + 13w$, S-pair$(f_2, f_3) = m_2 = -2y + 2z - 3w$, and S-pair$(f_1, f_3) = m_3 = 7y - 4z + 4w$. None of these S-pairs are divisible by f_1, f_2, f_3 and hence the current partial Gröbner basis consists of f_1, f_2, f_3, m_1, m_2 and m_3. Computing all S-pairs and their normal forms we see that the reduced Gröbner basis of the ideal is $x + 1/2w, y - 2/3w, z - 13/6w$. Check that the matrix of coefficient vectors of these three polynomials is a reduced row echelon form of the original matrix of coefficient vectors.
7. Let

$$A = \begin{pmatrix} 2 & 1 & 1 & 0 & 0 & 0 \\ 0 & 1 & 0 & 2 & 1 & 0 \\ 0 & 0 & 1 & 0 & 1 & 2 \end{pmatrix}.$$

Then one can see that the rows of

$$B = \begin{pmatrix} 0 & 2 & -2 & -1 & 0 & 1 \\ 1 & -1 & -1 & 0 & 1 & 0 \\ 1 & 0 & -2 & 0 & 0 & 1 \end{pmatrix}$$

form a basis for ker(A). Let $\tau = \{3, 4, 5, 6\}$. We compute C_τ, circuit corresponding to τ.

$$C_\tau = -\det\begin{pmatrix} -1 & 0 & 1 \\ 0 & 1 & 0 \\ 0 & 0 & 1 \end{pmatrix}\mathbf{e}_3 + \det\begin{pmatrix} -2 & 0 & 1 \\ -1 & 1 & 0 \\ -2 & 0 & 1 \end{pmatrix}\mathbf{e}_4$$

$$-\det\begin{pmatrix} -2 & -1 & 1 \\ -1 & 0 & 0 \\ -2 & 0 & 1 \end{pmatrix}\mathbf{e}_5 + \det\begin{pmatrix} -2 & -1 & 0 \\ -1 & 0 & 1 \\ -2 & 0 & 0 \end{pmatrix}\mathbf{e}_6$$

$$= (0, 0, 1, 0, 1, 2).$$

Therefore, $x_3 + x_5 + 2x_6$ is a circuit of I. By varying the set τ over all four element subsets of $\{1, \ldots, 6\}$, one can compute all circuits of I. They are $\{x_3 +$

$x_5+2x_6, x_2-x_3+2x_4-2x_6,\ 2x_1+x_2-x_5-2x_6,\ x_1+x_3-x_4+x_6,\ 2x_1+x_3-2x_4-x_5,\ x_1-x_4-x_5-x_6,\ x_2+2x_4+x_5,\ 2x_1+x_2+x_3,\ x_1+x_2+x_4-x_6\}$.

8. Without loss of generality assume $Y = \{x_1, \ldots, x_r\}$ for some $r \leq n$. Consider an elimination term order $\succ$ on S with the property that $\{x_{r+1}, \ldots, x_n\} \succ \{x_1, \ldots, x_r\}$. Since $\mathcal{U}$ is a universal Gröbner basis for I, $\mathcal{U}$ is a Gröbner basis of I with respect to the term order $\succ$. To show that $\mathcal{U} \cap \mathbf{k}[Y]$ generates $I \cap \mathbf{k}[Y]$ it suffices to show that $\mathcal{U} \cap \mathbf{k}[Y]$ contains a Gröbner basis for $I \cap \mathbf{k}[Y]$. Let $f \in I \cap \mathbf{k}[Y]$. Since $\mathcal{U}$ is a Gröbner basis for I with respect to $\succ$, there exists an element $u \in \mathcal{U}$ such that $\mathrm{in}_\succ(u)$ divides $\mathrm{in}_\succ(f) \subset \mathbf{k}[Y]$. This implies that $\mathrm{in}_\succ(u)$ and hence all of u lies in $\mathbf{k}[Y]$ since $\succ$ is the elimination order defined above. Therefore, we conclude that $\mathrm{in}_\succ(I \cap \mathbf{k}[Y]) \subseteq \langle \mathrm{in}_\succ(u) : u \in \mathcal{U} \cap \mathbf{k}[Y] \rangle$. Since the reverse containment is obviously true, we conclude that $\mathrm{in}_\succ(I \cap \mathbf{k}[Y]) = \langle \mathrm{in}_\succ(u) : u \in \mathcal{U} \cap \mathbf{k}[Y] \rangle$. By the definition of a Gröbner basis, $\mathcal{U} \cap \mathbf{k}[Y]$ is a Gröbner basis for $I \cap \mathbf{k}[Y]$ with respect to $\succ$ and hence generates $I \cap \mathbf{k}[Y]$.
9. We prove that $\mathrm{in}_\succ(I)\,\mathrm{in}_\succ(J) \subseteq \mathrm{in}_\succ(IJ)$. Let $f \in \mathrm{in}_\succ(I)$ and $g \in \mathrm{in}_\succ(J)$. Since $\mathrm{in}_\succ(I)$ and $\mathrm{in}_\succ(J)$ are generated by initial forms of elements in I and J respectively, we may assume that $f = \mathrm{in}_\succ(f_1)$ and $g = \mathrm{in}_\succ(g_1)$ for some polynomials $f_1, g_1 \in \mathbf{k}[x_1, \ldots, x_n]$. Then $f_1 g_1 \in IJ$ and by the first part $\mathrm{in}_\succ(f_1 g_1) = fg$. Therefore $fg \in \mathrm{in}_\succ(IJ)$. Hence $\mathrm{in}_\succ(I)\,\mathrm{in}_\succ(J) \subseteq \mathrm{in}_\succ(IJ)$.

   Let $S = \mathbf{k}[x, y]$ with grevlex monomial order and $I = (x^5, x^4y^2, x^2y^5(x+y), xy^8, y^{10})$. Then it can be seen that

$$\mathrm{in}(I) = (x^5, x^4y^2, x^3y^5, x^2y^7, xy^8, y^{10}),$$

$$\mathrm{in}(I)^2 = (x^{10}, x^9y^2, x^8y^4, x^7y^7, x^6y^8, x^5y^{10}, x^4y^{12}, x^2y^{16}, x^3y^{15}, xy^{18}, y^{20}),$$

$$\mathrm{in}(I^2) = (x^{10}, x^9y^2, x^8y^4, x^7y^6, x^6y^8, x^5y^{10}, x^4y^{12}, x^3y^{14}, x^2y^{16}, xy^{18}, y^{20}).$$

   Therefore $x^7y^6 \in \mathrm{in}(I^2)$ but $x^7y^6 \notin \mathrm{in}(I)^2$.

10. (a) If $g, h \in (I : f^\infty)$, then $f^m g, f^n h \in I$. Thus $f^{\max(m,n)}(g+h) \in I$. This implies $(g+h) \in (I : f^\infty)$. Now if $g \in (I : f^\infty)$, then $f^m g \in I$ for some $m > 0$. This implies that $h(f^m g) \in I$. But $h(f^m g) = f^m(hg)$. Thus $hg \in (I : f^\infty)$. Therefore $(I : f^\infty)$ is an ideal.
    (b) If $f^m g \in I$, then $f^{m+1} g \in I$. Therefore the chain of inequalities follows.
    (c) Let $g \in (I : f^\infty)$. Thus $f^m g \in I$ for some $m > 0$. If $m \leq N$, then $g \in (I : f^m) \subseteq (I : f^N)$. If $m > N$, then since $(I : f^N) = \cdots = (I : f^m)$, $g \in (I : f^N)$.
    (d) Clearly, if $(I : f^\infty) = (I : f^m)$, then $(I : f^m) = (I : f^{m+1})$. Now suppose that $(I : f^m) = (I : f^{m+1})$. Assume by induction that for any $0 \leq r < \ell$, $(I : f^{m+r}) = (I : f^{m+r+1})$. Then $(I : f^{m+l+1}) = ((I : f^{m+l}) : f) = ((I : f^{m+l-1}) : f) = (I : f^{m+l})$. Therefore, $(I : f^\infty) = (I : f^m)$.
    (e) Let $J = (I : x_n)$. To prove that G' is a Gröbner basis of J with respect to the given reverse lexicographic order, it suffices to show that $\{\mathrm{in}(g) : g \in G'\}$ generates $\mathrm{in}(J)$. Let $f \in J$ be a homogeneous element. Since G is a Gröbner basis for I and $x_n f \in I$, there exists homogeneous $r_i \in S$ such that $x_n f = \sum_i r_i g_i$ where $g_i \in G$. Since $x_n f$ and all the elements of G are homogeneous,

in fact, $\text{in}(x_n f) = \sum_i r_i \,\text{in}(g_i)$. Since the term order is reverse lex with x_n cheapest, $\text{in}(x_n f) = x_n \,\text{in}(f)$. Therefore $\text{in}(f) = \sum_{x_n \nmid \text{in}(g_i)} (r_i/x_n)\,\text{in}(g_i) + \sum_{x_n | \text{in}(g_i)} r_i (\text{in}(g_i)/x_n)$. Again due to the choice of term order, it is clear that if $x_n \mid \text{in}(g)$, then $x_n | g$, for any homogeneous element g. Therefore $\text{in}(g_i)/x_n = \text{in}(g_i/x_n)$ and hence we obtain that $\text{in}(J)$ is generated by $\{\text{in}(g) : g \in G'\}$.

(f) We prove by induction that the set $G_m = \bigcup_{i=0}^{m-1} \{(f/x_n^i) : f \in G$ and x_n^i is the largest power of x_n dividing $f\} \bigcup \{f/x_n^m : f \in G$ and x_n^m divides $f\}$ is a Gröbner basis for $(I : x_n^m)$, $m \geq 1$. The assertion for $m = 1$ is proved in (e). Assume by induction that $m > 1$ and the assertion holds for all $\ell < m$. Let $J = (I : x_n^{m-1})$. By induction, G_{m-1} is a Gröbner basis for J. Now using the fact $(I : x_n^m) = (J : x_n)$ and proceeding as in (e), we one proves the assertion.

2

The Gröbner Fan

2.1 Introduction

The main goal of this chapter is to introduce the Gröbner fan of an ideal in a polynomial ring, which is a polyhedral fan associated to the given ideal, with one top-dimensional cone for each initial ideal of the ideal. No familiarity with polyhedral theory is assumed. Throughout this chapter, the polynomial ring is written as $S = \mathbf{k}[x_1, \ldots, x_n]$, with no assumptions on the field $\mathbf{k}$. We denote by $\mathbb{R}^n_{\geq 0}$ the set of vectors $\mathbf{v} \in \mathbb{R}^n$ with $v_i \geq 0$ for $1 \leq i \leq n$.

2.2 Some more Gröbner facts

We begin by introducing some more Gröbner facts. Recall that a reduced Gröbner basis, and hence the corresponding initial ideal, for an ideal is determined by a term order on the monomials in the polynomial ring. Note that there are an uncountable number of different term orders. One way to see this is to observe that if $\mathbf{w} \in \mathbb{R}^n_{\geq 0}$ is a vector with algebraically independent transcendental entries, then the order defined by setting $\mathbf{x}^{\mathbf{u}} \prec \mathbf{x}^{\mathbf{v}}$ if $\mathbf{w} \cdot \mathbf{u} < \mathbf{w} \cdot \mathbf{v}$ is a total order that obeys the term order axioms. Different transcendental vectors with norm one give different term orders, so we conclude that there are an uncountable number of term orders. We will use these term orders extensively in this chapter. The vector $\mathbf{w}$ is called a *weight vector*.

In light of this infinite number of term orders, it is somewhat surprising that a fixed ideal has only a finite number of different initial ideals. The intuitive explanation is that most of the term orders constructed above only differ in very high degree, and so the Buchberger algorithm does not see the difference between them. We now present two proofs of this finiteness result.

Proposition 2.2.1. *Let I be an ideal in S. Then there are only finitely many distinct initial ideals of I.*

Both proofs of this proposition rely on the following lemma.

Lemma 2.2.2. *Let M be an initial ideal of an ideal $I \subseteq S$ with respect to a term order $\prec$. Then the monomials of S not in M form a vector space basis for S/I.*

Proof. To see that the monomials not in M are independent modulo I, note that any dependency relation would give a nonzero polynomial f in I none of whose terms lie in M. But this contradicts M being the initial ideal of I, as we must have $\text{in}_\prec(f) \in M$. To see that they span, note that any polynomial in S has a normal form g for which $f - g \in I$, and g is a linear combination of monomials not in M. □

Corollary 2.2.3. If $J = \text{in}_\prec(I)$ and $K = \text{in}_{\prec'}(I)$ are two initial ideals of an ideal $I \subseteq S$, with $J \subseteq K$, then $J = K$.

Proof. Suppose that $J \subsetneq K$, and let $\mathbf{x^u} \in K \setminus J$. By Lemma 2.2.2 we know that the monomials not in K form a basis for S/I, so there is some polynomial g none of whose terms lie in K for which $\mathbf{x^u} - g \in I$. But none of the terms of $\mathbf{x^u} - g$ lies in J, so in particular its leading term with respect to $\prec$ does not lie in J, contradicting the fact that J is the initial ideal of I with respect to $\prec$. From this contradiction we conclude that $J = K$. □

First proof of Proposition 2.2.1. Suppose I has an infinite number of initial ideals. Let Σ_0 be the set of all initial ideals of I. Since Σ_0 is infinite, I is not the zero ideal, so we can choose an element $f_1 \in I$. Since f_1 has only a finite number of terms, and each initial ideal $M \in \Sigma_0$ contains a term of f_1, there must be one term m_1 of f_1 that is contained in an infinite number of ideals in Σ_0. Let $\Sigma_1 = \{M \in \Sigma_0 : m_1 \in M\}$. Let $J_1 = \langle m_1 \rangle$. Since infinitely many initial ideals contain J_1, there is some initial ideal that properly contains J_1. Thus Lemma 2.2.2 implies that the monomials of S outside J_1 are linearly dependent modulo I, so there is some polynomial f_2 in I with no term lying in J_1. Again, there is a term m_2 of f_2 that is contained in an infinite number of ideals in Σ_1. Let $\Sigma_2 = \{M \in \Sigma_1 : m_2 \in M\}$, and let $J_2 = J_1 + \langle m_2 \rangle$. This procedure can now be iterated, at each stage finding a polynomial f_k none of whose terms are contained in J_{k-1}, and one of which, m_k, lies in infinitely many initial ideals in Σ_{k-1}. The new ideal $J_k = J_{k-1} + \langle m_k \rangle$ will be properly contained in some initial ideal, so the new f_{k+1} can be created.

In this fashion we get a properly increasing sequence of ideals

$$J_1 \subsetneq J_2 \subsetneq J_3 \subsetneq \dots.$$

Since S is Noetherian, this is impossible. We thus conclude that I has only finitely many initial ideals. □

The other proof relies on the following theorem in addition to Lemma 2.2.2.

Theorem 2.2.4 ([Mac01, Theorem 1.1]). *Let $\mathcal{I}$ be an infinite collection of monomial ideals in S. Then there exist $I, J \in \mathcal{I}$ such that $I \subsetneq J$.*

Second proof of Proposition 2.2.1. Suppose there are an infinite number of initial ideals of I. Then by Theorem 2.2.4 there are two distinct initial ideals J and K with $J \subseteq K$. This is impossible, by Corollary 2.2.3. □

Corollary 2.2.5. Let I be an ideal in S. Then there is a finite generating set for I that is a Gröbner basis for I with respect to any term order.

Proof. Note that if $\text{in}_{\prec_1}(I) = \text{in}_{\prec_2}(I) = J$, then the reduced Gröbner bases for I with respect to the term orders $\prec_1$ and $\prec_2$ are identical. To see this note that for each minimal generator $\mathbf{x^u}$ of J, there is a unique polynomial $p_{\mathbf{u}}(\mathbf{x}) = \mathbf{x^u} - q(\mathbf{x})$ in I where every monomial occurring in $q(\mathbf{x})$ does not lie in J. The existence of $p_{\mathbf{u}}$ follows from Lemma 2.2.2, since $\mathbf{x^u} \in J$, while if $p_{\mathbf{u}}$ were not unique the difference of two such polynomials would yield a polynomial in I whose leading term did not lie in J. The polynomial $q(\mathbf{x})$ is the normal form of $\mathbf{x^u}$ with respect to $\prec_1$ and $\prec_2$. The union of all $p_{\mathbf{u}}$ as $\mathbf{x^u}$ varies over all minimal generators of J is the reduced Gröbner basis of I for any term order $\prec$ for which $\text{in}_{\prec}(I) = J$. We can now take the union over the finitely many initial ideals of I of the corresponding reduced Gröbner bases to get the desired finite generating set. □

Remark 2.2.6. Corollary 2.2.5 shows that a universal Gröbner basis (see Definition 1.4.3) for an ideal I always exists. When we refer to the universal Gröbner basis of I, we will usually mean the union of the reduced Gröbner bases corresponding to each initial ideal of I.

2.3 Polytopal basics

In this section we review the basics of polytope theory.

Definition 2.3.1. A set $\mathcal{U} \subseteq \mathbb{R}^d$ is *convex* if $\lambda\mathbf{v} + (1-\lambda)\mathbf{w} \in \mathcal{U}$ for any $\mathbf{v}, \mathbf{w} \in \mathcal{U}$, and $0 \le \lambda \le 1$. The *convex hull* of a set $\mathcal{V} \subseteq \mathbb{R}^d$ is the intersection of all convex sets containing $\mathcal{V}$, and is itself convex. A *polytope* is the convex hull of a finite set of points in $\mathbb{R}^d$. We write $P = \text{conv}(\mathbf{v}_1, \ldots, \mathbf{v}_n) = \{\sum_{i=1}^n \lambda_i \mathbf{v}_i : \lambda_i \ge 0,\ \sum_{i=1}^n \lambda_i = 1\}$.

Example 2.3.2. Classic examples of polytopes in $\mathbb{R}^3$ include the cube, the tetrahedron and the octahedron.

Definition 2.3.3. A *face* of a polytope $P \subseteq \mathbb{R}^d$ is a set of the form $\text{face}_{\mathbf{c}}(P) := \{\mathbf{x} \in P : \mathbf{c}\cdot\mathbf{x} \ge \mathbf{c}\cdot\mathbf{y}\ \forall \mathbf{y} \in P\}$, where $\mathbf{c}$ is any vector in $\mathbb{R}^d$. The *affine span* of a set $V \subseteq \mathbb{R}^d$ is the affine subspace $\mathbf{v} + H$, where $\mathbf{v} \in V$, and H is the subspace of $\mathbb{R}^d$ spanned by $\{\mathbf{w} - \mathbf{v} : \mathbf{w} \in V\}$. The *dimension* of a face F of P is the dimension of its affine span.

Example 2.3.4. If P is the square $\text{conv}((0,0), (1,0), (0,1), (1,1))$, then P has one two-dimensional face, four one-dimensional faces (edges), and four zero-dimensional faces (vertices). The whole square is the unique two-dimensional face $\text{face}_{\mathbf{0}}(P)$. Note that the vertex $(1,0)$ is $\text{face}_{(1,-1)}(P)$. Similarly $\text{face}_{(0,1)}(P)$ is the the edge $\text{conv}((0,1),(1,1))$. It is straightforward to find vectors $\mathbf{c}$ for the other three vertices and three edges. Definition 2.3.10 will make it easier to check that we have not omitted any faces from the above list.

The $(d-1)$-dimensional faces of a d-dimensional polytope are called *facets*. A facet F is of the form $F = \{\mathbf{x} : \mathbf{c}\cdot\mathbf{x} = \mathbf{b}\} \cap P$ for some $\mathbf{c} \in \mathbb{R}^d$, $\mathbf{b} \in \mathbb{R}$. The vector $\mathbf{c} =: \mathbf{c}_F$ is called the *facet normal*, and the corresponding hyperplane is a *defining hyperplane*. A normal vector to a facet is only defined up to sign. We shall choose the sign that makes $F = \text{face}_{\mathbf{c}}(P)$. The facets determine the polytope in the following way.

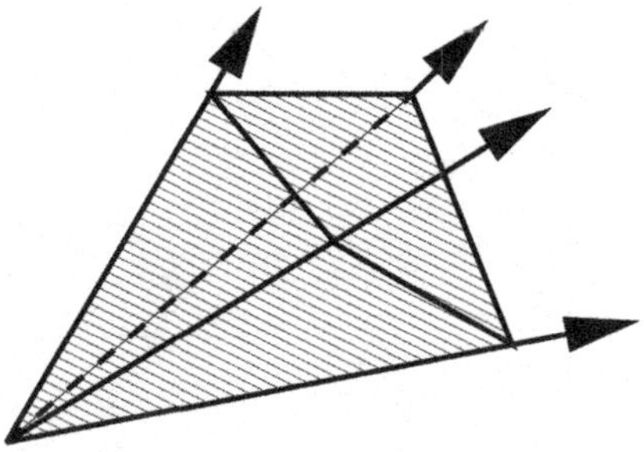

Figure 2.1. A polyhedral cone.

Proposition 2.3.5. *Let P be a polytope with facet normals $\{\mathbf{c}_F : F \text{ facet of } P\}$. Then $P = \{\mathbf{x} : \mathbf{c}_F \cdot \mathbf{x} \leq \mathbf{b}_F\}$, where $\mathbf{b}_F = \max_{\mathbf{y}\in P}\{\mathbf{c}_F \cdot \mathbf{y}\}$.*

A proof of this proposition can be found in [Zie95, Lecture 1].

Example 2.3.6. The square P of Example 2.3.4 has facets $\{\mathbf{x} \cdot \mathbf{c}_F = \mathbf{b}_F\} \cap P$ for $\mathbf{c}_F, \mathbf{b}_F = \{((0,-1),0), ((-1,0),0), ((1,0),1), ((0,1),1)\}$. Thus $P = \{(x,y) : 0 \leq x \leq 1, 0 \leq y \leq 1\}$.

Definition 2.3.7. A *cone* is a set $C \subseteq \mathbb{R}^d$ that is closed under addition and positive scalar multiplication, so if $\mathbf{v}, \mathbf{w} \in C$, then $\mathbf{v} + \mathbf{w} \in C$, and if $\mathbf{v} \in C$ and $\lambda > 0$, then $\lambda\mathbf{v} \in C$. If $\mathcal{V} = \{\mathbf{v}_1, \ldots, \mathbf{v}_n\} \subseteq \mathbb{R}^d$ then its *positive hull* is $\text{pos}(\mathcal{V}) = \{\sum_i \lambda_i \mathbf{v}_i : \lambda_i \geq 0\}$. We call $\text{pos}(\mathcal{V})$ the cone spanned by $\mathcal{V}$. A cone C is *polyhedral* if $C = \text{pos}(\mathcal{V})$ for a finite subset $\mathcal{V}$ of $\mathbb{R}^d$. As for polytopes, we define a face of a cone C to be $\text{face}_{\mathbf{c}}(C) = \{\mathbf{x} \in C : \mathbf{c} \cdot \mathbf{x} \geq \mathbf{c} \cdot \mathbf{y}\ \forall \mathbf{y} \in C\}$.

Example 2.3.8. An example of a polyhedral cone in $\mathbb{R}^2$ is shown in Figure 2.1.

Polytopes and cones are special cases of *polyhedra*, which are the intersection of finitely many halfspaces. Polytopes are bounded polyhedra, while cones are polyhedra for which the origin lies on every defining hyperplane. We are interested in particular families of polyhedral cones that fit together nicely.

If a set $V \subset \mathbb{R}^n$ is not full dimensional then it has no interior. However, every such V is full dimensional relative to its affine span. We call the interior of V in its affine span the *relative interior* . A set is *relatively open* if it is its own relative interior.

Definition 2.3.9. A *polyhedral fan* is a collection of polyhedral cones in $\mathbb{R}^n$ such that the intersection of any two cones is a face of each. The fan is called *complete* if in addition the union of the cones covers $\mathbb{R}^n$. A fan $\mathcal{F}$ is *simplicial* if every i-dimensional cone in $\mathcal{F}$ is the positive hull of i vectors for all i. It suffices to check this condition for the maximal cones in $\mathcal{F}$ (those not contained in any larger cone in $\mathcal{F}$).

We denote by $\overline{V}$ the closure (in the standard topology) of a set $V \subseteq \mathbb{R}^n$.

Definition 2.3.10. The *outer normal fan* of a polytope P is a polyhedral fan whose cones are indexed by the faces F of P, with the cone corresponding to the face F of P being the closure of the relatively open cone $C[F] = \{\mathbf{c} \in \mathbb{R}^n \text{ such that } \text{face}_{\mathbf{c}}(P) = F\}$.

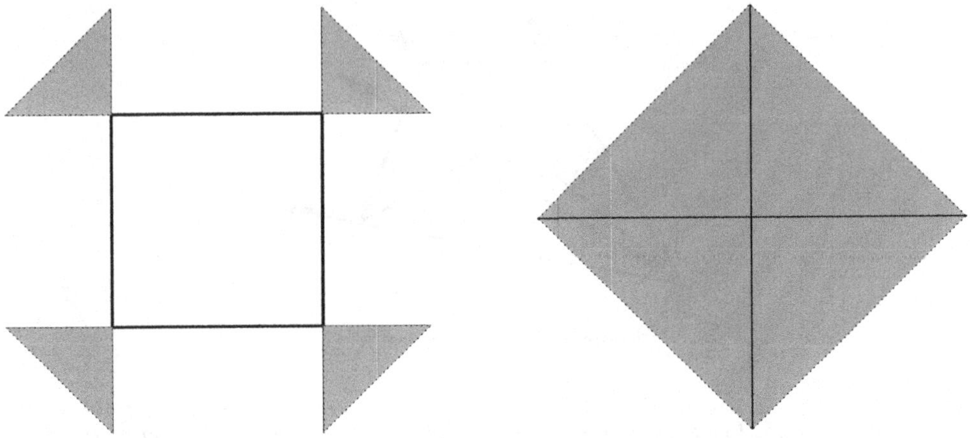

Figure 2.2. The outer normal fan for a square

Example 2.3.11. For the face $F = \{(1, 0)\}$ of the square from Example 2.3.4, we have $C[F] = \{\mathbf{v} \in \mathbb{R}^2 : v_1 > 0, v_2 < 0\}$. Its closure is the polyhedral cone $\overline{C[F]} = \{\mathbf{v} \in \mathbb{R}^2 : v_1 \geq 0, v_2 \leq 0\}$. The face $F = \text{conv}((0, 1), (1, 1))$ has $C[F] = \{\lambda(0, 1) : \lambda > 0\}$. The closure $\overline{C[F]}$ is the polyhedral cone $\text{pos}(\{(0, 1)\})$. Check that in both cases we have that $C[F]$ is the relative interior of $\overline{C[F]}$.

Figure 2.2 shows the outer normal fan for the square. The first picture shows the cones $C[\mathbf{p}]$ attached to each vertex $\mathbf{p}$, while the second is the standard picture of the fan.

Definition 2.3.12. A polyhedral fan $\mathcal{F}$ is *polytopal* if there is a polytope P for which $\mathcal{F}$ is the outer normal fan of P.

All complete fans in $\mathbb{R}^2$ are polytopal, but this is false in $\mathbb{R}^3$. See [Ful93, §1.5] for an example. We next describe a notion of sum for polytopes.

Definition 2.3.13. The *Minkowski sum* of two polytopes P and Q in $\mathbb{R}^n$ is the set $\{\mathbf{x} + \mathbf{y} : \mathbf{x} \in P, \mathbf{y} \in Q\}$. This is again a polytope.

Example 2.3.14. Figure 2.3 illustrates the Minkowski sum of a square and a triangle in $\mathbb{R}^2$. Note that the picture of the sum can be obtained by marking the points covered by the triangle as we slide it so its bottom vertex lies in the square.

We finish this section with a fundamental theorem in linear programming which we will use later.

Theorem 2.3.15. *Let $A \in \mathbb{R}^{d \times n}$ and $\mathbf{z} \in \mathbb{R}^d$. Either there exists $\mathbf{x} \in \mathbb{R}^n$ with $A\mathbf{x} \leq \mathbf{z}$, where the inequality is term-wise, or there exists a vector $\mathbf{c} \in \mathbb{R}^d$ with $\mathbf{c} \geq 0$ such that $\mathbf{c}^T A = 0$ and $\mathbf{c} \cdot \mathbf{z} < 0$, but not both.*

This theorem is known as the Farkas Lemma. A proof can be found in [Zie95, Section 1.4].

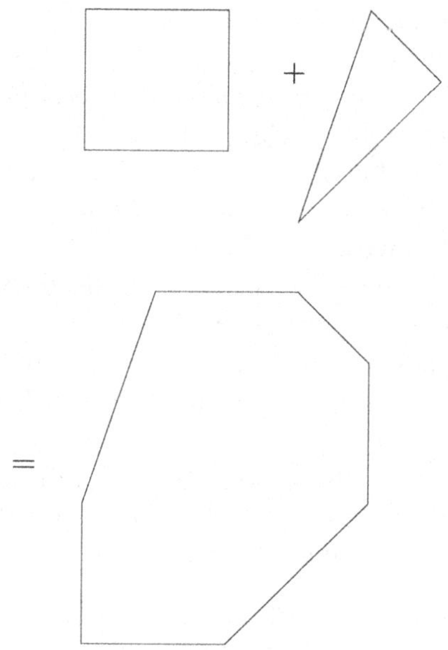

Figure 2.3. The Minkowski sum of a square and a triangle.

2.4 The Gröbner fan

In this section we associate a polyhedral fan to an ideal, each top-dimensional cone of which corresponds to a different initial ideal of the given ideal. We begin by noting that the term orders determined by weight vectors mentioned at the beginning of this chapter are the only ones that need to be considered.

Definition 2.4.1. For $\mathbf{w} \in \mathbb{R}^n_{\geq 0}$, and $f = \sum c_{\mathbf{u}} \mathbf{x}^{\mathbf{u}} \in S$, we define the *leading form* $\text{in}_{\mathbf{w}}(f)$ to be the sum of terms $c_{\mathbf{u}} \mathbf{x}^{\mathbf{u}}$ in f with $\mathbf{w} \cdot \mathbf{u}$ maximized. The initial ideal $\text{in}_{\mathbf{w}}(I) := \langle \text{in}_{\mathbf{w}}(f) : f \in I \rangle$. Note that $\text{in}_{\mathbf{w}}(I)$ need not be a monomial ideal. Given a term order $\prec$, we also define the term order $\prec_{\mathbf{w}}$, for which $\mathbf{x}^{\mathbf{u}} \prec_{\mathbf{w}} \mathbf{x}^{\mathbf{v}}$ if $\mathbf{w} \cdot \mathbf{u} < \mathbf{w} \cdot \mathbf{v}$ or if $\mathbf{w} \cdot \mathbf{u} = \mathbf{w} \cdot \mathbf{v}$ and $\mathbf{x}^{\mathbf{u}} \prec \mathbf{x}^{\mathbf{v}}$. If no specific term order is specified when referring to $\prec_{\mathbf{w}}$, we take $\prec$ to be the lexicographic term order with $x_1 \succ x_2 \succ \cdots \succ x_n$.

We note that the above definitions still make sense for a general $\mathbf{w} \in \mathbb{R}^n$, possibly with negative coordinates. However for general $\mathbf{w}$ there may be no relation between any invariants of $\text{in}_{\mathbf{w}}(I)$ and those of I (such as the Hilbert function, or dimension). This is because if $\mathbf{w}$ has negative coordinates $\prec_{\mathbf{w}}$ is not a term order, as 1 is no longer the smallest monomial.

Lemma 2.4.2. *Let $\prec$ be a term order. If $\mathbf{w} \in \mathbb{R}^n_{\geq 0}$ and $\mathcal{G}$ is a Gröbner basis for I with respect to $\prec_{\mathbf{w}}$, then $\{\text{in}_{\mathbf{w}}(g) : g \in \mathcal{G}\}$ is a Gröbner basis for $\text{in}_{\mathbf{w}}(I)$ with respect to $\prec$.*

Proof. For every polynomial $f \in I$ we have $\text{in}_{\prec}(\text{in}_{\mathbf{w}}(f)) = \text{in}_{\prec_{\mathbf{w}}}(f)$ by the definition of $\prec_{\mathbf{w}}$. Since every monomial in the ideal $\text{in}_{\prec_{\mathbf{w}}}(I)$ is of the form $\text{in}_{\prec_{\mathbf{w}}}(f)$ for some $f \in I$, this means that $\text{in}_{\prec_{\mathbf{w}}}(I) \subseteq \text{in}_{\prec}(\text{in}_{\mathbf{w}}(I))$.

To see the reverse inclusion, we first note that the vector $\mathbf{w}$ gives a $\mathbb{R}$-grading of S by $\deg(x_i) = w_i$. Since the group generated by the w_i is isomorphic to $\mathbb{Z}^k$ for some $k \le n$, this is a grading by a finitely-generated abelian group. The ideal $\mathrm{in}_{\mathbf{w}}(I)$ is homogeneous with respect to this grading, and thus so is the reduced Gröbner basis for $\mathrm{in}_{\mathbf{w}}(I)$ with respect to $\prec$. Let $\mathbf{x}^{\mathbf{u}}$ be a minimal generator of $\mathrm{in}_{\prec}(\mathrm{in}_{\mathbf{w}}(I))$, so $\mathbf{x}^{\mathbf{u}} = \mathrm{in}_{\prec}(g)$ for some $\mathbf{w}$-homogeneous $g \in \mathrm{in}_{\mathbf{w}}(I)$. We first show that $g = \mathrm{in}_{\mathbf{w}}(f)$ for some $f \in I$. Indeed, we can write $g = \sum h_i \, \mathrm{in}_{\mathbf{w}}(g_i)$, where $g_i \in I$, h_i is a monomial for all i, and the sum has as few terms as possible. Since multiplying by a monomial multiplies the leading form by that monomial, $g = \sum \mathrm{in}_{\mathbf{w}}(h_i g_i)$. Since we have as few terms as possible, the polynomials $\mathrm{in}_{\mathbf{w}}(h_i g_i)$ must be $\mathbf{w}$-homogeneous of the same degree, since g is $\mathbf{w}$-homogeneous, so $g = \mathrm{in}_{\mathbf{w}}(\sum h_i g_i)$. This proves the claim, since $f = \sum h_i g_i \in I$.

Now this means that $\mathrm{in}_{\prec}(g) = \mathrm{in}_{\prec}(\mathrm{in}_{\mathbf{w}}(f)) = \mathrm{in}_{\prec_{\mathbf{w}}}(f)$, and therefore $\mathrm{in}_{\prec}(\mathrm{in}_{\mathbf{w}}(I)) \subseteq \mathrm{in}_{\prec_{\mathbf{w}}}(I)$. Thus $\mathrm{in}_{\prec}(\mathrm{in}_{\mathbf{w}}(I)) = \mathrm{in}_{\prec_{\mathbf{w}}}(I)$, and so $\langle \mathrm{in}_{\prec}(\mathrm{in}_{\mathbf{w}}(g)) : g \in \mathcal{G}\rangle = \langle \mathrm{in}_{\prec_{\mathbf{w}}}(g) : g \in \mathcal{G}\rangle = \mathrm{in}_{\prec}(\mathrm{in}_{\mathbf{w}}(I))$, which completes the proof. □

Example 2.4.3. For $\mathbf{w} = (10, 1, 11, 12)$, $\mathrm{in}_{\mathbf{w}}(a^2c - b^2) = a^2c$, and $\mathrm{in}_{\mathbf{w}}(a^2b^2 - c^2) = a^2b^2 - c^2$. Let $I = \langle ad - bc, ac - b^2, bd - c^2\rangle \subseteq \mathbf{k}[a, b, c, d]$. The initial ideal $\mathrm{in}_{\mathbf{w}}(I) = \langle ad, ac, c^2\rangle$. The ideal $\mathrm{in}_{\mathbf{w}}(I)$ can be computed using Lemma 2.4.2 and Macaulay 2.

Proposition 2.4.4. *Let I be a fixed ideal contained in S. For every term order $\prec$ there is a weight vector $\mathbf{w} \in \mathbb{R}^n_{\ge 0}$ for which $\mathrm{in}_{\prec}(I) = \mathrm{in}_{\mathbf{w}}(I)$.*

Proof. Let $\mathcal{G} = \{g_1, \dots, g_r\}$ be the reduced Gröbner basis of I with respect to $\prec$. Write $g_i = \sum_j c_{ij}\mathbf{x}^{\mathbf{u}_{ij}}$, where $\mathrm{in}_{\prec}(g_i) = c_{i1}\mathbf{x}^{\mathbf{u}_{i1}}$. Let $C_{\prec} = \{\mathbf{w} \in \mathbb{R}^n_{\ge 0} : \mathbf{w}\cdot\mathbf{u}_{i1} > \mathbf{w}\cdot\mathbf{u}_{ij}$ for all $j \ge 2, 1 \le i \le r\}$. For any weight vector $\mathbf{w} \in C_{\prec}$ we have $\mathrm{in}_{\prec}(I) \subseteq \mathrm{in}_{\prec_{\mathbf{w}}}(I)$, so since we cannot have a proper inclusion of initial ideals by Corollary 2.2.3, we conclude that $\mathrm{in}_{\prec}(I) = \mathrm{in}_{\mathbf{w}}(I)$. It thus remains to show that $C_{\prec}$ is nonempty.

Suppose that $C_{\prec} = \emptyset$. Form the matrix U with n columns whose s rows are the vectors $\mathbf{u_{i1}} - \mathbf{u_{ij}}$ for $1 \le i \le r$, and $j \ge 2$. The fact that $C_{\prec} = \emptyset$ means that there is no vector $\mathbf{w} \in \mathbb{R}^n_{\ge 0}$ with $U\mathbf{w} > 0$, where the inequality is term-wise. It follows that there is no $\mathbf{w}' \in \mathbb{R}^n$ with $U'\mathbf{w}' \le (-\mathbf{1}, \mathbf{0})^T$, where U' is the $(s+n) \times n$ matrix with first s rows $-U$, and last n rows $-I$, $\mathbf{1}$ is the vector of length s all of whose components are one, and $\mathbf{0}$ is the zero vector of length n. The Farkas Lemma (Theorem 2.3.15) now implies that there is some vector $\mathbf{c} \in \mathbb{R}^{s+n}$ with $\mathbf{c} \ge 0$, $\mathbf{c} \ne 0$, and $\mathbf{c}^T U' = 0$. Since U' has integral entries, we can choose $\mathbf{c} \in \mathbb{N}^{s+n}$. Let $c_{i,m} \ge 0$ be the component of $\mathbf{c}$ corresponding to the row $\mathbf{u}_{im} - \mathbf{u}_{i1}$ of U'. Then we have $\sum_{i,m} c_{im}(\mathbf{u_{im}} - \mathbf{u_{i1}}) \ge 0$, because when this sum is subtracted from $\mathbf{c}U' = 0$ the result has all nonpositive coordinates. Thus

$$\prod_{i,m} (\mathbf{x}^{\mathbf{u_{i1}}})^{c_{im}} \text{ divides } \prod_{i,m} (\mathbf{x}^{\mathbf{u_{im}}})^{c_{im}},$$

so

$$\prod_{i,m} (\mathbf{x}^{\mathbf{u_{i1}}})^{c_{im}} \preceq \prod_{i,m} (\mathbf{x}^{\mathbf{u_{im}}})^{c_{im}}.$$

But for all i, m we have $\mathbf{x}^{\mathbf{u}_{im}} \prec \mathbf{x}^{\mathbf{u}_{i1}}$, so

$$\prod_{i,m}(\mathbf{x}^{\mathbf{u_{im}}})^{c_{im}} \prec \prod_{i,m}(\mathbf{x}^{\mathbf{u_{i1}}})^{c_{im}}.$$

From this contradiction we conclude that $C_{\prec}$ is nonempty, so the proposition follows. □

Remark 2.4.5. A *rational cone* is the positive hull of vectors in $\mathbb{Q}^n$, or equivalently one whose facet normals all lie in $\mathbb{Q}^n$. Note that $C_{\prec}$ is the interior of an n-dimensional rational cone in $\mathbb{R}^n_{\geq 0}$, so it contains a nonnegative rational vector, and thus a vector in $\mathbb{N}^n$. Hence we can take $\mathbf{w} \in \mathbb{N}^n$. By interior we mean here in the induced topology on $\mathbb{R}^n_{\geq 0}$.

Also, we emphasize that the weight vector $\mathbf{w}$ assigned to $\prec$ in Proposition 2.4.4 depends on I. For example there is no weight vector $\mathbf{w}$ for which $\text{in}_{\mathbf{w}}(I) = \text{in}_{\text{lex}}(I)$ for every ideal $I \subseteq S$, where lex denotes the lexicographic term order.

Proposition 2.4.6. *Let I be a fixed ideal, let $\mathbf{w} \in \mathbb{R}^n_{\geq 0}$ and let $C[\mathbf{w}]$ be the set $\{\mathbf{w}' \in \mathbb{R}^n_{\geq 0} : \text{in}_{\mathbf{w}}(I) = \text{in}_{\mathbf{w}'}(I)\}$. Then $C[\mathbf{w}]$ is the relative interior of a polyhedral cone inside $\mathbb{R}^n_{\geq 0}$.*

Proof. For a given $\mathbf{w} \in \mathbb{R}^n_{\geq 0}$, let $J = \text{in}_{\mathbf{w}}(I)$, and let $\mathcal{G} = \{g_1, \ldots, g_r\}$ be the reduced Gröbner basis for I with respect to $\prec_{\mathbf{w}}$. We note that J need not be a monomial ideal. For $g_i \in \mathcal{G}$, write $g_i = \sum_j c_{ij}\mathbf{x}^{\mathbf{a}_{ij}} + \sum_j c'_{ij}\mathbf{x}^{\mathbf{b}_{ij}}$ where $\text{in}_{\mathbf{w}}(g_i) = \sum_j c_{ij}\mathbf{x}^{\mathbf{a}_{ij}}$. It suffices to show that

$$C[\mathbf{w}] = \{\mathbf{w}' \in \mathbb{R}^n_{\geq 0} : \text{in}_{\mathbf{w}'}(g) = \text{in}_{\mathbf{w}}(g) \text{ for all } g \in \mathcal{G}\} \tag{2.1}$$

The proposition then follows because the set on the right in this equation is

$$\{\mathbf{w}' \in \mathbb{R}^n_{\geq 0} : \mathbf{w}' \cdot \mathbf{a}_{ij} = \mathbf{w}' \cdot \mathbf{a}_{ik}, \mathbf{w}' \cdot \mathbf{a}_{ij} > \mathbf{w}' \cdot \mathbf{b}_{ik} \text{ for } i = 1, \ldots, r \text{ and all } j, k\}.$$

This is the relative interior of a polyhedral cone by definition.

Let $\mathbf{w}'$ lie in the set on the right-hand side of Equation 2.1. Then $\text{in}_{\mathbf{w}}(I) \subseteq \text{in}_{\mathbf{w}'}(I)$. These ideals may not be monomial ideals, but the containment (and whether it is proper) must be preserved by taking any initial ideal with respect to an arbitrary order $\prec$. So we have $\text{in}_{\prec}(\text{in}_{\mathbf{w}}(I)) \subseteq \text{in}_{\prec}(\text{in}_{\mathbf{w}'}(I))$. By Corollary 2.2.3 we know that this inclusion cannot be proper, so we conclude that $\text{in}_{\mathbf{w}}(I) = \text{in}_{\mathbf{w}'}(I)$. This shows that the set on the right is contained in $C[\mathbf{w}]$.

Now consider $\mathbf{w}' \in C[\mathbf{w}]$. By Lemma 2.4.2 we know that $\text{in}_{\mathbf{w}}(\mathcal{G}) = \{\text{in}_{\mathbf{w}}(g) : g \in \mathcal{G}\}$ is a Gröbner basis for $\text{in}_{\mathbf{w}}(I) = \text{in}_{\mathbf{w}'}(I)$ with respect to $\prec$. Fix $g \in \mathcal{G}$. Then $\text{in}_{\mathbf{w}'}(g)$ reduces to zero with respect to $\text{in}_{\mathbf{w}}(\mathcal{G})$ using $\prec$. Now $m := \text{in}_{\prec_{\mathbf{w}}}(g)$ is the only monomial occurring in g which is divisible by the leading term with respect to $\prec$ of a polynomial in $\text{in}_{\mathbf{w}}(\mathcal{G})$, so it must occur in $\text{in}_{\mathbf{w}'}(g)$ for the reduction to be possible. Write $\text{in}_{\mathbf{w}}(g) = m + h$, and $\text{in}_{\mathbf{w}'}(g) = m + h'$. By the choice of m we know that h and h' both are sums of terms not in $\text{in}_{\prec_{\mathbf{w}}}(I)$. However $\text{in}_{\mathbf{w}}(g) - \text{in}_{\mathbf{w}'}(g) = h - h' \in \text{in}_{\mathbf{w}}(I)$, so $\text{in}_{\prec}(h - h') \in \text{in}_{\mathbf{w}}(\text{in}_{\prec_{\mathbf{w}}}(I)) = \text{in}_{\prec_{\mathbf{w}}}(I)$. This is only possible if $h - h' = 0$, so $\text{in}_{\mathbf{w}}(g) = \text{in}_{\mathbf{w}'}(g)$, and thus $\mathbf{w}'$ lies in the right-hand side of Equation 2.1. From this we conclude that Equation 2.1 holds, and so the proposition follows. □

Remark 2.4.7. If I is homogeneous with respect to some positive grading $\deg(x_i) = p_i > 0$, then we can define $\prec_{\mathbf{w}}$ for all $\mathbf{w} \in \mathbb{R}^n$ by setting $\mathbf{x}^{\mathbf{u}} \prec_{\mathbf{w}} \mathbf{x}^{\mathbf{v}}$ if $\deg(\mathbf{x}^{\mathbf{u}}) < \deg(\mathbf{x}^{\mathbf{v}})$ or if $\deg(\mathbf{x}^{\mathbf{u}}) = \deg(\mathbf{x}^{\mathbf{v}})$ and $\mathbf{w} \cdot \mathbf{u} < \mathbf{w} \cdot \mathbf{v}$, or finally if $\deg(\mathbf{x}^{\mathbf{u}}) = \deg(\mathbf{x}^{\mathbf{v}})$, $\mathbf{w} \cdot \mathbf{u} = \mathbf{w} \cdot \mathbf{v}$ and $\mathbf{x}^{\mathbf{u}} \prec \mathbf{x}^{\mathbf{v}}$, where deg is with respect to the grading by p_i. While this definition differs from the previous one, the term $\mathrm{in}_{\prec_{\mathbf{w}}}(f)$ is unchanged for homogeneous f. The same leading terms will also be achieved if we use $\prec_{\mathbf{w}'}$, where $\mathbf{w}' = \mathbf{w} + N\mathbf{p}$, where $\mathbf{p} = (p_i)$, and $N > 0$ has been chosen to be sufficiently large so $\mathbf{w}' \in \mathbb{R}^n_{\geq 0}$.

With this new definition the result of Lemma 2.4.2 still holds, and thus we can define $\mathrm{in}_{\mathbf{w}}(I)$ for any $\mathbf{w} \in \mathbb{R}^n$. This means that the conclusion of Proposition 2.4.6 holds for the cone $C_h[\mathbf{w}] = \{\mathbf{w}' \in \mathbb{R}^n : \mathrm{in}_{\mathbf{w}}(I) = \mathrm{in}_{\mathbf{w}'}(I)\}$. Note that $C_h[\mathbf{w}] \supseteq C[\mathbf{w}] + \mathrm{span}(\mathbf{p})$. In fact the lineality space of $C_h[\mathbf{w}]$ (the largest subspace contained in $C_h[\mathbf{w}]$) consists of all degree vectors $\mathbf{p}$ for which I is homogeneous. We leave this fact as an exercise to the reader.

Example 2.4.8. Let $S = \mathbf{k}[a,b,c,d,e]$ and let I be the ideal $\langle ac - b^2, ae - bd, be - cd\rangle$. Let $\mathbf{w}$ be the weight-vector $(0,2,0,1,0)$. The initial ideal $\mathrm{in}_{\prec_{\mathbf{w}}}(I)$ is then $\langle b^2, bd, be, cd^2\rangle$. The reduced Gröbner basis for $\prec_{\mathbf{w}}$ is $\{b^2 - ac, bd - ae, be - cd, cd^2 - ae^2\}$, so by Lemma 2.4.2 we see that $\mathrm{in}_{\mathbf{w}}(I) = \mathrm{in}_{\prec_{\mathbf{w}}}(I)$. Thus $C[\mathbf{w}] = \{\mathbf{w}' \in \mathbb{R}^5_{\geq 0} : 2w'_2 > w'_1 + w'_3, w'_2 + w'_4 > w'_1 + w'_5, w'_2 + w'_5 > w'_3 + w'_4, w'_3 + 2w'_4 > w'_1 + 2w'_5\}$. To obtain the closure $\overline{C[\mathbf{w}]}$ we turn the strict inequalities in the above set into non-strict inequalities.

Note that I is homogeneous with respect to the standard degree $\deg(a) = \deg(b) = \deg(c) = \deg(d) = \deg(e) = 1$, so we can also consider $C_h[\mathbf{w}]$. The vectors $\{(1,1,1,1,1), (0,0,0,1,1), (0,1,2,0,1)\}$ are all contained in the lineality space of $C_h[\mathbf{w}]$, so we can project onto the orthogonal subspace spanned by $(1,-1,0,-1,1)$ and $(1,-2,1,0,0)$. In these coordinates we have $C_h[\mathbf{w}] = \{(s,t) : s > 3t, 0 > s\}$. To see this, note that only the last two inequalities above are facet-defining for $\overline{C_h[\mathbf{w}]}$. The first of these then turns into $(-s-2t)+s > t+(-s)$, which simplifies to $s > 3t$, and similarly for the second inequality.

We next note that these cones fit together to form a polyhedral fan.

Proposition 2.4.9. *The set* $\{\overline{C[\mathbf{w}]} : \mathbf{w} \in \mathbb{R}^n_{\geq 0}\}$ *forms a polyhedral fan.*

Proof. We first show that if $\mathbf{w}'$ lies in a face of $\overline{C[\mathbf{w}]}$ with $\mathbf{w}' \notin C[\mathbf{w}]$, then $\overline{C[\mathbf{w}']}$ is a face of $\overline{C[\mathbf{w}]}$. Fix a monomial term order $\prec$ and let $\mathcal{G}$ be the reduced Gröbner basis for I with respect to $\prec_{\mathbf{w}}$. Let $J = \langle \mathrm{in}_{\mathbf{w}'}(g) : g \in \mathcal{G}\rangle$. Since $\mathbf{w}'$ lies in a face of $C[\mathbf{w}]$, we know that $\mathrm{in}_{\mathbf{w}}(g) = \mathrm{in}_{\mathbf{w}}(\mathrm{in}_{\mathbf{w}'}(g))$ for all $g \in \mathcal{G}$. This means $\mathrm{in}_{\prec}(\mathrm{in}_{\mathbf{w}}(I)) \subseteq \mathrm{in}_{\prec}(\mathrm{in}_{\mathbf{w}}(J))$ By Lemma 2.4.2 we have $\mathrm{in}_{\prec_{\mathbf{w}}}(I) = \mathrm{in}_{\prec}(\mathrm{in}_{\mathbf{w}}(I))$. Suppose that $J \subsetneq \mathrm{in}_{\mathbf{w}'}(I)$. Then $\mathrm{in}_{\prec_{\mathbf{w}}}(J) \subsetneq \mathrm{in}_{\prec_{\mathbf{w}}}(\mathrm{in}_{\mathbf{w}'}(I)) = \mathrm{in}_{\prec_{\mathbf{w},\mathbf{w}'}}(I)$, where $\prec_{\mathbf{w},\mathbf{w}'}$ is the term order that first compares monomials using $\mathbf{w}'$, and then breaks ties with $\prec_{\mathbf{w}}$. But this means $\mathrm{in}_{\prec_{\mathbf{w}}}(I) \subsetneq \mathrm{in}_{\prec_{\mathbf{w},\mathbf{w}'}}(I)$ is a proper inclusion of initial ideals, which is impossible by Corollary 2.2.3. Thus $J = \mathrm{in}_{\mathbf{w}'}(I)$. Since $\mathrm{in}_{\prec_{\mathbf{w}}}(I) \subseteq \mathrm{in}_{\prec_{\mathbf{w}}}(J) = \mathrm{in}_{\prec_{\mathbf{w},\mathbf{w}'}}(I)$, we conclude, again by Corollary 2.2.3, that they are equal, and so $\mathcal{G}$ is the reduced Gröbner basis for $\prec_{\mathbf{w},\mathbf{w}'}$. This implies that $C[\mathbf{w}'] = \{\mathbf{w}'' \in \mathbb{R}^n_{\geq 0} : \mathrm{in}_{\mathbf{w}''}(g) = \mathrm{in}_{\mathbf{w}'}(g)$ for all $g \in \mathcal{G}\}$ is a face of $C[\mathbf{w}]$.

Now suppose that $\overline{C[\mathbf{w}_1]}$ and $\overline{C[\mathbf{w}_2]}$ are two cones with neither closure contained in the other. Then by the above argument we know that the intersection, which at least contains 0, is a union of common faces of these two closed cones. Since this intersection is convex, it must in fact be one face, so $\overline{C[\mathbf{w}_1]} \cap \overline{C[\mathbf{w}_2]}$ is a face of each. □

Definition 2.4.10. The polyhedral fan $\{\overline{C[\mathbf{w}]} : \mathbf{w} \in \mathbb{R}^n_{\geq 0}\}$ is called the *Gröbner fan* of the ideal I. If I is homogeneous with respect to some positive grading, as in Remark 2.4.7, then the homogeneous Gröbner fan is the fan $\{\overline{C_h[\mathbf{w}]} : \mathbf{w} \in \mathbb{R}^n_{\geq 0}\}$. This is a complete fan.

From now until the end of this section we assume that I is homogeneous with respect to a positive grading, which allows us to use the forms of the definitions in Remark 2.4.7.

Proposition 2.4.11. *Suppose I is homogeneous with respect to some positive grading $\deg(x_i) = p_i > 0$. Then the homogeneous Gröbner fan of I is polytopal.*

Proof. For a monomial ideal M, we define $(\sum M)_d$ to be the vector $\sum \mathbf{u}$ where the sum is over all vectors $\mathbf{u}$ with $\mathbf{x}^{\mathbf{u}} \in M$ with $\sum_i p_i u_i = d$. The hypothesis that the grading is positive is used here to guarantee that this sum is finite. Define

$$\mathrm{State}_d(I) = \mathrm{conv}\left(\left\{\left(\sum J\right)_d : J \text{ is a monomial initial ideal of } I\right\}\right).$$

Set D to be the largest degree (using the weights p_i) of any element of a universal Gröbner basis for I. We then set

$$\mathrm{State}(I) = \sum_{d=1}^{D} \mathrm{State}_d(I),$$

where the sum is the Minkowski sum of polytopes.

To prove the proposition it suffices to prove that the outer normal fan to $\mathrm{State}(I)$ is the Gröbner fan of I. To show that two polyhedral fans are the same it suffices to show that their maximal open cones are the same. So we need only prove the claim that for *generic* weight vectors $\mathbf{w}, \mathbf{w}' \in \mathbb{R}^n$ we have $\mathrm{in}_{\mathbf{w}}(I) = \mathrm{in}_{\mathbf{w}'}(I)$ if and only if the vertices of $\mathrm{State}(I)$ maximizing $\mathbf{w}$ and $\mathbf{w}'$ are the same. The notion of genericity we are using here is that $\mathbf{w}$ is generic if both $\mathrm{in}_{\mathbf{w}}(I)$ is a monomial ideal, and $\mathrm{face}_{\mathbf{w}}(\mathrm{State}(I))$ is a vertex of $\mathrm{State}(I)$.

For the "only-if" direction, suppose that $\mathrm{in}_{\mathbf{w}}(I) = \mathrm{in}_{\mathbf{w}'}(I)$. This means that $(\mathrm{in}_{\mathbf{w}}(I))_d = (\mathrm{in}_{\mathbf{w}'}(I))_d$ for $d = 1 \dots D$. Lemma 2.4.12, which follows this proof, thus says that the faces of $\mathrm{State}_d(I)$ maximizing $\mathbf{w}$ and $\mathbf{w}'$ are the same. This direction now follows from the fact that the faces of $\mathrm{State}(I)$ maximizing $\mathbf{w}$ and $\mathbf{w}'$ are the Minkowski sums of the faces of $\mathrm{State}_d(I)$ maximizing $\mathbf{w}$ and $\mathbf{w}'$ respectively. Indeed, if $\mathbf{v}_d \in \mathrm{State}_d(I)$ satisfies $\mathbf{w} \cdot \mathbf{v}_d \geq \mathbf{w} \cdot \mathbf{y}_d$ for all $\mathbf{y}_d \in \mathrm{State}_d(I)$, for $1 \leq d \leq D$, then $\mathbf{w} \cdot \sum_{d=1}^{D} \mathbf{v}_d \geq \mathbf{w} \cdot \mathbf{y}$, where $\mathbf{y} = \sum_{d=1}^{D} \mathbf{y}_d$ for $\mathbf{y}_d \in \mathrm{State}_d(I)$ is an arbitrary point of $\mathrm{State}(I)$.

For the "if" direction, suppose that the vertices of $\mathrm{State}(I)$ maximizing $\mathbf{w}$ and $\mathbf{w}'$ coincide. Properties of Minkowski sums now imply that the vertices of each $\mathrm{State}_d(I)$

maximizing $\mathbf{w}$ must coincide. Since $\mathbf{w}, \mathbf{w}'$ were chosen to be generic, $\mathrm{in}_{\mathbf{w}}(I)$ and $\mathrm{in}_{\mathbf{w}'}(I)$ are honest monomial initial ideals of I. Corollary 2.4.13, which follows this proof, now implies that $\mathrm{in}_{\mathbf{w}}(I) = \mathrm{in}_{\mathbf{w}'}(I)$, which completes the proof. □

Lemma 2.4.12. *For all* $\mathbf{w} \in \mathbb{R}^n$ *we have* $\mathrm{face}_{\mathbf{w}}(\mathrm{State}_d(I)) = \mathrm{State}_d(\mathrm{in}_{\mathbf{w}}(I))$.

Proof. We first prove the lemma in the case where $\mathbf{w}$ is generic, so $\mathrm{in}_{\mathbf{w}}(I)$ is a monomial ideal, and the face of $\mathrm{State}_d(I)$ is a vertex. We denote by I_d the vector space of homogeneous polynomials of degree d in I. Let $\mathbf{x}^{\mathbf{a}_1}, \ldots, \mathbf{x}^{\mathbf{a}_m}$ be the monomials of degree d in S, and let $r = \dim_{\mathbf{k}}(I_d) \leq m$. Let $\prec$ be a term order for which $\mathrm{in}_{\prec}(I) = \mathrm{in}_{\mathbf{w}}(I)$. We may assume that $\mathbf{x}^{\mathbf{a}_1}, \ldots, \mathbf{x}^{\mathbf{a}_r}$ are the monomials in $\mathrm{in}_{\prec}(I)$. Since every initial ideal of a monomial ideal is itself, we have $\mathrm{State}_d(\mathrm{in}_{\mathbf{w}}(I)) = \mathbf{a}_1 + \cdots + \mathbf{a}_r$. Lemma 2.2.2 implies the existence of polynomials $\mathbf{x}^{\mathbf{a}_i} - \sum_{j=r+1}^{m} c_{ij}\mathbf{x}^{\mathbf{a}_j} \in I_d$ for $1 \leq i \leq r$. In each of these polynomials $\mathbf{x}^{\mathbf{a}_i}$ is the leading term with respect to the weight vector $\mathbf{w}$.

By the construction of $\mathrm{State}_d(I)$, we know that there is some term order $\prec'$ for which the face of $\mathrm{State}_d(I)$ maximized by $\mathbf{w}$ is $(\sum \mathrm{in}_{\prec'}(I))_d$. Let $\mathbf{x}^{\mathbf{a}_{i_1}}, \ldots, \mathbf{x}^{\mathbf{a}_{i_r}}$ be the monomials of degree d in $\mathrm{in}_{\prec'}(I)$.

If the lemma were not true we would have $\mathbf{w} \cdot (\mathbf{a}_{i_1} + \cdots + \mathbf{a}_{i_r}) > \mathbf{w} \cdot (\mathbf{a}_1 + \cdots + \mathbf{a}_r)$. This would mean that $\mathbf{w} \cdot (\sum_{k \neq i_j} \mathbf{a}_k) < \mathbf{w} \cdot (\sum_{k>r} \mathbf{a}_k)$ for the two bases $\mathcal{B}_1 = \{\mathbf{x}^{\mathbf{a}_k} : \mathbf{x}^{\mathbf{a}_k} \notin \mathrm{in}_{\prec'}(I)\}$ and $\mathcal{B}_2 = \{\mathbf{x}^{\mathbf{a}_k} : k > r\}$ of $(S/I)_d$. We will obtain a contradiction to this statement by constructing a path of bases for this vector space from $\mathcal{B}_1$ to $\mathcal{B}_2$ each of which differs from the previous one by one element, and consists of elements from $\mathcal{B}_1 \cup \mathcal{B}_2$. These bases will have the property that the dot product with $\mathbf{w}$ of their sum decreases at each step, which will give the desired contradiction.

Suppose we have constructed a path from $\mathcal{B}_1$ to a basis $\mathcal{B}$. Let $\mathbf{x}^{\mathbf{a}_k}$ be an element of $\mathcal{B} \setminus \mathcal{B}_2$ (so $k \leq r$), and let $f = \mathbf{x}^{\mathbf{a}_k} - \sum_{j=r+1}^{m} c_{kj}\mathbf{x}^{\mathbf{a}_j}$ be the corresponding element of I_d. Let $\mathcal{C} = \{\mathbf{x}^{\mathbf{a}_j} : c_{kj} \neq 0\}$. The collection $\mathcal{C}$ is linearly independent modulo I, since it is a subset of $\mathcal{B}_2$. Each element $\mathbf{x}^{\mathbf{a}_j}$ of $\mathcal{C}$ can be written as a linear combination of elements of $\mathcal{B}$. If none of these linear combinations involved the element $\mathbf{x}^{\mathbf{a}_k}$, then $\mathcal{C} \cup \{\mathbf{x}^{\mathbf{a}_k}\}$ would be a linearly independent set, which is not the case since f is a linear dependency relation. Thus there is some $\mathbf{x}^{\mathbf{a}_i}$ that equals $c\mathbf{x}^{\mathbf{a}_k} + \sum_{\mathbf{x}^j \in \mathcal{B}, j \neq k} b_j \mathbf{x}^j$ where $c \neq 0$. This means that $\mathcal{B}' = \mathcal{B} \setminus \{\mathbf{x}^{\mathbf{a}_k}\} \cup \{\mathbf{x}^{\mathbf{a}_i}\}$ has the same span as $\mathcal{B}$, and thus is also a basis. Note that $\mathbf{w} \cdot (\sum_{\mathbf{x}^{\mathbf{a}_j} \in \mathcal{B}'} \mathbf{a}_j) < \mathbf{w} \cdot (\sum_{\mathbf{x}^{\mathbf{a}_j} \in \mathcal{B}} \mathbf{a}_j)$. Thus comparing along the constructed path from $\mathcal{B}_1$ to $\mathcal{B}_2$ we see that $\mathbf{w} \cdot (\sum_{\mathbf{x}^{\mathbf{a}_k} \in \mathcal{B}_1} \mathbf{a}_k) < \mathbf{w} \cdot (\sum_{\mathbf{x}^{\mathbf{a}_k} \in \mathcal{B}_2} \mathbf{a}_k)$. From this contradiction we conclude that the two bases are the same, and so the face of $\mathrm{State}_d(I)$ maximizing $\mathbf{w}$ is $\mathrm{State}_d(\mathrm{in}_{\mathbf{w}}(I))$ for generic $\mathbf{w}$.

To complete the proof for nongeneric $\mathbf{w}$ it suffices to show that the face of $\mathrm{State}_d(I)$ maximizing $\mathbf{w}$ has the same vertices as $\mathrm{State}_d(\mathrm{in}_{\mathbf{w}}(I))$. This follows from the generic case by considering the face of each polytope with respect to a generic weight vector $\mathbf{w}'$. □

Corollary 2.4.13. If $\prec$, $\prec'$ are two distinct term orders then $\mathrm{in}_{\prec}(I)_d \neq \mathrm{in}_{\prec'}(I)_d$ implies that $(\sum \mathrm{in}_{\prec}(I))_d \neq (\sum \mathrm{in}_{\prec'}(I))_d$.

Proof. Let the monomials of degree d in $\mathrm{in}_{\prec}(I)$ be $\{\mathbf{x}^{\mathbf{a}_1}, \ldots, \mathbf{x}^{\mathbf{a}_r}\}$ and the monomials of degree d in $\mathrm{in}_{\prec'}(I)$ be $\{\mathbf{x}^{\mathbf{a}_{i_1}}, \ldots, \mathbf{x}^{\mathbf{a}_{i_r}}\}$, and pick $\mathbf{w} \in \mathbb{R}^n_{\geq 0}$ with $\mathrm{in}_{\mathbf{w}}(I) = \mathrm{in}_{\prec}(I)$. In

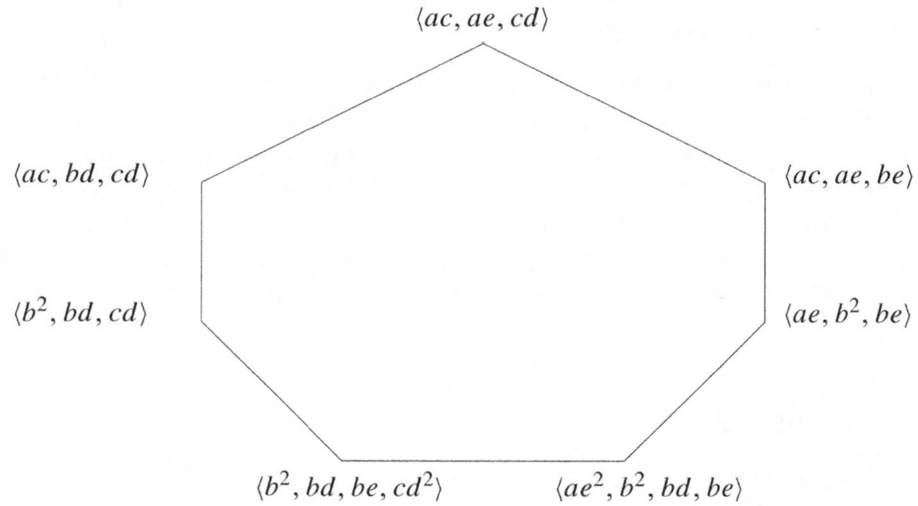

Figure 2.4. A state polytope for the ideal of Example 2.4.15

the proof of Lemma 2.4.12 we showed that $\mathbf{w}\cdot(\mathbf{a}_1+\cdots+\mathbf{a}_r) < \mathbf{w}\cdot(\mathbf{a}_{i_1}+\cdots+\mathbf{a}_{i_r})$. Since $(\sum \mathrm{in}_{\prec}(I))_d = \mathbf{a}_1+\cdots+\mathbf{a}_r$ and $(\sum \mathrm{in}_{\prec'}(I))_d = \mathbf{a}_{i_1}+\cdots+\mathbf{a}_{i_r}$, this shows that they are not the same. □

Definition 2.4.14. A polytope P whose normal fan is the Gröbner fan of an ideal $I \subseteq S$ is called a *state polytope* of I.

Example 2.4.15. Let $S = \mathbf{k}[a, b, c, d, e]$, and let I be the ideal $\langle ac-b^2, ae-bd, be-cd\rangle$.

The ideal I has seven initial ideals.

1. $\langle ac, ae, cd\rangle$
2. $\langle ac, bd, cd\rangle$
3. $\langle b^2, bd, cd\rangle$
4. $\langle b^2, bd, be, cd^2\rangle$
5. $\langle ae^2, b^2, bd, be\rangle$
6. $\langle ae, b^2, be\rangle$
7. $\langle ac, ae, be\rangle$

This list can be computed using Gfan [Jenb]. A state polytope for I is shown in Figure 2.4. As in Example 2.4.8 the Gröbner fan has a three-dimensional lineality space, so we draw a two-dimensional polytope.

Remark 2.4.16. We note that while the Gröbner fan is defined for any ideal in the polynomial ring, the hypothesis in Proposition 2.4.11 that the ideal is homogeneous with respect to some positive grading cannot be removed. In [Jen05] Anders Jensen gives an example of an ideal with three generators in the polynomial ring $\mathbb{Q}[a, b, c, d]$ whose Gröbner fan is not the normal fan of any polyhedron. Positive grading is not strictly a necessary condition for the Gröbner fan to be the normal fan of a polyhedron, as there are examples of non-positively graded ideals with with this property.

2.5 Further reading

The material covered in this chapter is taken from Chapters 2 and 3 of [Stu96]. An excellent introduction to polytope theory is [Zie95]. Term orders in the polynomial ring were first classified by Robbiano in [Rob86]. The existence of the state polytope and the corresponding Gröbner fan was shown in two papers by Bayer and Morrison [BM88] and Mora and Robbiano [MR88] which both appeared in the same issue of the Journal of Symbolic Computation. Anders Jensen has new code Gfan [Jenb] to compute the list of all reduced Gröbner bases of a given ideal, and thus the Gröbner fan.

2.6 Tutorial 2

2.6.1 Polytopes and cones

The exercises in this section are designed to provide practice working with polytopes, polyhedral cones, and normal fans.

1. Consider the collection of points in $\mathbb{R}^2$:

$$S = \{(0,0), (-2,1), (0,2), (1,2), (2,4), (-1,1), (0,3), (1,0)\}.$$

   a) Draw the convex hull of S. What are the vertices?
   b) Find generators for all of the cones in the normal fan of the polygon you drew in part a).

2. Consider the convex hull of the points

$$\{(0,0,1), (1,0,0), (0,1,0), (0,-1,0), (-1,0,0), (0,0,-1)\} \subset \mathbb{R}^3.$$

   How many faces are there of dimension $0, 1, 2$, and 3 respectively? Describe the normal fan.
3. Prove that there is an inclusion reversing correspondence between the faces of a polytope $P \subset \mathbb{R}^n$ and the cones in its normal fan. Specifically, if $F_1 \subset F_2$ are two faces of P, then $\overline{C[F_1]} \supset \overline{C[F_2]}$. More generally for any two faces F_1 and F_2 show that $C[F_1 \cap F_2]$ is the smallest cone whose closure contains both $C[F_1]$ and $C[F_2]$.
4. Prove that if P is a polytope of dimension d in $\mathbb{R}^n$, then the closures of all the n-dimensional cones in the normal fan of P share a lineality space of dimension $n-d$.

2.6.2 Gröbner fan and state polytope

5. Given a non-zero polynomial $f = \sum_{i=1}^{m} c_i \mathbf{x}^{\mathbf{a}_i} \in S$, recall that $\mathrm{supp}(f) = \{\mathbf{a}_i \mid c_i \neq 0\}$. We call $\mathrm{New}(f) := \mathrm{conv}(\mathrm{supp}(f))$, the *Newton polytope* of f. Let $\mathrm{Vert}(f)$ be the subset of $\mathrm{supp}(f)$ consisting of the vertices of $\mathrm{New}(f)$.

(a) Prove that if $\mathbf{a} \in \text{supp}(f) \setminus \text{Vert}(f)$, then there is no weight vector $\mathbf{w}$ such that $\text{in}_{\mathbf{w}}(f) = \mathbf{x}^{\mathbf{a}}$.

(b) Let $f = 3x^6y^2 + 2x^3y^3 - xy + 5x^3y^5 \in \mathbf{k}[x, y]$.

1) Find a weight vector $\mathbf{w}$ such that $\text{in}_{\mathbf{w}}(f) = x^6y^2$.
2) Find a weight vector $\mathbf{w}$ such that $\text{in}_{\mathbf{w}}(f) = x^3y^5$.
3) Show that $\text{in}_{\succ}(f) \neq xy$ for every term ordering $\succ$.
4) Show that $\text{in}_{\succ}(f) \neq x^3y^3$ for every term ordering $\succ$.

(c) If $f \in S$ is a homogeneous polynomial, then what is the state polytope of the principal ideal $\langle f \rangle$?

In the next two exercises you will need to compute the cones of Gröbner fans. Let $\mathcal{G} = \{g_1, \ldots, g_r\}$ be a reduced Gröbner basis of an ideal I with respect to some term order $\prec$. We can write $g_i = \mathbf{x}^{\mathbf{a}_i} + \sum c_{ij}\mathbf{x}^{\mathbf{b}_{ij}}$ where $\mathbf{x}^{\mathbf{a}_i}$ is the leading term of g_i and the various $\mathbf{x}^{\mathbf{b}_{ij}}$ are the other non-leading monomials in g_i. In the proof of Proposition 2.4.4 it is shown that these g_i suffice to determine the corresponding cone in the Gröbner fan. We can write the corresponding inequalities in the homogeneous setting as:

$$\begin{aligned} C[\prec] &= \{\mathbf{w} \in \mathbb{R}^n \ : \ \text{in}_{\mathbf{w}}(g_i) = \mathbf{x}^{\mathbf{a}_i} \text{ for } i = 1, \ldots, r\} \\ &= \{\mathbf{w} \in \mathbb{R}^n \ : \mathbf{w} \cdot \mathbf{b}_{ij} < \mathbf{w} \cdot \mathbf{a}_i \text{ for } i = 1, \ldots, r \text{ and all } j\}. \end{aligned}$$

The set of inequalities above are called *defining inequalities* for the open polyhedral cone $C[\prec]$. The non-redundant inequalities in this set define the *facets* of the cone $C[\prec]$.

6. Let I be the ideal of Example 2.4.15.

a) For each of the seven given initial ideals find the corresponding reduced Gröbner basis.
b) What is the three dimensional lineality space of the Gröbner fan? Write down a vector space basis.
c) Compute the defining inequalities for each of the cones in the Gröbner fan.
d) Find weight vectors $\mathbf{w} \in \mathbb{N}^5$ such that $\text{in}_{\mathbf{w}}(I)$ is equal to each of the seven initial ideals.

7. Let $I = \langle x^2 + yz, xy + z^2 \rangle \subset \mathbb{Q}[x, y, z]$.

a) Compute the grevlex and lex reduced Gröbner bases of I with $x > y > z$.
b) Find defining inequalities for the corresponding cones in the Gröbner fan.
c) Some of these inequalities are redundant. For each cone find a non-redundant set. (Hint: There is a one-dimensional lineality space and each cone will have two non-redundant inequalities).
d) Find all of the remaining initial ideals and cones in the Gröbner fan. One way to proceed is to compute lex and grevlex Gröbner bases with respect to different variable orderings then check if the corresponding cones cover. In section 2.6.3 of this tutorial we will see a more systematic way to compute the Gröbner fan.

e) Draw the 2-dimensional state polytope of I with vertices labeled by initial ideals, as in Figure 2.4.

2.6.3 The Gröbner walk

One of the most important applications of the Gröbner fan is the *Gröbner walk* [CKM97] (see also [Stu96, Chapter 3]), which is a general algorithm for converting a Gröbner basis from one term order to another. The lexicographic order is very useful when eliminating variables but its Gröbner bases are notoriously difficult to compute. The Gröbner walk allows us to start with an easier order such as graded reverse lex and transform the corresponding Gröbner basis.

Suppose we have a Gröbner basis $\mathcal{G}$ for I with respect to some starting order $\prec_s$. Let $\mathbf{w}_s$ be a weight vector realizing this order. The goal will be to compute a new Gröbner basis with respect to a target order $\prec_t$ represented by $\mathbf{w}_t$. To that end we pick a path from $\mathbf{w}_s$ to $\mathbf{w}_t$, which in sufficiently nice situations can be taken to be a straight line. We will assume that our path only passes through codimension-one cones of the Gröbner fan, and never enters lower-dimensional cones.

The Gröbner walk has two basic steps.

- Cross from one open cone C_i in the Gröbner fan to another C_{i+1} along the chosen path.
- Modify the current Gröbner basis to respect the new cone.

For the first step, suppose we are in some open cone C_i for which we know the Gröbner basis $\mathcal{G}_i$. Let $\mathbf{w}_{new}$ be the last point in $\overline{C_i}$ along the path, computed by intersecting the path with each facet of C_i.

For the second step consider the ideal $\langle \mathrm{in}_{\mathbf{w}_{new}}(\mathcal{G}_i)\rangle$ of initial forms of $\mathcal{G}_i$ with respect to the new weight vector. Since $\mathbf{w}_{new}$ is on the boundary of C_i, $\langle \mathrm{in}_{\mathbf{w}_{new}}(\mathcal{G}_i)\rangle$ will not be a monomial ideal, but for nice cases there will be mostly monomials with a couple of other polynomials (ideally just one binomial).

Define a new term order $\prec_{i+1}$ using the weight vector $\mathbf{w}_{new}$ but breaking ties with our target order $\prec_t$. The remarkable fact is that to compute the Gröbner basis $\mathcal{G}_{i+1}$ of I with respect to $\prec_{i+1}$ it is enough to compute the Gröbner basis of the far simpler ideal $\langle \mathrm{in}_{\mathbf{w}_{new}}(\mathcal{G}_i)\rangle = \mathrm{in}_{\mathbf{w}_{new}}(I)$.

Proposition 2.6.1 ([Stu96, page 23]). *Let $H = \{h_1, \ldots, h_s\}$ be a Gröbner basis of $\langle \mathrm{in}_{\mathbf{w}_{new}}(G_i)\rangle$ with respect to $\prec_{i+1}$. Write*

$$h_i = \sum_{g \in \mathcal{G}_i} p_g \, \mathrm{in}_{\mathbf{w}_{new}}(g),$$

then the set of all polynomials

$$\overline{h}_i = \sum_{g \in \mathcal{G}_i} p_g g$$

is a Gröbner basis for I with respect to $\prec_{i+1}$.

8. Use the Gröbner walk to convert the grevlex Gröbner basis to lex in Exercise 7a).

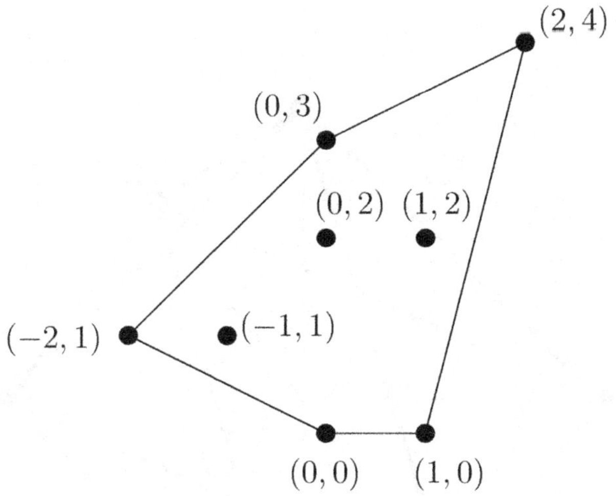

Figure 2.5.

2.7 Solutions to Tutorial 2

2.7.1 Polytopes and cones

1. (a) The convex hull is the pentagon shown in Figure 2.5.

   The points $(0, 0)$, $(1, 0)$, $(2, 4)$, $(0, 3)$, and $(-2, 1)$ are the vertices and the remaining three points $(0, 2)$, $(1, 2)$, and $(-1, 1)$ are in the interior.

   (b) There are five one-dimensional cones (rays) in the normal fan. They are spanned in counterclockwise order by $(-1, 2)$, $(-1, 1)$, $(-1, -2)$, $(0, -1)$, and $(4, -1)$. The five two-dimensional cones are generated by consecutive pairs of rays. A picture of the normal fan is shown in Figure 2.6

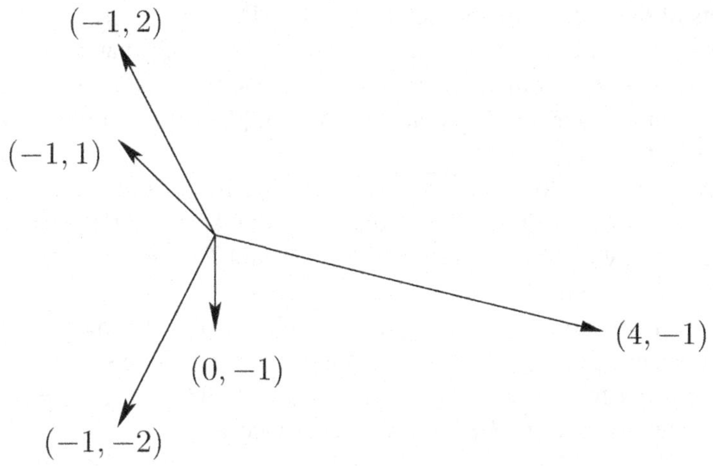

Figure 2.6.

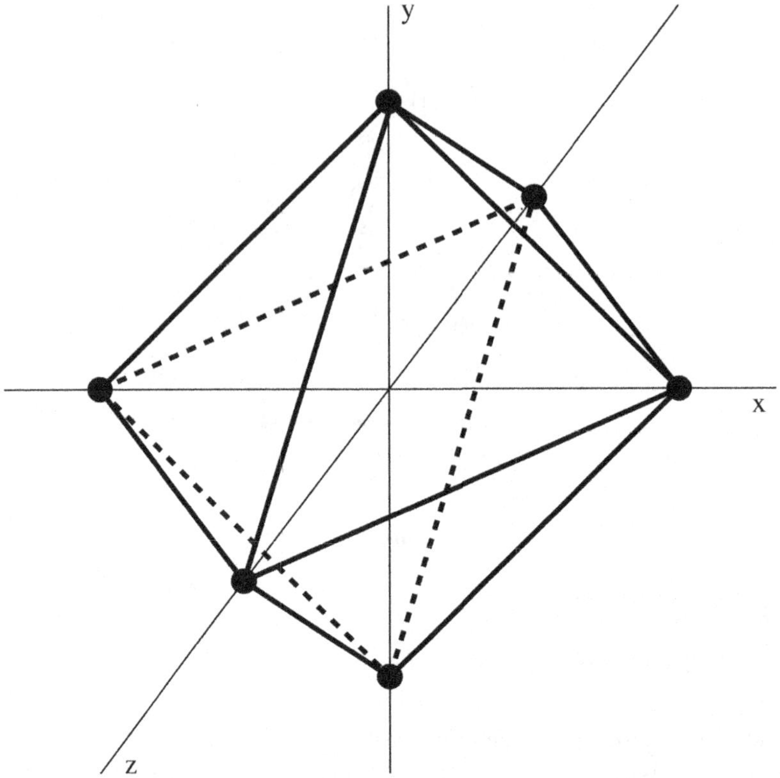

Figure 2.7.

2. The convex hull of the given points is the octahedron shown in Figure 2.7. There are 6 vertices, 12 edges, 8 two-dimensional faces, and 1 three-dimensional face (the whole polytope). The normal fan has three-dimensional cones the cones over the facets of the cube with vertices $(\pm 1, \pm 1, \pm 1)$.
3. The cone $C[F]$ was defined as the set of $\mathbf{c} \in \mathbb{R}^n$ such that $\mathbf{c} \cdot \mathbf{x}$ is maximized for exactly $\mathbf{x} \in F$. To make this a closed set we must allow $\mathbf{c} \cdot \mathbf{x}$ to achieve its maximum on a larger set. Thus the closure of $C[F]$ is the set of $\mathbf{c} \in \mathbb{R}^n$ such that $\text{face}_{\mathbf{c}}(P) \supseteq F$.

   It immediately follows that $\overline{C[F_1]} \supset \overline{C[F_2]}$ if and only if $F_1 \subset F_2$. More generally the closure of $C[F_1 \cap F_2]$ contains both $C[F_1]$ and $C[F_2]$, and conversely any other $C[F]$ with this property has $F \subset F_1$ and $F \subset F_2$, thus $F \subset F_1 \cap F_2$ and $\overline{C[F]} \supset \overline{C[F_1 \cap F_2]}$.
4. If $P \subset \mathbb{R}^n$ is of dimension d, it spans an affine space of dimension d. Let V be the linear space of dimension $n - d$ orthogonal to this affine space. For any $\mathbf{v} \in V$, $\mathbf{v} \cdot \mathbf{p}$ is constant for $\mathbf{p} \in P$. Hence, $\text{face}_{\mathbf{v}}(P) = P$. Therefore, $V = C[P]$. By the previous exercise $V \subset C[F]$ for any smaller face F.

2.7.2 Gröbner fan and state polytope

5. (a) Let $V = \{\mathbf{v}_1, \mathbf{v}_2, \ldots \mathbf{v}_r\} \subset \mathbb{Q}^n$ and $P = \text{conv}(\mathbf{v}_i : 1 \leq i \leq r)$. We use the fact that every element $\mathbf{v}$ of $\mathcal{P} \cap \mathbb{Q}^n$ has a representation $\mathbf{v} = \sum_{i=1}^{r} \lambda_i \mathbf{v}_i$ with $\lambda_1, \lambda_2, \ldots, \lambda_r \in \mathbb{Q}_{\geq 0}$ and $\lambda_1 + \lambda_2 + \ldots + \lambda_r = 1$. Let $U = \text{Vert (New } (f))$ and let $\mathbf{a} \in \text{supp } (f) \setminus U$. Then we can write $\mathbf{a} = \sum_{\mathbf{v} \in U} \lambda_{\mathbf{v}} \mathbf{v}$ with $\lambda_{\mathbf{v}} \in \mathbb{Q}_{\geq 0}$ and $\sum_{\mathbf{v} \in U} \lambda_{\mathbf{v}} = 1$. Let the least common multiple of the denominators be L. Then we have $L\mathbf{a} = \sum_{\mathbf{v} \in U} l_{\mathbf{v}} \mathbf{v}$, where $l_{\mathbf{v}} \in \mathbb{N}$ and $\sum_{\mathbf{v}} l_{\mathbf{v}} = L$. If $\mathbf{x}^{\mathbf{a}}$ is the leading term then $\mathbf{x}^{\mathbf{a}} \succ \mathbf{x}^{\mathbf{v}}$ for any $\mathbf{v} \in V$. This implies $\mathbf{x}^{L\mathbf{a}} \succ \mathbf{x}^{\sum_{\mathbf{v} \in V} l_{\mathbf{v}} \mathbf{v}}$ (by the multiplicative axiom), which is a contradiction. Thus $\mathbf{x}^{\mathbf{a}}$ cannot be the leading term under any term order.

   (b) 1) This is true for the lex ordering with $x > y$, so we take $\mathbf{w} = (10, 1)$.
   2) This is true for the lex ordering with $y > x$, so we take $\mathbf{w} = (1, 10)$.
   3) For any order $\prec$ we must have $1 \prec x^2y^2$, so $xy \prec x^3y^3$.
   4) For any order $\prec$ we must have $1 \prec y^2$, so $x^3y^3 \prec x^3y^5$.

   (c) Write $f = \sum c_i \mathbf{x}^{\mathbf{a}_i}$. We will show that if f is homogeneous of degree d, the initial ideals of f correspond exactly to the vertices of the Newton polytope. If that is the case then the Newton polytope is equal to $\text{State}_d(\langle f \rangle)$ which is equal to $\text{State}(\langle f \rangle)$ since the set consisting of the generator f of degree d is itself a universal Gröbner basis.

   The vertices of the Newton polytope of f are those $\mathbf{a}_i$ such that there exists $\mathbf{w} \in \mathbb{R}^n$ such that $\mathbf{w} \cdot \mathbf{a}_i > \mathbf{w} \cdot \mathbf{a}_j$ for all $j \neq i$. That is to say $\text{in}_{\mathbf{w}}(f) = \mathbf{x}^{\mathbf{a}_i}$. Therefore, any leading monomial of f (corresponding to $\mathbf{w}$ nonnegative) is a vertex of the Newton polytope. Conversely, since f is homogeneous of degree d, $(1, \ldots, 1) \cdot a_j = d$ for all j. Therefore, we can take $\mathbf{w}' = \mathbf{w} + (k, \ldots, k)$ and still have $\text{in}_{\mathbf{w}'}(f) = \mathbf{x}^{\mathbf{a}_i}$. Taking k sufficiently large, $\mathbf{w}'$ will be nonnegative. This results fails if f is not homogeneous as can be seen by considering the above nonhomogeneous polynomial.

6. a) The seven Gröbner bases, with leading terms underlined, are:

   (a) $\{\underline{ac} - b^2, \underline{ae} - bd, \underline{cd} - be\}$
   (b) $\{\underline{ac} - b^2, \underline{bd} - ae, \underline{cd} - be\}$
   (c) $\{\underline{b^2} - ac, \underline{bd} - ae, \underline{cd} - be\}$
   (d) $\{\underline{b^2} - ac, \underline{bd} - ae, \underline{be} - cd, \underline{cd^2} - ae^2\}$
   (e) $\{\underline{ae^2} - cd^2, \underline{b^2} - ac, \underline{bd} - ae, \underline{be} - cd\}$
   (f) $\{\underline{b^2} - ac, \underline{ae} - bd, \underline{be} - cd\}$
   (g) $\{\underline{ac} - b^2, \underline{ae} - bd, \underline{be} - cd\}$

   These can be computed using the software package Gfan [Jenb], or can be done by hand by finding polynomials in I with the given leading terms.

   b) The lineality space consists of those $\mathbf{w}$ such that $\text{in}_{\mathbf{w}}(I) = I$ (do you see why?). Thus, for each binomial generator of I, the two monomials have equal weight. So, $w_1 + w_3 = 2w_2$, $w_1 + w_5 = w_2 + w_4$, and $w_3 + w_4 = w_2 + w_5$. The solution

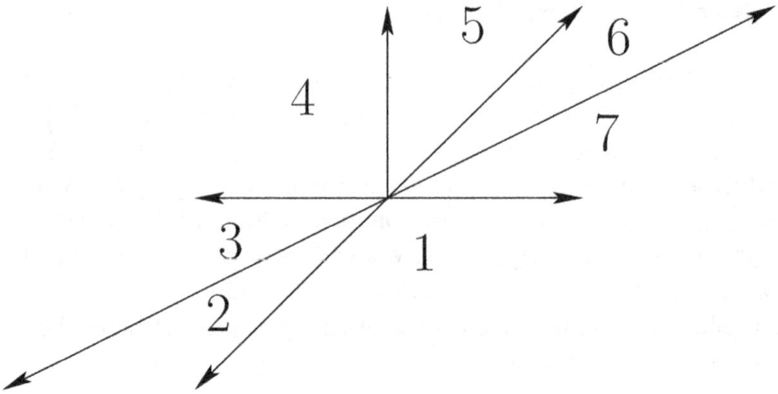

Figure 2.8.

space is w_3, w_4, w_5 arbitrary, $w_1 = w_3+2w_4-2w_5$, and $w_2 = w_3+w_4-w_5$. A vector space basis is $\{(1,1,1,0,0), (2,1,0,1,0), (-2,-1,0,0,1)\}$.

c) To simplify computation we quotient out by the lineality space by making the change of coordinates $w_1' = w_1 - w_3 - 2w_4 + 2w_5$ and $w_2' = w_2 - w_3 - w_4 + w_5$. Therefore, any point $(w_1, w_2, w_3, w_4, w_5)$ can be transformed into $(w_1', w_2', 0, 0, 0)$ by adding something in the lineality space.

The defining inequalities for the first initial ideal $\langle ac, ae, cd\rangle$ are

$$w_1' > 2w_2',\ w_1' > w_2',\ \text{and } 0 > w_2'.$$

The first inequality is redundant (the sum of the second and third) so the cone is defined by $w_1' > w_2'$ and $w_2' < 0$. Similarly, for each of the other 6 initial ideals, removing redundant inequalities, we get

(b) $w_1' > 2w_2',\ w_2' > w_1'$
(c) $2w_2' > w_1',\ w_2' < 0$
(d) $w_1' < 0,\ w_2' > 0$
(e) $w_1' > 0,\ w_2' > w_1'$
(f) $w_1' < 2w_2',\ w_1' > w_2'$
(g) $w_1' > 2w_2',\ w_2' > 0.$

A picture of the Gröbner fan in the plane with coordinates w_1' and w_2' is shown in Figure 2.8

d) Using the inequalities found above we can find weight vectors of the form $(w_1', w_2', 0, 0, 0)$ in each region. These may not be nonnegative so we can add an appropriate multiple of $(1,1,1,1,1)$ to get the following **w**:

(a) $(2, 0, 1, 1, 1)$
(b) $(0, 1, 3, 3, 3)$
(c) $(0, 2, 3, 3, 3)$
(d) $(0, 2, 1, 1, 1)$
(e) $(1, 2, 0, 0, 0)$

(f) (3, 2, 0, 0, 0)
(g) (3, 1, 0, 0, 0)

7. a) The grevlex basis is $\langle x^2+yz, xy+z^2, y^2z-xz^2\rangle$. The lex basis is $\langle x^2+yz, xy+z^2, xz^2-y^2z, y^3z+z^4\rangle$. These are computed by the following Macaulay 2 code:

```
i1 : R = QQ[x,y,z];
i2 : I = ideal(x^2 + y*z, x*y + z^2);
o2 : Ideal of R
i3 : gb(I)

o3 = | xy+z2 x2+yz y2z-xz2 |

i4 : R1 = QQ[x,y,z, MonomialOrder=>Lex];
i5 : I1 = substitute(I, R1);
i6 : gb(I1)

o6 = | xy+z2 x2+yz xz2-y2z y3z+z4 |
```

b) For the grevlex basis the defining inequalities are:

$$w_1 + w_2 > 2w_3, \quad 2w_1 > w_2 + w_3, \quad 2w_2 > w_1 + w_3$$

For the lex basis the inequalities are:

$$w_1 + w_2 > 2w_3, \quad 2w_1 > w_2 + w_3, \quad w_1 + w_3 > 2w_2, \quad w_2 > w_3$$

c) Make a change of variables $w_1' = w_1 - w_3$, and $w_2' = w_2 - w_3$ to account for the one-dimensional lineality space arising from the homogeneity. The inequalities for the grevlex cone become

$$w_1' + w_2' > 0, \quad 2w_1' > w_2', \quad 2w_2' > w_1',$$

and the first one is redundant as it is the sum of the second and third. The corresponding cone is generated by the vectors (1, 2) and (2, 1).

	Gröbner basis	Initial ideal
1)	$x^2 + yz, xy + z^2, y^2z - xz^2$	x^2, xy, y^2z
2)	$x^2 + yz, xy + z^2, xz^2 - y^2x, y^3z + z^4$	x^2, xy, xz^2, y^3z
3)	$x^2 + yz, xy + z^2, xz^2 - y^2x, z^4 + y^3z$	x^2, xy, xz^2, z^4
4)	$x^2 + yz, z^2 + xy$	x^2, z^2
5)	$yz + x^2, z^2 + xy, x^2z - xy^2, x^4 + xy^3$	yz, z^2, x^2z, x^4
6)	$yz + x^2, z^2 + xy, x^2z - xy^2, xy^3 + x^4$	yz, z^2, x^2z, xy^3
7)	$yz + x^2, z^2 + xy, xy^2 - x^2z$	yz, z^2, xy^2
8)	$yz + x^2, xy + z^2, z^3 - x^3$	yz, xy, z^3
9)	$yz + x^2, xy + z^2, x^3 - z^3$	yz, xy, x^3

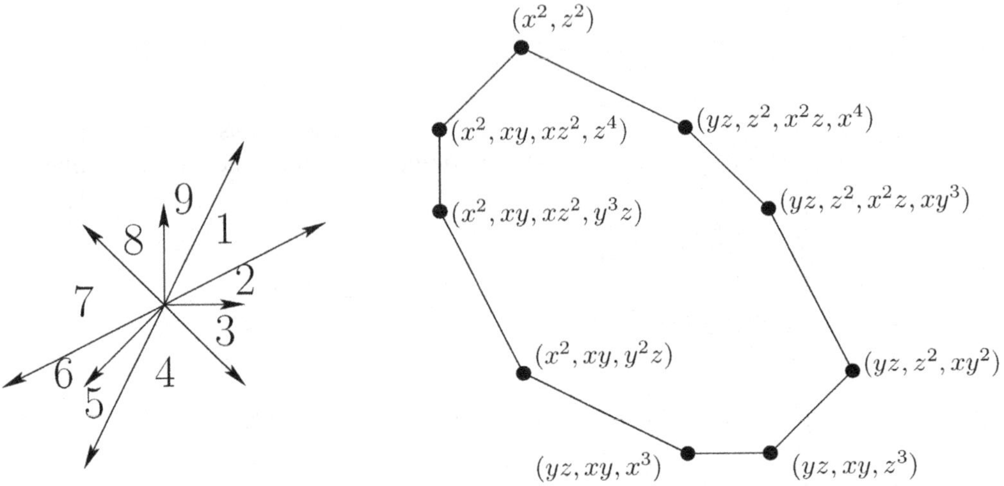

Figure 2.9. Gröbner fan and state polytope

For the lex cone we get:

$$w_1' + w_2' > 0, \quad 2w_1' > w_2', \quad w_1' > 2w_2', \quad w_2' > 0$$

A little thought gives that the last two are nonredundant and the corresponding cone is generated by (2, 1) and (1, 0).

d) and e) There are 9 total reduced Gröbner bases all of which can be found as grevlex or lex orders with respect to some variable ordering. We list the 9 Gröbner bases and corresponding initial ideals. We leave it up to the reader to determine which ideals are determined by which orders.

Computing all the cones yields the Gröbner fan in Figure 2.9. The state polytope is also shown with vertices labeled by the corresponding initial ideals.

2.7.3 The Gröbner walk

8. Weight vectors realizing the grevlex and lex term orders are $(1, 1, 0)$ and $(3, 1, 0)$ respectively. Computing the grevlex cone as before, we find that we hit the boundary of the grevlex cone at $\mathbf{w}_{new} = (2, 1, 0)$. Our starting Gröbner basis is $\mathcal{G} = \{x^2 - yz, xy - z^2, y^2z - xz^2\}$. The ideal of initial forms with respect to $\mathbf{w}_{new}$ is $\langle x^2, xy, y^2z - xz^2\rangle$. We next compute a Gröbner basis of this ideal of initial forms with respect to the new lex order, so the leading term of the binomial $y^2z - xz^2$ is now xz^2. There is one non-trivial S-pair:

$$z^2(xy) + y(-xz^2 + y^2z) = y^3z.$$

Lifting this up to the original ideal yields

$$z^2(xy - z^2) + y(-xz^2 + y^2z) = y^3z - z^4.$$

It is not difficult to check that all other S-pairs reduce to zero.

So, our new Gröbner basis is

$$\{x^2 + yz, xy - z^2, xz^2 - y^2z, y^3z - z^4\}$$

which is already the desired reduced lex basis.

3

Toric Ideals

3.1 Introduction

In this chapter we focus on a special class of polynomial ideals called toric ideals which allow interactions among algebra, geometry and combinatorics. They are the defining ideals of toric varieties which are a rich but fairly accessible class of varieties in algebraic geometry. See Chapter 6 for the connection to algebraic geometry. They also encode the combinatorics of polytopes and vector configurations, and have several applications. This chapter is based on Chapters 4, 5 and 12 of [Stu96].

To motivate the definition of a toric ideal, consider the following problem. Suppose you had an unlimited supply of coins in a certain currency, of denominations 5, 10, 25 and 50, and you would like to answer the following questions for any positive number b.

1. Is there a combination of coins that adds up to b?
2. How many coins of each type would you have to use if you wanted to make the combination in Problem 1 with as few coins as possible?

Problem 2 is the following integer program:

$$\begin{aligned} \text{minimize} \quad & x_1 + x_2 + x_3 + x_4 \\ \text{subject to} \quad & 5x_1 + 10x_2 + 25x_3 + 50x_4 = b \\ & x_1, x_2, x_3, x_4 \in \mathbb{N} \end{aligned}$$

while Problem 1 is asking whether this program has a *feasible* solution, i.e., is $\{(x_1, x_2, x_3, x_4) \in \mathbb{N}^4 : 5x_1 + 10x_2 + 25x_3 + 50x_4 = b\}$ nonempty?

Let us examine the possible ways of making change for $b = 100$ using the four types of coins, or equivalently all the feasible solutions to the above integer program with $b = 100$. There are 40 feasible solutions:

$$\begin{array}{llllll} (0,0,0,2) & (0,0,2,1) & (0,0,4,0) & (1,2,1,1) & (1,2,3,0) & (0,5,0,1) \\ (0,5,2,0) & (3,1,1,1) & (3,1,3,0) & (2,4,0,1) & (2,4,2,0) & (1,7,1,0) \\ (0,10,0,0) & (5,0,1,1) & (5,0,3,0) & (4,3,0,1) & (4,3,2,0) & (3,6,1,0) \end{array}$$

(2, 9, 0, 0)	(6, 2, 0, 1)	(6, 2, 2, 0)	(5, 5, 1, 0)	(4, 8, 0, 0)	(8, 1, 0, 1)
(8, 1, 2, 0)	(7, 4, 1, 0)	(6, 7, 0, 0)	(10, 0, 0, 1)	(10, 0, 2, 0)	(9, 3, 1, 0)
(8, 6, 0, 0)	(11, 2, 1, 0)	(10, 5, 0, 0)	(13, 1, 1, 0)	(12, 4, 0, 0)	(15, 0, 1, 0)
(14, 3, 0, 0)	(16, 2, 0, 0)	(18, 1, 0, 0)	(20, 0, 0, 0)		

In this example, it is easy to see that (0, 0, 0, 2) is the combination that minimizes the number of coins used. This chapter lays the foundation for a general strategy to solve integer programs which will be explained in detail in the tutorial. The method uses toric ideals which is the main topic of this chapter.

Continuing the example, we record each combination as a monomial in the polynomial ring $\mathbf{k}[x_1, x_2, x_3, x_4]$. For instance, (6, 2, 0, 1) is recorded as $x_1^6 x_2^2 x_4$. Letting $\mathcal{A} = \{5, 10, 25, 50\}$, we define the $\mathcal{A}$-degree of a monomial $x_1^{m_1} x_2^{m_2} x_3^{m_3} x_4^{m_4}$ to be $5m_1 + 10m_2 + 25m_3 + 50m_4$. Thus the 40 monomials gotten from the above vectors are precisely all the monomials in $\mathbf{k}[x_1, x_2, x_3, x_4]$ of $\mathcal{A}$-degree $100 = b$. In general, Problem 1 is asking whether the bth graded part of $\mathbf{k}[x_1, x_2, x_3, x_4]$ is nontrivial. Theoretically, the answer is "yes" if and only if b lies in the semigroup $\mathbb{N}\mathcal{A} = \{5x_1 + 10x_2 + 25x_3 + 50x_4 \,:\, x_1, x_2, x_3, x_4 \in \mathbb{N}\}$. How do we check this in practice? More ambitiously, one could ask to enumerate all monomials in $\mathbf{k}[x_1, x_2, x_3, x_4]$ of $\mathcal{A}$-degree b. We have listed the exponents of all such monomials for $b = 100$. How can we do this in practice?

We will see in the exercises that toric ideals and their Gröbner bases give algorithms to answer both Problem 1 and 2.

3.2 Toric ideals

Fix a subset $\mathcal{A} = \{\mathbf{a}_1, \mathbf{a}_2, \ldots, \mathbf{a}_n\} \subset \mathbb{Z}^d \setminus \{\mathbf{0}\}$. We assume that the matrix $A = [\mathbf{a}_1\ \mathbf{a}_2\ \ldots\ \mathbf{a}_n] \in \mathbb{Z}^{d \times n}$ has rank d. Consider the following semigroup homomorphism:

$$\pi \,:\, \mathbb{N}^n \to \mathbb{Z}^d, \quad \mathbf{u} = (u_1, \ldots, u_n) \mapsto \sum_{i=1}^{n} \mathbf{a}_i u_i = A\mathbf{u}.$$

Then $\pi(\mathbb{N}^n) = \{A\mathbf{u} \,:\, \mathbf{u} \in \mathbb{N}^n\} =: \mathbb{N}\mathcal{A}$ is called the *monoid (semigroup)* generated by $\mathcal{A}$. The semigroup ring of $\mathbb{N}^n$ is $\mathbf{k}[\mathbf{x}] = \mathbf{k}[x_1, \ldots, x_n]$, and that of $\mathbb{Z}^d$ is the *Laurent polynomial ring* $\mathbf{k}[\mathbf{t}^{\pm 1}] := \mathbf{k}[t_1^{\pm 1}, \ldots, t_d^{\pm 1}]$. The map π lifts to the ring homomorphism:

$$\hat{\pi} \,:\, \mathbf{k}[\mathbf{x}] \to \mathbf{k}[\mathbf{t}^{\pm 1}], \quad x_j \mapsto \mathbf{t}^{\mathbf{a}_j} := t_1^{a_{1j}} t_2^{a_{2j}} \cdots t_d^{a_{dj}}.$$

Definition 3.2.1. The *toric ideal* of $\mathcal{A}$, denoted as $I_{\mathcal{A}}$, is the kernel of the map $\hat{\pi}$.

In this situation it is natural to grade the polynomial ring $\mathbf{k}[\mathbf{x}]$ by setting $\deg(x_i) = \mathbf{a}_i$ for $i = 1, \ldots, n$. Then the set of all degrees of polynomials in $\mathbf{k}[\mathbf{x}]$ is $\mathbb{N}\mathcal{A}$. A polynomial in $\mathbf{k}[\mathbf{x}]$ is $\mathcal{A}$-homogeneous if it is homogeneous under this multigrading.

Proposition 3.2.2.

1. *The toric ideal $I_{\mathcal{A}}$ is a prime ideal in* $\mathbf{k}[\mathbf{x}]$.

2. *[Stu96, Lemma 4.1, Corollary 4.3] The ideal $I_\mathcal{A}$ is generated as a* **k***-vector space by the infinitely-many binomials* $\{\mathbf{x}^\mathbf{u} - \mathbf{x}^\mathbf{v} : \pi(\mathbf{u}) = \pi(\mathbf{v}), \ \mathbf{u}, \mathbf{v} \in \mathbb{N}^n\}$, *and hence* $I_\mathcal{A} = \langle \mathbf{x}^\mathbf{u} - \mathbf{x}^\mathbf{v} : \pi(\mathbf{u}) = \pi(\mathbf{v})\rangle$.
3. *[Stu96, Lemma 4.2] The ring* $\mathbf{k}[\mathbf{x}]/I_\mathcal{A}$ *has Krull dimension* d.
4. *[Stu96, Corollary 4.4] For every term order* $\succ$ *the reduced Gröbner basis of* $I_\mathcal{A}$ *with respect to* $\succ$ *consists of a finite set of binomials of the form* $\mathbf{x}^\mathbf{u} - \mathbf{x}^\mathbf{v} \in I_\mathcal{A}$.

Proof.

1. Since $\mathbf{k}[\mathbf{x}]/I_\mathcal{A} \cong \hat{\pi}(\mathbf{k}[\mathbf{x}]) = \mathbf{k}[\mathbf{t}^{\mathbf{a}_1}, \ldots, \mathbf{t}^{\mathbf{a}_n}]$, which is an integral domain, $I_\mathcal{A}$ is a prime ideal.
2. First note that a binomial $\mathbf{x}^\mathbf{u} - \mathbf{x}^\mathbf{v} \in I_\mathcal{A}$ if and only if it is $\mathcal{A}$-homogeneous, so $\pi(\mathbf{u}) = \pi(\mathbf{v})$. It therefore suffices to show that each polynomial $f \in I_\mathcal{A}$ is a **k**-linear combination of these binomials. Fix a term order $\succ$ on $\mathbf{k}[\mathbf{x}]$, and suppose f cannot be written as a **k**-linear combination of $\mathcal{A}$-homogeneous binomials. Choose such an f for which $\text{in}_\succ(f) = \mathbf{x}^\mathbf{u}$ is minimal with respect to $\succ$, which we call a "minimal criminal". Since $f \in I_\mathcal{A} = \ker(\hat{\pi})$, $f(\mathbf{t}^{\mathbf{a}_1}, \ldots, \mathbf{t}^{\mathbf{a}_n}) = 0$. In particular, $\hat{\pi}(\mathbf{x}^\mathbf{u}) = \mathbf{t}^{\pi(\mathbf{u})}$ must cancel in this expansion. Hence there is a monomial $\mathbf{x}^\mathbf{v} \prec \mathbf{x}^\mathbf{u}$ in f such that $\pi(\mathbf{u}) = \pi(\mathbf{v})$. Then the polynomial $f' = f - \mathbf{x}^\mathbf{u} + \mathbf{x}^\mathbf{v}$ cannot be written as a **k**-linear combination of binomials in $I_\mathcal{A}$. However since $\text{in}_\succ(f') \prec \text{in}_\succ(f)$, this contradicts the assumption. Minimal criminal arguments are a staple of Gröbner bases proofs.
3. Theorem A in Chapter 8 of [Eis94] states that if R is an affine domain over a field **k** then the Krull dimension of R is the transcendence degree of R over **k**. Thus the Krull dimension of $\mathbf{k}[\mathbf{x}]/I_\mathcal{A} \cong \mathbf{k}[\mathbf{t}^{\mathbf{a}_1}, \ldots, \mathbf{t}^{\mathbf{a}_n}]$ is the maximum number of algebraically independent monomials $\mathbf{t}^{\mathbf{a}_i}$ which in turn is equal to the maximum number of linearly independent vectors in $\mathcal{A}$ which is d by assumption.
4. By the Hilbert basis theorem and part 2 there exists a finite list of binomials from the list $\{\mathbf{x}^\mathbf{u} - \mathbf{x}^\mathbf{v} : \pi(\mathbf{u}) = \pi(\mathbf{v})\}$ that generate $I_\mathcal{A}$. We use this generating set as the input for Buchberger's algorithm. It can be checked that the S-pair of two binomials is again a binomial, and that the normal form of a homogeneous binomial with respect to a set of homogeneous binomials with marked leading terms is also a homogeneous binomial. Thus each reduced Gröbner basis of $I_\mathcal{A}$ computed from the binomial generating set mentioned above consists of a finite set of $\mathcal{A}$-homogeneous binomials. □

Example 3.2.3. For $\mathcal{A} = \{5, 10, 25, 50\}$, $I_\mathcal{A} = \langle x_3^2 - x_4, x_1x_2^2 - x_3, x_1^2 - x_2\rangle$. This toric ideal has 12 distinct reduced Gröbner bases that one can compute using the software package CaTS [Jena] (using the algorithm found in [HT00]). Each reduced Gröbner basis is numbered as a vertex of the state polytope of $I_\mathcal{A}$. In each case, the first list of binomials give the facets of the Gröbner cone and the second list is the reduced Gröbner basis. CaTS uses $a, b, c, \ldots$ for variables in a ring. The following computation requires that the matrix A be stored in a file. See the CaTS homepage for various acceptable formats. The command needed to obtain the following output is `cats -p1 -i filename`.

```
Vtx: 0 (3 facets/4 binomials/degree 3)
{# c^2-d, # b^3-a*c, # a^2-b}
{c^2-d, b^3-a*c, a*b^2-c, a^2-b}
Vtx: 1 (3 facets/4 binomials/degree 3)
{# a^2-b, # b^3-a*c, # d-c^2}
{a^2-b, a*b^2-c, b^3-a*c, d-c^2}
Vtx: 2 (3 facets/3 binomials/degree 5)
{# a^5-c, # b-a^2, # d-c^2}
{a^5-c, b-a^2, d-c^2}
Vtx: 3 (3 facets/5 binomials/degree 5)
{# b^5-c^2, # a*c-b^3, # d-c^2}
{a^2-b, a*b^2-c, b^5-c^2, a*c-b^3, d-c^2}
Vtx: 4 (3 facets/5 binomials/degree 3)
{# a*b^2-c, # c^2-b^5, # d-b^5}
{a^2-b, a*b^2-c, a*c-b^3, c^2-b^5, d-b^5}
Vtx: 5 (3 facets/3 binomials/degree 2)
{# a^2-b, # c-a*b^2, # d-b^5}
{a^2-b, c-a*b^2, d-b^5}
Vtx: 6 (3 facets/3 binomials/degree 1)
{# b-a^2, # c-a^5, # d-a^10}
{b-a^2, c-a^5, d-a^10}
Vtx: 7 (3 facets/6 binomials/degree 5)
{# a*b^2-c, # a*c-b^3, # b^3*c-a*d}
{a^2-b, a*b^2-c, b^5-d, a*c-b^3, b^3*c-a*d, c^2-d}
Vtx: 8 (3 facets/3 binomials/degree 5)
{# a^2-b, # b^5-d, # c-a*b^2}
{a^2-b, b^5-d, c-a*b^2}
Vtx: 9 (3 facets/3 binomials/degree 10)
{# a^10-d, # b-a^2, # c-a^5}
{a^10-d, b-a^2, c-a^5}
Vtx: 10 (3 facets/6 binomials/degree 5)
{# b^5-d, # c^2-d, # a*d-b^3*c}
{a^2-b, a*b^2-c, b^5-d, a*c-b^3, c^2-d, a*d-b^3*c}
Vtx: 11 (3 facets/3 binomials/degree 5)
{# a^5-c, # b-a^2, # c^2-d}
{a^5-c, b-a^2, c^2-d}
```

3.3 Algorithms for toric ideals

Since $I_{\mathcal{A}}$ is a prime ideal that does not contain any monomials, if $\mathbf{x}^{\mathbf{w}}(\mathbf{x}^{\mathbf{u}} - \mathbf{x}^{\mathbf{v}}) \in I_{\mathcal{A}}$ then $\mathbf{x}^{\mathbf{u}} - \mathbf{x}^{\mathbf{v}} \in I_{\mathcal{A}}$. Thus every minimal generating set and reduced Gröbner basis of $I_{\mathcal{A}}$ consists of $\mathcal{A}$-homogeneous binomials $\mathbf{x}^{\mathbf{u}} - \mathbf{x}^{\mathbf{v}}$ with $\text{supp}(\mathbf{u}) \cap \text{supp}(\mathbf{v}) = \emptyset$. We may record $\mathbf{x}^{\mathbf{u}} - \mathbf{x}^{\mathbf{v}}$ as the vector $\mathbf{u} - \mathbf{v}$. Conversely, given any vector $\mathbf{p} = (p_1, \ldots, p_n) \in \mathbb{Z}^n$, write $\mathbf{p} = \mathbf{p}^+ - \mathbf{p}^-$ where $\mathbf{p}^+, \mathbf{p}^- \in \mathbb{N}^n$ are the unique vectors obtained as follows: $(\mathbf{p}^+)_i = p_i$ when $p_i > 0$ and $(\mathbf{p}^+)_i = 0$ otherwise while $(\mathbf{p}^-)_i = -p_i$ when $p_i < 0$ and $(\mathbf{p}^-)_i = 0$ otherwise. For example, if $\mathbf{p} = (3, -4, 0, 5, 1)$, then $\mathbf{p}^+ = (3, 0, 0, 5, 1)$ and $\mathbf{p}^- = (0, 4, 0, 0, 0)$. If $\mathbf{p} \in \mathbb{Z}^n$ such that $A\mathbf{p} = \mathbf{0}$, then the binomial $\mathbf{x}^{\mathbf{p}^+} - \mathbf{x}^{\mathbf{p}^-}$ is $\mathcal{A}$-homogeneous and lies in $I_\mathcal{A}$.

We now outline algorithms for computing reduced Gröbner bases of the toric ideal $I_\mathcal{A}$ starting with the configuration $\mathcal{A}$. We compute reduced Gröbner bases with respect to a weight vector $\mathbf{w}$ which, as we saw in Chapter 2, is more general than computing reduced Gröbner bases with respect to term orders.

The Conti-Traverso algorithm [CT91]

Input: A vector configuration $\mathcal{A} \subset \mathbb{Z}^d$ and a term order given by the weight vector $\mathbf{w}$.
Output: The reduced Gröbner basis of $I_\mathcal{A}$ with respect to $\mathbf{w}$.

1. Introduce $n + d + 1$ indeterminates $t_0, t_1, \ldots, t_d, x_1, \ldots, x_n$. Let $\succ$ be any elimination order such that $\{t_i\} \succ \{x_j\}$ and the x variables are ordered by $\mathbf{w}$.
2. Compute the reduced Gröbner basis $\mathcal{G}_\succ(J)$ of the ideal
$$J = \langle t_0 t_1 \cdots t_d - 1,\ x_j \mathbf{t}^{\mathbf{a}_j^-} - \mathbf{t}^{\mathbf{a}_j^+},\ j = 1, \ldots, n \rangle.$$
3. Output the set $\mathcal{G}_\succ(J) \cap \mathbf{k}[\mathbf{x}]$ which is the reduced Gröbner basis of $I_\mathcal{A}$ with respect to $\mathbf{w}$.

Proof. The correctness of this algorithm follows from Theorem 2 in Section 3.3 of [CLO97] on computing equations of the smallest variety containing a rationally parametrized set of points. Note that $I_\mathcal{A}$ is such a variety given by the rational parametrization $\hat{\pi}$. □

Buchberger's algorithm is very sensitive to the number of variables used, and in that light the above algorithm is not ideal as it requires $d + 1$ extra variables beyond the variables $x_1, \ldots, x_n$ of $I_\mathcal{A}$. However, it has the advantage that a generating set for $I_\mathcal{A}$ is not needed to begin the algorithm. There are examples of toric ideals for which a minimal generating set is also a minimal universal Gröbner basis, and hence the problem of finding a minimal generating set is not an easy task. Several fast algorithms for computing a generating set for $I_\mathcal{A}$ that do not require additional variables are known. Many of these are implemented in CoCoA [COC], Singular [GPS05], and 4ti2 [Hem]. The 4ti2 code is currently the fastest for computing toric ideals and their Gröbner bases. We describe one of these algorithms here. See [Stu96, Chapter 12] for more details.

The Hoşten-Sturmfels algorithm [Stu96, Algorithm 12.3]

Input: $\mathcal{A}$ and a term order $\mathbf{w}$.
Output: The reduced Gröbner basis of $I_\mathcal{A}$ with respect to $\mathbf{w}$.

1. Find a lattice basis $\mathcal{B}$ for the lattice $\ker_\mathbb{Z}(\mathcal{A}) := \{\mathbf{u} \in \mathbb{Z}^n : A\mathbf{u} = 0\}$.
2. (optional) Replace $\mathcal{B}$ by a *reduced* basis in the sense of Lenstra, Lenstra and Lovász [Sch86, Chapter 6.2]. Call it $\mathcal{B}$ as well.
3. Let $J_0 = \langle \mathbf{x}^{\mathbf{u}^+} - \mathbf{x}^{\mathbf{u}^-} : \mathbf{u} \in \mathcal{B} \rangle$.

4. For $i = 1, 2, \ldots, n$: Compute $J_i := (J_{i-1} : x_i^\infty)$.
5. Compute the reduced Gröbner basis of $J_n = I_\mathcal{A}$ with respect to $\mathbf{w}$.

Proof. The proof of the correctness of this algorithm follows from Lemma 3.3.1 which was proved originally in [HS95]. □

Lemma 3.3.1 ([Stu96, Lemma 12.2]). *A set $\mathcal{B}$ is a lattice basis for* $\ker_\mathbb{Z}(\mathcal{A})$ *if and only if* $(J_\mathcal{B} : (x_1 \cdots x_n)^\infty) = I_\mathcal{A}$*, where* $J_\mathcal{B} := \langle \mathbf{x}^{\mathbf{b}^+} - \mathbf{x}^{\mathbf{b}^-} \ : \mathbf{b} \in \mathcal{B}\rangle$.

Proof. Clearly $J_\mathcal{B} \subseteq I_\mathcal{A}$. Suppose $\mathcal{B} = \{\mathbf{b}_1, \ldots, \mathbf{b}_r\}$ is a basis for $\ker_\mathbb{Z}(\mathcal{A})$. Let $\mathbf{u} \in \ker_\mathbb{Z}(\mathcal{A})$. Then $\mathbf{u} = \sum_{i=1}^r \lambda_i \mathbf{b}_i$ for some $\lambda_i \in \mathbb{Z}$. This implies that

$$\frac{\mathbf{x}^{\mathbf{u}^+}}{\mathbf{x}^{\mathbf{u}^-}} - 1 = \prod_{i=1}^{r} \left(\frac{\mathbf{x}^{\mathbf{b}_i^+}}{\mathbf{x}^{\mathbf{b}_i^-}} \right)^{\lambda_i} - 1.$$

Clearing denominators we get that

$$\prod_{i=1}^{r} (\mathbf{x}^{\mathbf{b}_i^-})^{\lambda_i} (\mathbf{x}^{\mathbf{u}^+} - \mathbf{x}^{\mathbf{u}^-}) = \mathbf{x}^{\mathbf{u}^-} \left(\prod_{i=1}^{r} (\mathbf{x}^{\mathbf{b}_i^+})^{\lambda_i} - \prod_{i=1}^{r} (\mathbf{x}^{\mathbf{b}_i^-})^{\lambda_i} \right).$$

If we show that the right hand side lies in $J_\mathcal{B}$ then we will get that a monomial multiple of $\mathbf{x}^{\mathbf{u}^+} - \mathbf{x}^{\mathbf{u}^-}$ lies in $J_\mathcal{B}$ which will finish the "only-if" direction of the argument. Note that if $\mathbf{x}^{\mathbf{p}^+} - \mathbf{x}^{\mathbf{p}^-}$ and $\mathbf{x}^{\mathbf{q}^+} - \mathbf{x}^{\mathbf{q}^-}$ lie in an ideal then $\mathbf{x}^{\mathbf{q}^+}(\mathbf{x}^{\mathbf{p}^+} - \mathbf{x}^{\mathbf{p}^-}) + \mathbf{x}^{\mathbf{p}^-}(\mathbf{x}^{\mathbf{q}^+} - \mathbf{x}^{\mathbf{q}^-}) = \mathbf{x}^{\mathbf{q}^+}\mathbf{x}^{\mathbf{p}^+} - \mathbf{x}^{\mathbf{p}^-}\mathbf{x}^{\mathbf{q}^-}$ also lies in the ideal. Applying this argument to $\{\mathbf{x}^{\mathbf{b}_i^+} - \mathbf{x}^{\mathbf{b}_i^-} \ : i = 1, \ldots, r\} \subset J_\mathcal{B}$ we get that $\prod_{i=1}^r (\mathbf{x}^{\mathbf{b}_i^+})^{\lambda_i} - \prod_{i=1}^r (\mathbf{x}^{\mathbf{b}_i^-})^{\lambda_i}$ lies in $J_\mathcal{B}$.

To argue the "if" direction, we have to show that if $\mathbf{u} \in \ker_\mathbb{Z}(\mathcal{A})$, then $\mathbf{u} = \sum_{i=1}^r \lambda_i \mathbf{b}_i$ for some integers λ_i. Now $\mathbf{u} \in \ker_\mathbb{Z}(\mathcal{A})$ implies that $\mathbf{x}^{\mathbf{u}^+} - \mathbf{x}^{\mathbf{u}^-} \in I_\mathcal{A} = (J_\mathcal{B} : (x_1 \cdots x_n)^\infty)$. Hence $\mathbf{x}^{\mathbf{a}}(\mathbf{x}^{\mathbf{u}^+} - \mathbf{x}^{\mathbf{u}^-}) \in J_\mathcal{B}$. Hence, we can connect the lattice points $\mathbf{a} + \mathbf{u}^+$ and $\mathbf{a} + \mathbf{u}^-$ by a sequence of vectors in $\mathcal{B}$ which shows that $\mathbf{u} = \mathbf{a} + \mathbf{u}^+ - \mathbf{a} + \mathbf{u}^-$ is an integer combination of vectors in $\mathcal{B}$. □

In Tutorial 1 we saw how to compute $(I \ : \ x_i^\infty)$ for a homogeneous ideal I and hence we can compute $(J_\mathcal{B} \ : \ (x_1 \cdots x_n)^\infty)$. Lattice bases can be computed using a computer algebra package such as Maple or Macaulay 2.

3.4 Universal Gröbner bases for toric ideals

In the rest of this chapter we construct a universal Gröbner basis for $I_\mathcal{A}$ called the Graver basis of $I_\mathcal{A}$, and derive a bound on the maximum degree of elements in this universal Gröbner basis. This basis plays a fundamental role in the study and applications of toric ideals.

Recall from Chapter 2 that a particular choice of universal Gröbner basis of $I_\mathcal{A}$ is the union (up to sign) of all the reduced Gröbner bases of $I_\mathcal{A}$. We denote this set by $\mathcal{U}_\mathcal{A}$.

Example 3.4.1. Taking the union over all reduced Gröbner bases of $I_{\{5,10,25,50\}}$ from Example 3.2.3, we get $\mathcal{U}_\mathcal{A} = \{x_3^2 - x_4, x_2^3 - x_1x_3, x_1x_2^2 - x_3, x_1^2 - x_2, x_1^5 - x_3, x_2^5 - x_3^2, x_4 - x_2^5, x_4 - x_1^{10}, x_2^3x_3 - x_1x_4\}$. This is computed easily from the CaTS output from earlier.

From Chapter 2 and Proposition 3.2.2(4) we know that $\mathcal{U}_{\mathcal{A}}$ is finite and consists of $\mathcal{A}$-homogeneous binomials. A binomial $\mathbf{x}^{\mathbf{u}^+} - \mathbf{x}^{\mathbf{u}^-} \in I_{\mathcal{A}}$ is *primitive* if there is no binomial $\mathbf{x}^{\mathbf{v}^+} - \mathbf{x}^{\mathbf{v}^-} \in I_{\mathcal{A}}$ such that $\mathbf{x}^{\mathbf{v}^+}$ properly divides $\mathbf{x}^{\mathbf{u}^+}$ and $\mathbf{x}^{\mathbf{v}^-}$ properly divides $\mathbf{x}^{\mathbf{u}^-}$. The following is an easy fact.

Lemma 3.4.2 (Lemma 4.6 [Stu96]). *Every binomial in $\mathcal{U}_{\mathcal{A}}$ is primitive.*

Definition 3.4.3. The *Graver basis* of $I_{\mathcal{A}}$, denoted by $Gr_{\mathcal{A}}$, is the set of all primitive binomials in $I_{\mathcal{A}}$.

There are many examples of $\mathcal{A}$ for which $Gr_{\mathcal{A}}$ is a proper superset of $\mathcal{U}_{\mathcal{A}}$. This is the case for our running example where $Gr_{\mathcal{A}} = \{x_1^2 - x_2, x_1x_2^2 - x_3, x_1^3x_2 - x_3, x_1^5 - x_3, x_2^3 - x_1x_3, x_3^2 - x_4, x_1x_2^2x_3 - x_4, x_1^3x_2x_3 - x_4, x_1^5x_3 - x_4, x_2^5 - x_3^2, x_2^5 - x_4, x_1^2x_2^4 - x_4, x_1^4x_2^3 - x_4, x_1^6x_2^2 - x_4, x_1^8x_2 - x_4, x_1^{10} - x_4, x_2^3x_3 - x_1x_4\}$. Graver bases can be computed using the software package 4ti2.

For a collection of vectors $\mathcal{P} = \{\mathbf{p}_1, \ldots, \mathbf{p}_n\} \subset \mathbb{Z}^d$, recall that $\mathrm{pos}(\mathcal{P})$ is the cone generated by $\mathcal{P}$.

Definition 3.4.4. Let D denote the semigroup $\mathrm{pos}(\mathcal{P}) \cap \mathbb{Z}^d$ for $\mathcal{P} \subset \mathbb{Z}^d$. A *Hilbert basis* of $\mathrm{pos}(\mathcal{P})$ is a finite generating set for the semigroup D.

Example 3.4.5. The Hilbert basis of $\mathrm{pos}(\{(1,0),(1,3),(1,4),(1,6)\})$ is

$$\{(1,0),(1,1),(1,2),(1,3),(1,4),(1,5),(1,6)\}.$$

The following lemma shows that Definition 3.4.4 makes sense.

Lemma 3.4.6. *Every rational polyhedral cone has a Hilbert basis.*

Proof. Assume that $\mathbf{p}_1, \ldots, \mathbf{p}_k$ are primitive vectors generating the cone $\mathrm{pos}(\mathcal{P})$. Consider the finite set $D' = \{\mathbf{x} \in \mathbb{Z}^d : \mathbf{x} = \sum_i \lambda_i \mathbf{p}_i \text{ with } 0 \leq \lambda_i < 1\}$. Note that if $\mathbf{y} \in D$, then $\mathbf{y} = \sum_j \mu_j \mathbf{p}_j$ for some $\mu_j \in \mathbb{R}_{\geq 0}$. Write this as $\mathbf{y} = \sum_i \lfloor \mu_i \rfloor \mathbf{p}_i + \sum_i (\mu_i - \lfloor \mu_i \rfloor)\mathbf{p}_i$. The second sum belongs to D', so, since $\mathbf{y}$ was arbitrary, this shows that $\{\mathbf{p}_1, \ldots, \mathbf{p}_k\} \cup D'$ generate D. We may delete redundant elements in this set to minimalize the generating set and obtain a Hilbert basis for $\mathrm{pos}(\mathcal{P})$. □

When $\mathrm{pos}(\mathcal{P})$ is *pointed*, which means that there is no line in the cone, then $\mathrm{pos}(\mathcal{P})$ has a unique minimal Hilbert basis.

We use Hilbert bases to give an explicit (theoretical) construction of $Gr_{\mathcal{A}}$. For each sign pattern $\sigma \in \{+,-\}^n$, consider the pointed polyhedral cone $C_\sigma := \ker(\mathcal{A}) \cap \mathbb{R}^n_\sigma$, where $\ker(\mathcal{A})$ is the vector space $\{\mathbf{u} \in \mathbb{R}^n : A\mathbf{u} = 0\}$. Here $\mathbb{R}^n_\sigma$ is the orthant of $\mathbb{R}^n$ with sign pattern σ. Let $\mathcal{H}_\sigma$ be the Hilbert basis of C_σ which is a pointed cone. The Graver basis $Gr_{\mathcal{A}} = \{\mathbf{x}^{\mathbf{u}^+} - \mathbf{x}^{\mathbf{u}^-} : \mathbf{u} \in \bigcup_\sigma \mathcal{H}_\sigma \backslash \{\mathbf{0}\}\}$.

Lemma 3.4.7. *The above construction yields the Graver basis $Gr_{\mathcal{A}}$. By construction it is finite.*

Proof. If $\mathbf{u} \in C_\sigma$, then $\mathbf{u}$ is a $\mathbb{N}$-linear combination of elements of $\mathcal{H}_\sigma$ all of which are sign-compatible to $\mathbf{u}$. If $\mathbf{u} \notin \mathcal{H}_\sigma$, then every element $\mathbf{v} \in \mathcal{H}_\sigma$ involved in the combination has the property that $\mathbf{v}^+ \leq \mathbf{u}^+$ and $\mathbf{v}^- \leq \mathbf{u}^-$ which implies that $\mathbf{x}^{\mathbf{u}^+} - \mathbf{x}^{\mathbf{u}^-}$ is not primitive. Thus if $\mathbf{x}^{\mathbf{u}^+} - \mathbf{x}^{\mathbf{u}^-} \in I_{\mathcal{A}}$ is primitive then $\mathbf{u} \in \cup_\sigma \mathcal{H}_\sigma \backslash \{\mathbf{0}\}$. On the other hand, if $\mathbf{x}^{\mathbf{u}^+} - \mathbf{x}^{\mathbf{u}^-}$ is such that $\mathbf{u} \in \mathcal{H}_\sigma$, then it is primitive by construction. □

Algorithm 7.2 in [Stu96] is a more practical method for constructing $Gr_{\mathcal{A}}$. See Tutorial 4 for an example.

Our next goal is to provide bounds for the maximum degree of an element in the Graver basis. In order to do this, we isolate a special subset of the Graver basis known as the set of circuits of $I_{\mathcal{A}}$. A bound on the maximum degree of a circuit can be derived from linear algebra. This in turn yields a bound on the maximum degree of an element in $\mathcal{G}r_{\mathcal{A}}$.

Definition 3.4.8. A primitive binomial $\mathbf{x}^{\mathbf{u}^+} - \mathbf{x}^{\mathbf{u}^-} \in I_{\mathcal{A}}$ is called a *circuit* if its support is minimal among all binomials in $I_{\mathcal{A}}$. The set of circuits of $I_{\mathcal{A}}$ is denoted as $\mathcal{C}_{\mathcal{A}}$.

We refer to the set of vectors $\{\mathbf{u} : \mathbf{x}^{\mathbf{u}^+} - \mathbf{x}^{\mathbf{u}^-} \in \mathcal{C}_{\mathcal{A}}\}$ as the circuits of $\mathcal{A}$. This set will also be denoted as $\mathcal{C}_{\mathcal{A}}$. Recall that circuits of a matrix were also defined in Chapter 1 in the context of linear ideals. The same notion is being used here. We now state a few facts about circuits. See [Stu96, Lemmas 4.8, 4.9, 4.10 and 4.11] for details.

Lemma 3.4.9.

1. *The circuits $\mathcal{C}_{\mathcal{A}}$ are precisely the generators of the pointed rational polyhedral cones C_σ in the construction of the Graver basis.*
2. *Every vector $\mathbf{v} \in \ker(\mathcal{A})$ can be written as a non-negative rational combination of $n-d$ circuits, each of which is sign-compatible to $\mathbf{v}$.*
3. *The support of a circuit has cardinality at most $d+1$.*
4. *Every circuit $\mathbf{x}^{\mathbf{u}^+} - \mathbf{x}^{\mathbf{u}^-}$ of $I_{\mathcal{A}}$ appears in some reduced Gröbner basis of $I_{\mathcal{A}}$. Thus $\mathcal{C}_{\mathcal{A}} \subseteq \mathcal{U}_{\mathcal{A}} \subseteq Gr_{\mathcal{A}}$.*
5. *Let $D(\mathcal{A}) := \max\{|\det[\mathbf{a}_{i_1} \ldots \mathbf{a}_{i_d}]| : 1 \leq i_1 < i_2 < \cdots < i_d \leq n\}$. If $\mathbf{u} \in \mathcal{C}_{\mathcal{A}}$ then $\|\mathbf{u}\|_1 \leq (d+1)D(\mathcal{A})$.*

Proof (Sketch of the main ideas.).

1. Recall that circuits of $\mathcal{A}$ are vectors in $\ker_{\mathbb{Z}}(\mathcal{A})$ of minimal support. Thus they lie at the intersection of as many coordinate hyperplanes as possible. The extreme rays of the cones C_σ are cut out by a maximal number of coordinate hyperplanes.
2. Caratheodory's Theorem for polyhedral cones says that every vector in a d-dimensional rational polyhedral cone is a non-negative rational combination of d generators of the cone [Zie95].
3. The circuits of $\mathcal{A}$ are the minimal dependencies among the columns of A. Recall that $\text{rank}(A) = d$.
4. This fact follows from an argument very similar to the minimality argument in Proposition 4.3 (2) of Chapter 1.
5. By part 3 it suffices to show that $|u_i| \leq D(\mathcal{A})$ for each $i = 1, \ldots, n$. Let $\text{supp}(\mathbf{u}) = \{i_1, \ldots, i_r\}$. Then the $d \times r$ matrix $(\mathbf{a}_{i_1} \ldots \mathbf{a}_{i_r})$ has rank $r-1$ since $\mathbf{u}$ is a minimal dependency on the columns of A. Choose columns $\mathbf{a}_{i_{r+1}}, \ldots, \mathbf{a}_{i_{d+1}}$ of A such that $B = (\mathbf{a}_{i_1} \ldots \mathbf{a}_{i_{d+1}})$ has rank d. Hence the kernel of B is one-dimensional, and by Cramer's rule is spanned by the vector

$$\sum_{j=1}^{d+1} (-1)^j \det(\mathbf{a}_{i_1} \ldots \mathbf{a}_{i_{j-1}} \mathbf{a}_{i_{j+1}} \ldots \mathbf{a}_{i_{d+1}}) \mathbf{e}_{i_j} \qquad (*).$$

Thus $\mathbf{u}$ is a rational multiple of $(*)$ (extended by zeros to be a vector in $\mathbb{R}^n$). However, since $\mathbf{u}$ is a circuit and $(*)$ is integral, $(*)$ is in fact an integer multiple of $\mathbf{u}$ which proves the claim. □

Using Lemma 3.4.9 we obtain a bound on the maximum degree of a Graver basis element.

Theorem 3.4.10. *Let* $D(\mathcal{A}) := \max\{|\det[\mathbf{a}_{i_1} \ldots \mathbf{a}_{i_d}]| : 1 \leq i_1 < i_2 < \cdots < i_d \leq n\}$. *The total degree of any element of* $Gr_{\mathcal{A}}$ *is less than* $(d+1)(n-d)D(\mathcal{A})$.

Proof. Note that the ideal $I_{\mathcal{A}}$ need not be homogeneous in the total degree grading. The "total degree" of an element $\mathbf{x}^{\mathbf{u}^+} - \mathbf{x}^{\mathbf{u}^-}$ is taken to be $\max\{\|\mathbf{u}^+\|_1, \|\mathbf{u}^-\|_1\}$.

If $\mathbf{x}^{\mathbf{v}^+} - \mathbf{x}^{\mathbf{v}^-} \in Gr_{\mathcal{A}}$, then apply Lemma 3.4.9 (2) to find $n-d$ circuits sign-compatible to $\mathbf{v}$ and non-negative rational numbers $\lambda_1, \ldots, \lambda_{n-d}$ such that

$$\mathbf{v} = \lambda_1 \mathbf{u}_1 + \cdots + \lambda_{n-d}\mathbf{u}_{n-d}.$$

The sign-compatibility implies that $\mathbf{v}^+ = \lambda_1 \mathbf{u}_1^+ + \cdots + \lambda_{n-d}\mathbf{u}_{n-d}^+$ and $\mathbf{v}^- = \lambda_1 \mathbf{u}_1^- + \cdots + \lambda_{n-d}\mathbf{u}_{n-d}^-$. Further, since $\mathbf{v}$ is primitive, $\lambda_i < 1$ for each i. Assuming that the total degree of $\mathbf{x}^{\mathbf{v}^+} - \mathbf{x}^{\mathbf{v}^-}$ equals $\|\mathbf{v}^+\|_1$, we get from Lemma 3.4.9 (5) that

$$\|\mathbf{v}^+\|_1 \leq \sum_{j=1}^{n-d} \lambda_j \|\mathbf{u}_j{}^+\|_1 < (n-d)(d+1)D(\mathcal{A}).$$

□

It has been conjectured by Sturmfels that the factor $(n-d)$ in the bound proved above is unnecessary.

Conjecture 3.4.11. The total degree of any element of $Gr_{\mathcal{A}}$ is less than or equal to $(d+1)D(\mathcal{A})$.

Apart from intrinsic mathematical interest, tighter bounds on the degree of Graver basis elements have important consequences in integer programming. Several bounds in this field appear as functions of this maximum degree. The maximum degree of elements in a universal Gröbner basis of an ideal is a measure of the complexity of the ideal and the complexity of Buchberger's algorithm working on this ideal. The single-exponential upper bound on the degree of elements in $Gr_{\mathcal{A}}$ should be contrasted with the double-exponential lower bounds on the elements in a universal Gröbner basis of a class of binomial ideals found by Mayr and Meyer [MM82]. These ideals have been used in the literature to show that Gröbner bases calculations can be difficult from the view of computational complexity.

3.5 Tutorial 3

In Tutorial 1 the program Macaulay 2 was introduced. In this tutorial we will create our own functions in Macaulay 2 to implement the algorithms in Lecture 3. A function in Macaulay 2 has the syntax

function-name = (*names for input(s) separated by commas*) ->
(*computations to execute separated by semi-colons*)

Below are two examples. The first function `toBinomial` creates a binomial from an integer vector. The second function `HSAlg` encodes the Hoşten-Sturmfels algorithm described in Lecture 3. The package "LLL.m2" allows us to do the optional reduction of the basis in the sense of Lenstra, Lenstra and Lovász from step two of the Hoşten-Sturmfels algorithm.

```
load "LLL.m2" -- load a package for doing LLL reduction

toBinomial = (b,R) -> (
  -- take a vector b of the form {*,*,...,*}, and
  -- a ring R having length of b number of variables
  -- return binomial x^(b-) - x^(b+)
  pos := 1_R;
  neg := 1_R;
  scan(#b, i-> if b_i > 0 then pos = pos * R_i^(b_i)
       else if b_i < 0 then neg = neg * R_i^(-b_i));
  pos - neg);

HSAlg = (A,w) -> (
  -- take list of rows of A written {{*},{*},...,{*}}
  -- and take a weight vector w
  -- return a reduced GB for I_A with respect to w
  n := #(A_0);    -- NOTE: A is not a matrix
  R = QQ[x_1..x_n,Degrees=>transpose A,MonomialSize=>16,
        Weights=>w];
  B := transpose LLL syz matrix A;
  J := ideal apply(entries B, b -> toBinomial(b,R));
  scan(gens ring J, f -> J = saturate(J,f));
  gens gb J)
```

Notice that the prefix '`--`' is used to denote the start of a comment; the entire rest of the line is ignored by the program. A list in Macaulay 2 is denoted by curly braces and its elements are separated by commas. In the function `HSAlg` instead of using one of the predefined orders and the command `MonomialOrder=>`*name-of-order*, we use `Weights=>w` which tells Macaulay 2 to use a weight order determined by the vector `w`. The command `saturate(J,f)` computes $(J : f^\infty)$. The command `#L` returns the size of the list `L`. More help with commands is available by typing `help` *name-of-command* in Macaulay 2, or better, by using the html-based help either locally or at `http://www.math.uiuc.edu/Macaulay2/Manual`.

In one of the functions above, we used a monomial order defined by a weight vector. Another useful order is an elimination order. For an ideal $I \subset \mathbf{k}[\mathbf{t}, \mathbf{x}]$, generators for the ideal $I \cap \mathbf{k}[\mathbf{x}]$ are found by calculating a Gröbner basis of I with respect to an elimination order for the t_i variables and then throwing out any generators which contain t_i's. In Macaulay 2 including the setting `MonomialOrder => Eliminate n` in the definition of a ring will force the use of an elimination order eliminating the first n variables. Note that an elimination order need not be a total order, just one where the variables to be eliminated are bigger than those that are not to be eliminated. Macaulay 2 refines this partial order by using the graded reverse lexicographic order whenever necessary to attain a total order on the monomials.

1. (a) Working by hand, use the Conti-Traverso algorithm to find a reduced Gröbner basis $\mathcal{G}_{(3,2,1)}$ of $I_{\mathcal{A}}$ for $A = [1\ 2\ 3]$. Use grevlex as a tiebreaking term order.
   (b) Now, using Macaulay 2, implement the Conti-Traverso algorithm by replacing the question marks in the following function by your own code to complete the algorithm.

```
CTAlg = (A,w) -> (
  -- take a list of rows of matrix A of the form
  --    {{*},{*},...,{*}}
  --    and a weight vector w
  -- return a reduced GB of I_A
  n := #(A_0);
  d := #A;
  R := QQ[t_0..t_d,x_1..x_n, MonomialSize=>16,
       MonomialOrder=>     ???     ];
  J := ideal (     ???     );
  I := selectInSubring(1, gens gb J); -- select entries
    -- without vars from 1st 'part' of mon. order
  S :=    ???     ; -- the ring I should end up in
  gens gb substitute(I, S))
```

2. Recall that while defining the toric ideal, we defined the map $\pi : \mathbb{N}^n \rightarrow \mathbb{Z}^d$ by $\mathbf{u} \mapsto A\mathbf{u}$, and its image was $\mathbb{N}\mathcal{A}$. For any $\mathbf{b} \in \mathbb{N}\mathcal{A}$, we define the *fiber* of π over $\mathbf{b}$ to be the set $\pi^{-1}(\mathbf{b}) = \{\mathbf{u} \in \mathbb{N}^n : \pi(\mathbf{u}) = \mathbf{b}\}$. We also sometimes use the term *fiber* to refer to the set of monomials having the same $\mathcal{A}$-degree, that is, $\{\mathbf{x}^{\mathbf{u}} : \pi(\mathbf{u}) = \mathbf{b}\}$ which is also a basis for the vector space $\mathbf{k}[\mathbf{x}]_{\mathbf{b}}$.

   For any subset $H \subset \ker(\pi)$, we may define a graph on the fiber $\pi^{-1}(\mathbf{b})$ by taking the vertices to be the vectors in the fiber, and connecting two vertices $\mathbf{u}$ and $\mathbf{v}$ when $\mathbf{u} - \mathbf{v} \in H$. Furthermore, given a term order $\succ$ on $\mathbb{N}^n$, we can create a directed graph from the undirected one by orienting an edge from $\mathbf{u}$ to $\mathbf{v}$ when $\mathbf{u} \succ \mathbf{v}$.

   (a) Let $H \subset \ker(\pi)$. Prove that $\{\mathbf{x}^{\mathbf{u}^+} - \mathbf{x}^{\mathbf{u}^-} : \mathbf{u} \in H\}$ is a generating set of $I_{\mathcal{A}}$ if and only if for every $\mathbf{b} \in \mathbb{N}\mathcal{A}$ the graph on the fiber $\pi^{-1}(\mathbf{b})$ defined by H is connected.
   (b) Let $H \subset \ker(\pi)$, and let $\succ$ be a term order on $\mathbb{N}^n$. Prove that $\mathcal{G} = \{\mathbf{x}^{\mathbf{u}^+} - \mathbf{x}^{\mathbf{u}^-} : \mathbf{u} \in H\}$ is a Gröbner basis of $I_{\mathcal{A}}$ with respect to $\succ$ if and only if for every $\mathbf{b} \in \mathbb{N}\mathcal{A}$ the directed graph on the fiber $\pi^{-1}(\mathbf{b})$ defined by H and $\succ$ is connected as an undirected graph and has a unique sink (a vertex with only incoming edges) at the unique minimal vector in the fiber under the ordering by $\succ$. Note that a term order $\succ$ on $\mathbb{N}^n$ also defines a term order on monomials where $\mathbf{x}^{\mathbf{u}} \succ \mathbf{x}^{\mathbf{v}}$ if $\mathbf{u} \succ v$.

3. (a) Describe an algorithm using toric ideals to solve the integer program $\min\{\mathbf{c}\cdot\mathbf{x} : A\mathbf{x} = \mathbf{b},\ \mathbf{x} \in \mathbb{Z}^n,\ \mathbf{x} \geq 0\}$ where $\mathbf{c}$ is the cost (weight) vector and $\mathbf{b} \in \mathbb{N}\mathcal{A} =$

$\{A\mathbf{u} \ : \ \mathbf{u} \in \mathbb{Z}^n, \ \mathbf{u} \geq 0\}$. That is, find a vector or set of vectors $\mathbf{x}$ that minimize $\mathbf{c} \cdot \mathbf{x}$ subject to the constraints above.

(b) In order to do the calculations from your algorithm in part a using Macaulay 2 it may be helpful to know that you can reduce a monomial m modulo an ideal I using the command `m % I`.

Use your algorithm to solve the integer program about currency from the introduction to Lecture 3.

4. The programs 4ti2 [Hem] and CoCoA [COC] both have specialized packages for toric computations. To use the program 4ti2, create a file *matrix-file* for each A matrix with the following lines in the file:

line 1: *number-of-rows-in-A number-of-columns-in-A*
remaining lines: *entries of row of A, each separated by spaces*

A Gröbner basis may be calculated by running `groebner` *matrix-file*. This process will return a file *matrix-file.gro* in the same format as the input. The output file can be formatted with the command `output`. To list the output as binomials, for example, type `output bin` *matrix-file.gro*. Caution: you must include the extension *.gro*, or else you will get binomials from the input matrix A. For further information, see the web site `http://www.4ti2.de`.

In CoCoA, `Toric(`*matrix*`)` returns the toric ideal corresponding to the matrix. CoCoA also has an integer programming package. Details on using CoCoA may be found at `http://cocoa.dima.unige.it`.

(a) Use one or both of the above programs to find a Gröbner basis for the toric ideal generated by the matrix

$$\begin{pmatrix} 3 & 2 & 2 & 1 & 1 & 0 & 0 & 0 & 0 \\ 0 & 1 & 0 & 2 & 0 & 3 & 2 & 1 & 0 \\ 0 & 0 & 1 & 0 & 2 & 0 & 1 & 2 & 3 \end{pmatrix}.$$

(b) For more practice, check your previous calculations using these programs.

5. (a) Show that the result in Exercise 2b can be used to enumerate all the elements of a fiber.

(b) The program CaTS will enumerate all lattice points in a fiber using the command `cats_fiber`. The input file is a list of the columns of A in the form {(col 1)(col2) $\cdots$ (col n)} followed by a vector of the fiber in the form $(d_1, d_2, \ldots, d_n)$. For more information see `http://www.soopadoopa.dk/anders/cats/cats.html`.

Use CaTS to find all the lattice points in the fiber necessary to solve the currency problem in the introduction to Lecture 3.

6. [*From [CLO98]*, p. 359] Suppose a small local trucking firm has two customers, A and B, that generate shipments to the same location. Each shipment from A is a pallet weighing exactly 400 kilos and taking up 2 cubic meters of volume. Each pallet from B weighs 500 kilos and takes up 3 cubic meters of volume. The

shipping firm uses small trucks that can carry any load up to 3700 kilos, and up to 20 cubic meters. B's product is more perishable, though, and they are willing to pay a higher price for on-time delivery: \$15 per pallet verses \$11 per pallet from A. How many pallets from each of the two companies should be included in each truckload to maximize the revenues generated?

7. For each of the following matrices, calculate, by hand, a Hilbert basis of the cone spanned by the columns of A.

   (a) $A = \begin{pmatrix} 1 & 1 & 1 & 1 \\ 0 & 1 & 3 & 4 \end{pmatrix}$

   (b) $A = \begin{pmatrix} 1 & 2 & 2 & 3 \\ 0 & 1 & 3 & 4 \end{pmatrix}$

8. As we saw in Exercise 4, the program 4ti2 can compute a generating set and Gröbner basis for $I_{\mathcal{A}}$. It can also calculate Graver bases by running `graver` *matrix-file*.

   (a) For each of the following matrices, use 4ti2 to calculate the Graver basis for the lattice ideal defined by A.

      i. $A = \begin{pmatrix} 1 & 1 & 1 & 1 \\ 0 & 1 & 3 & 4 \end{pmatrix}$

      ii. $A = \begin{pmatrix} 1 & 2 & 2 & 3 \\ 0 & 1 & 3 & 4 \end{pmatrix}$

   (b) For the above matrices, check, by hand or using Macaulay 2, that the conjectured bound on the total degree of any element of the Graver basis (Conjecture 3.4.11) holds, that is, the total degree is always less than $(d+1)D(A)$ where

$$D(A) = \max\{|\det[\mathbf{a}_{i_1} \cdots \mathbf{a}_{i_d}]| \ : \ 1 \leq i_1 \leq i_2 \leq \ldots \leq i_d \leq n\}.$$

3.6 Solutions to Tutorial 3

1. (a) A reduced Gröbner basis of $I_{\mathcal{A}}$ is $\{x_1^2 - x_2,\ x_1x_2 - x_3,\ x_2^2 - x_1x_3\}$.
   (b) Here is one way to fill in the gaps to create a program which implements the Conti-Traverso algorithm.

```
CTAlg = (A,w) -> (
  -- take a list of rows of matrix A of the form
  --    {{*},{*},...,{*}}
  --    and a weight vector w
  -- return a reduced GB of I_A
  n := # (A_0);
  d := #A;
  R := QQ[t_0..t_d,x_1..x_n, MonomialSize=>16,
        MonomialOrder=>Eliminate(d+1)];
  J := ideal (append(
        apply(n, j -> (
```

```
        firstmon = x_(j+1);
        secondmon = 1_R;
        scan(d, i-> (
            if A_i_j < 0
            then firstmon = firstmon*(t_(i+1))^(-A_i_j)
            else secondmon = secondmon*(t_(i+1))^(A_i_j)));
        firstmon - secondmon)),
        product(toList (t_0..t_d))-1));
  I := selectInSubring(1, gens gb J);
  S := QQ[x_1..x_n, Degrees => transpose A, Weights=>w];
  gens gb substitute(I, S))
```

2. (a) Assume $\{\mathbf{x}^{\mathbf{u}^+} - \mathbf{x}^{\mathbf{u}^-} : \mathbf{u} \in H\}$ generates $I_{\mathcal{A}}$. Fix a vector $\mathbf{b}$. Then we want to show the graph on $\pi^{-1}(\mathbf{b})$ defined by H is connected. Consider any two vertices p, and q in the graph $\pi^{-1}(\mathbf{b})$. We want to show that $\mathbf{p} - \mathbf{q} \in H$. Proposition 3.2.2 gives us that $\mathbf{x}^{\mathbf{p}} - \mathbf{x}^{\mathbf{q}} \in I_{\mathcal{A}}$, so we can write this binomial in terms of our generating set for $I_{\mathcal{A}}$ as $\mathbf{x}^{\mathbf{p}} - \mathbf{x}^{\mathbf{q}} = \sum_{i=1}^{m} \mathbf{x}^{\mathbf{w}_i}(\mathbf{x}^{\mathbf{u}_i^+} - \mathbf{x}^{\mathbf{u}_i^-})$, where the vectors $\mathbf{u}_i$ are not necessarily unique. We proceed by induction. If $m = 1$ then up to relabeling $\mathbf{p} = \mathbf{w}_1 + \mathbf{u}_1{}^+$ and $\mathbf{q} = \mathbf{w}_1 + \mathbf{u}_1{}^-$. So $\mathbf{p} - \mathbf{q} = \mathbf{u} \in H$, and hence there is an edge between $\mathbf{p}$ and $\mathbf{q}$. If $m > 1$, then by relabeling, we may assume $\mathbf{x}^{\mathbf{p}} = \mathbf{x}^{\mathbf{w}_1}\mathbf{x}^{\mathbf{u}_1{}^+}$. Since $\mathbf{u} \in H$, this means that $\mathbf{p}$ and $(\mathbf{w}_1 + \mathbf{u}_1{}^-)$ are connected by an edge in the graph. So we have reduced the problem to showing that $(\mathbf{w}_1 + \mathbf{u}_1{}^-)$ and $\mathbf{q}$ are connected. However, $\mathbf{x}^{\mathbf{w}_1}\mathbf{x}^{\mathbf{u}_1{}^-} - \mathbf{x}^{\mathbf{q}} = \sum_{i=2}^{m} \mathbf{x}^{\mathbf{w}_i}(\mathbf{x}^{\mathbf{u}_i^+} - \mathbf{x}^{\mathbf{u}_i^-})$ is a smaller sum, so by induction $\mathbf{w}_1 + \mathbf{u}_1{}^-$ and $\mathbf{q}$ are connected.

For the other direction, assume that H defines a connected graph on $\pi^{-1}(\mathbf{b})$ for every $\mathbf{b} \in \mathbb{N}\mathcal{A}$. Take any $\mathbf{x}^{\mathbf{p}} - \mathbf{x}^{\mathbf{q}} \in I_{\mathcal{A}}$. Set $\mathbf{b} = \pi(\mathbf{p}) = \pi(\mathbf{q})$. By the hypothesis, the graph on $\pi^{-1}(\mathbf{b})$, the vertices $\mathbf{p}$ and $\mathbf{q}$ must have a path connecting them, say $\mathbf{p}, \mathbf{v}_1, \mathbf{v}_2, \ldots, \mathbf{v}_m, \mathbf{q}$. So the vectors $\mathbf{p} - \mathbf{v}_1, \mathbf{v}_1 - \mathbf{v}_2, \ldots, \mathbf{v}_m - \mathbf{q}$ are elements of H. But then we can write $\mathbf{x}^{\mathbf{p}} - \mathbf{x}^{\mathbf{q}} = (\mathbf{x}^{\mathbf{p}} - \mathbf{x}^{\mathbf{v}_1}) + (\mathbf{x}^{\mathbf{v}_1} - \mathbf{x}^{\mathbf{v}_2}) + \cdots + (\mathbf{x}^{\mathbf{v}_m} - \mathbf{x}^{\mathbf{q}})$, hence H generates $I_{\mathcal{A}}$.

(b) Assume that $\mathcal{G}$ is a Gröbner basis of $I_{\mathcal{A}}$ with respect to some ordering $\succ$. We can think of this as an ordering on the monomials or the exponent vectors. By part a, each fiber graph is connected. So we want to show there is a unique sink when $\succ$ is used to orient the graph. Suppose there are two sinks $\mathbf{p}$ and $\mathbf{q}$. Since they are in the same fiber, the binomial $\mathbf{x}^{\mathbf{p}} - \mathbf{x}^{\mathbf{q}} \in I_{\mathcal{A}}$ by Proposition 3.2.2. We may assume $\text{in}(\mathbf{x}^{\mathbf{p}} - \mathbf{x}^{\mathbf{q}}) = \mathbf{x}^{\mathbf{p}}$. Since $\mathcal{G}$ is a reduced Gröbner basis of $I_{\mathcal{A}}$, $\mathbf{x}^{\mathbf{p}} = \mathbf{x}^{\mathbf{w}}\,\text{in}(g)$ for some $g = \mathbf{x}^{\mathbf{u}^+} - \mathbf{x}^{\mathbf{u}^-} \in \mathcal{G}$. Without loss of generality $\text{in}(g) = \mathbf{x}^{\mathbf{u}^+}$ so $\mathbf{p} - (\mathbf{w} + \mathbf{u}^-) \in H$ and $\mathbf{p} \succ (\mathbf{w} + \mathbf{u}^-)$. Therefore there an edge from p to $(\mathbf{w} + \mathbf{u}^-)$, and hence $\mathbf{p}$ cannot be a sink.

For the other direction, assume that the directed graph on each fiber defined by H and $\succ$ has a unique sink at the minimal vector in that fiber. We want to show that $\mathcal{G}$ is a Gröbner basis of $I_{\mathcal{A}}$ with respect to $\succ$. By part a, the connectedness of each fiber graph implies that $\mathcal{G}$ generates $I_{\mathcal{A}}$, so $\langle \text{in}(g) : g \in \mathcal{G}\rangle \subset \text{in}(I_{\mathcal{A}})$. To verify equality, take a monomial $\mathbf{x}^{\mathbf{p}} \in \text{in}(I_{\mathcal{A}})$. Either $\mathbf{p}$ is the unique sink of the graph on $\pi^{-1}(\pi(\mathbf{p}))$ or there is an edge leaving $\mathbf{p}$ and going to some other vertex $\mathbf{q}$. If it is the sink, then it is minimal in the term order and hence cannot have been the initial term of some polynomial in $I_{\mathcal{A}}$. If it is not the sink, then we have $\mathbf{p}-\mathbf{q} \in H$ and $\mathbf{x}^{\mathbf{p}} \succ \mathbf{x}^{\mathbf{q}}$. So there is some element of $g = \{\mathbf{x}^{\mathbf{u}^+} - \mathbf{x}^{\mathbf{u}^-} \in G$ such that $\mathbf{u}^+ - \mathbf{u}^- = \mathbf{p} - \mathbf{q}$, hence $\text{in}(g)$ divides $\mathbf{x}^{\mathbf{p}}$. This proves that $\mathbf{x}^{\mathbf{p}} \in \langle\{\text{in}(g) \ : \ g \in \mathcal{G}\}\rangle$.

3. (a) First calculate $I_{\mathcal{A}}$ in the polynomial ring using a term order defined by the weight vector $\mathbf{c}$. Using the grading defined by $\deg(x_i) = \mathbf{a}_i$, find some monomial of degree $\mathbf{b}$. Now reduce the monomial modulo $I_{\mathcal{A}}$ using the weight vector term order, and take the exponent vector to get the $\mathbf{x}$ which minimizes the integer program.
   (b) For this problem $\mathbf{c} = (1, 1, 1, 1)$, $A = [5\ 10\ 25\ 50]$, and $b = 100$. To find $I_{\mathcal{A}}$ in Macaulay 2 using the Hoşten-Sturmfels algorithm type `I = ideal HSAlg({{5,10,25,50}},{1,1,1,1})`. A monomial of degree 100 in $I_{\mathcal{A}}$ is x_1^{20}. If you didn't know this, you could find such a monomial in Macaulay 2 by typing `basis(100, ring I)` and taking any entry of the resulting matrix. To reduce the monomial modulo $I_{\mathcal{A}}$ in Macaulay 2, type `x_1^(20)= % I`. The result is x_4^2, so the solution to the integer program is the vector $\mathbf{x} = (0, 0, 0, 2)$. In terms of coins, this solution vector means that the smallest number of coins adding up to a value of 100 is two coins each of value 50.

4. Using `4ti2`, the output of the command `output bin exercise.gro`, where *exercise* is the file containing the matrix and *exercise.gro* is the result of running `groebner exercise`, is

```
[
x[5]*x[6]-x[4]*x[8],
x[7]^2-x[6]*x[8],
x[3]*x[6]-x[2]*x[7],
x[4]^2-x[2]*x[6],
x[8]^2-x[7]*x[9],
x[2]*x[4]-x[1]*x[6],
x[2]*x[5]-x[1]*x[8],
x[3]*x[7]-x[2]*x[8],
x[3]*x[4]-x[1]*x[7],
x[2]^2-x[1]*x[4],
x[5]^2-x[3]*x[9],
x[5]*x[7]-x[4]*x[9],
x[3]*x[5]-x[1]*x[9],
x[4]*x[5]-x[2]*x[8],
x[7]*x[8]-x[6]*x[9],
x[3]*x[8]-x[2]*x[9],
x[3]^2-x[1]*x[5],
x[2]*x[3]*x[9]-x[1]*x[5]*x[8]
]
```

5. (a) Given a reduced Gröbner basis $\mathcal{G}_{\succ}$ of $I_{\mathcal{A}}$, Exercise 2b shows that the exponent vectors form connected directed graphs on the elements of each fiber. The idea is to start with some element of the fiber and then wander around the graph to find all the other elements of the fiber. Computationally, start with a fiber element $\mathbf{u} \in \pi^{-1}(\mathbf{b})$. Reduce $\mathbf{x}^{\mathbf{u}}$ modulo $\mathcal{G}_{\succ}$ to get a monomial whose exponent is the unique minimal element of the fiber. Now the elements of $\mathcal{G}_{\succ}$ can be used to move backward from the unique sink to find all vertices of the graph.

   One method for effectively finding all the vertices in a graph without using large amounts of memory to keep track of which vertices have already been visited is the reverse search algorithm by Avis and Fukuda [AF92]. This reverse search has been implemented in CaTS.

   (b) Create a file containing the following line:

   ```
   {(5)(10)(25)(50)} (20,0,0,0)
   ```

   Then run the program `cats_fiber` on it to get the elements of the fiber. The output should be the same as the fiber elements listed in the beginning of Lecture 3.

6. Let a be the number of pallets of A's product shipped and b be the number of pallets of B's product shipped. We must have $a, b \geq 0$ and they both must be integers. We want to maximize $15b + 11a$ subject to the constraints $400a + 500b \leq 3700$ and $2a + 3b \leq 20$. Solving this integer program using the method of Exercise 4 or Exercise 5, you should find the revenues will be maximized with 4 pallets of each. Note that it is helpful to add "slack variables" in order to change the inequalities to equalities before solving.
7. To find the Hilbert basis for a set of vectors $\mathcal{P}$, we draw the cone $\text{pos}(\mathcal{P})$ and the lattice $\mathbb{Z}\mathcal{P}$. Then we find a minimal set of vectors whose $\mathbb{N}$-linear combinations give the vectors of $\text{pos}(\mathcal{P}) \cap \mathbb{Z}^2$.

   (a) $\mathcal{P} = \{(1,0), (1,1), (1,3), (1,4)\}$ can be drawn as the black dots in Figure 3.1. The cone $\text{pos}(\mathcal{P})$ is the gray shaded region.

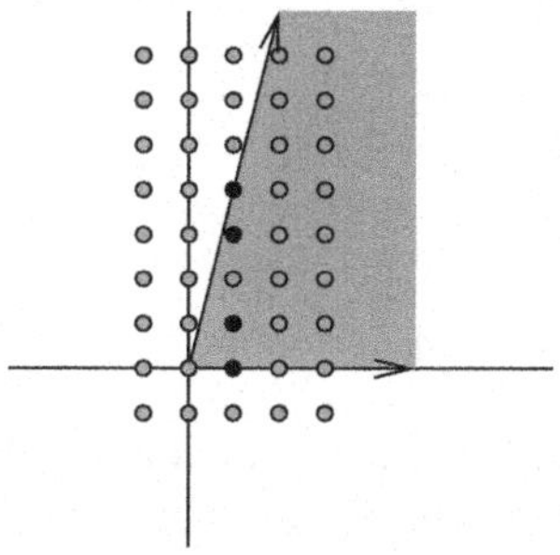

Figure 3.1. The cone $\text{pos}(\mathcal{P})$ for $\mathcal{P}$ in 7a.

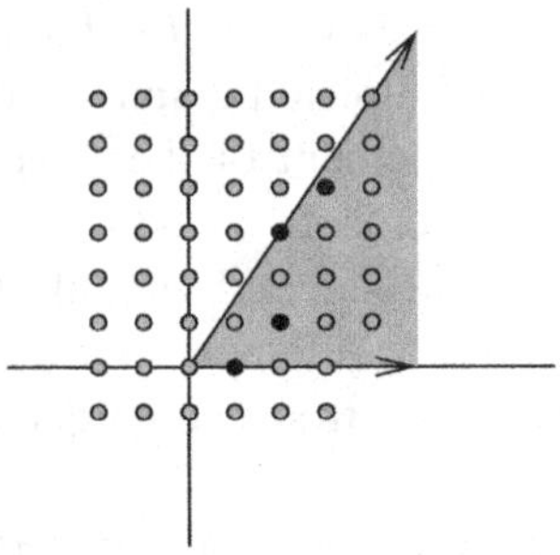

Figure 3.2. The cone pos($\mathcal{P}$) for $\mathcal{P}$ in 7b.

The intersection of the cone with $\mathbb{Z}^2$ is the semigroup of all integer lattice points within the cone. Therefore the Hilbert basis is $\{(1,0), (1,1), (1,2), (1,3), (1,4)\}$.

(b) $\mathcal{P} = \{(1,0), (2,1), (2,3), (3,4)\}$ can be drawn as the black dots in Figure 3.2. The cone pos($\mathcal{P}$) is the gray shaded region.

Again the intersection of the cone with $\mathbb{Z}^2$ is the semigroup of all integer lattice points within the cone, so the Hilbert basis is $\{(1,0), (1,1), (2,3)\}$.

8. (a) i. Run the command `graver` *filename* from 4ti2 on the matrix A and use `output bin`*filename.gra* to format the output file. The result is that the Graver basis is $\{x_2^3 - x_1^2x_3,\ x_2^4 - x_1^3x_4,\ x_2x_3 - x_1x_4,\ x_2^2x_4 - x_1x_3^2,\ x_2x_4^2 - x_3^3,\ x_3^4 - x_1x_4^3\}$.
   ii. Again, using 4ti2, we find that the Graver basis is $\{x_2^3 - x_1^4x_3,\ x_2^4 - x_1^5x_4,\ x_2x_3 - x_1x_4,\ x_2^2x_4 - x_1^3x_3^2,\ x_2x_4^2 - x_1^2x_3^3,\ x_1x_3^4 - x_4^3,\ x_2x_3^5 - x_4^4\}$.

(b) i. First we calculate the total degrees of the elements of the Graver basis.

element	total degree
$x_2^3 - x_1^2x_3$	3
$x_2^4 - x_1^3x_4$	4
$x_2x_3 - x_1x_4$	2
$x_2^2x_4 - x_1x_3^2$	3
$x_2x_4^2 - x_3^3$	3
$x_3^4 - x_1x_4^3$	4

Now we calculate the determinants of the two by two minors of A. This can be done by hand, or using Macaulay 2 with the command `minors(2,matrix {{1,1,1,1},{0,1,3,4}})`. Either way, we find that $D(A) = 4$. So now we check that indeed $4 \leq (3)(4)$.

ii. First we calculate the total degrees of the elements of the Graver basis.

element	total degree
$x_2^3 - x_1^4 x_3$	5
$x_2^4 - x_1^5 x_4$	6
$x_2 x_3 - x_1 x_4$	2
$x_2^2 x_4 - x_1^3 x_3^2$	5
$x_2 x_4^2 - x_1^2 x_3^3$	5
$x_1 x_3^4 - x_4^3$	5
$x_2 x_3^5 - x_4^4$	6

Now we calculate the determinants of the two by two minors of A and find that $D(A) = 5$. So now we check that indeed $6 \leq (3)(5)$.

4

Triangulations

The main point of this chapter is to discuss the relationship between initial ideals of toric ideals and triangulations of the vector configurations.

4.1 Background on triangulations

Let $\mathcal{A} = \{\mathbf{a}_1, \ldots, \mathbf{a}_n\}$ be a collection of nonzero integral vectors in $\mathbb{Z}^d$. We use the polyhedral language defined in Chapter 2. We will always assume that $\text{pos}(\mathbf{a}_i : 1 \leq i \leq n)$ is d-dimensional. We denote by $[n]$ the set $\{1, \ldots, n\}$. A *simplicial complex* is a collection $\{\sigma \subseteq [n]\}$ that is closed under taking subsets.

A *triangulation* of $\mathcal{A}$ is a simplicial complex Δ of subsets of $[n]$ such that the cones $\{\text{pos}(\mathbf{a}_i : i \in \sigma) : \sigma \in \Delta\}$ form a simplicial fan whose support is $\text{pos}(\mathbf{a}_i : 1 \leq i \leq n)$. A geometric realization of a triangulation has vertices at some of the points of $\mathcal{A}$.

Example 4.1.1. Let $\mathcal{A} \subset \mathbb{Z}^3$ be the collection $\{(1,0,0), (1,1,0), (1,2,0), (1,0,1), (1,1,1)\}$. These five vectors all have first coordinate one, so a cross-section of $\text{pos}(\mathbf{a}_i : 1 \leq i \leq 5)$ is obtained by taking the convex hull of the points $\{(0,0), (1,0), (2,0), (0,1), (1,1)\}$. Two triangulations of $\mathcal{A}$ are shown in cross-section in Figure 4.1, where the points are labeled $1, \ldots, 5$ is the order listed here. The first consists of all subsets of sets in $\{\{1,2,4\}, \{2,4,5\}, \{2,3,5\}\}$, while the second is all subsets of sets in $\{\{1,2,4\}, \{2,3,4\}, \{3,4,5\}\}$.

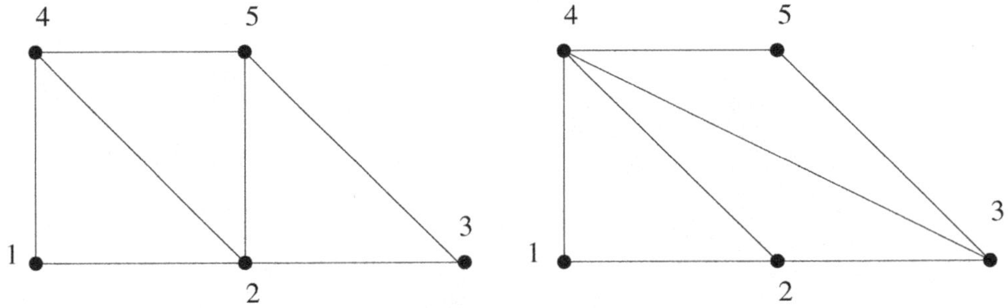

Figure 4.1. Triangulations of the configuration in Example 4.1.1

We focus on a subclass of triangulations known as regular or coherent triangulations. These subclasses were first introduced by Gelfand, Kapranov, and Zelevinsky [GKZ94].

Definition 4.1.2. A triangulation Δ of $\mathcal{A} = \{\mathbf{a}_1, \ldots, \mathbf{a}_n\}$ is *regular* or *coherent* if there is a weight vector $\mathbf{w} \in \mathbb{R}^n$ for which the following condition holds: a subset $\{i_1, \ldots, i_k\}$ is a face of Δ if and only if there exists a vector $\mathbf{c} \in \mathbb{R}^d$ with $\mathbf{a}_j \cdot \mathbf{c} = w_j$ for $j \in \{i_1, \ldots, i_k\}$, and $\mathbf{a}_j \cdot \mathbf{c} < w_j$ otherwise. In this case we denote by $\Delta_\mathbf{w}$ the triangulation Δ.

An alternative description of $\Delta_\mathbf{w}$ is that it is the triangulation of $\mathcal{A}$ induced by the lower faces of the cone in $\mathbb{R}^{d+1}$ formed by taking the polyhedral cone generated by the vectors $(\mathbf{a}_j, w_j)$. A *lower face* of a cone in $\mathbb{R}^{d+1}$ is one visible from the point $-N\mathbf{e}_{d+1}$ for $N \gg 0$. To see the equivalence, note that $\operatorname{pos}((\mathbf{a}_j, w_j) : j \in \sigma)$ is a lower face of this cone if and only if there is a vector $\mathbf{c} \in \mathbb{R}^d$ with $(\mathbf{c}, -1) \cdot (\mathbf{a}_j, w_j) = 0$ for $j \in \sigma$ and $(\mathbf{c}, -1) \cdot (\mathbf{a}_j, w_j) < 0$ for $j \notin \sigma$. But this means $\mathbf{c} \cdot \mathbf{a}_j = w_j$ for $j \in \sigma$ and $\mathbf{c} \cdot \mathbf{a}_j < w_j$ for $j \notin \sigma$, so $\operatorname{pos}(\mathbf{a}_j, w_j)$ is a face of $\Delta_\mathbf{w}$.

Note that for some $\mathbf{w}$ the lower faces will not be simplicial cones, so we will not get a triangulation of $\mathcal{A}$, but rather a subdivision. The requirement that $\Delta_\mathbf{w}$ be a triangulation of $\mathcal{A}$ places some conditions on $\mathbf{w}$, which are satisfied for most $\mathbf{w} \in \mathbb{R}^n$.

Example 4.1.3. The two triangulations shown in Figure 4.1 are both regular. The triangulation on the left is $\Delta_\mathbf{w}$ for $\mathbf{w} = (1, 0, 1, 0, 0)$. To check this, note that a vector $\mathbf{c}$ corresponding to the face $\{1, 2, 4\}$ is $(1, -1, -1)$, a $\mathbf{c}$ for $\{2, 4, 5\}$ is $(0, 0, 0)$, and a $\mathbf{c}$ for $\{2, 3, 5\}$ is $(-1, 1, 0)$. The triangulation on the left is $\Delta_\mathbf{w}$ for $\mathbf{w} = (1, 0, 0, 0, 1)$.

A non-regular triangulation is shown in Figure 4.2. This is the smallest such example in terms of dimension d, number of vectors n, and codimension $n - d$. Checking that this triangulation is not regular involves writing down a list of inequalities on the

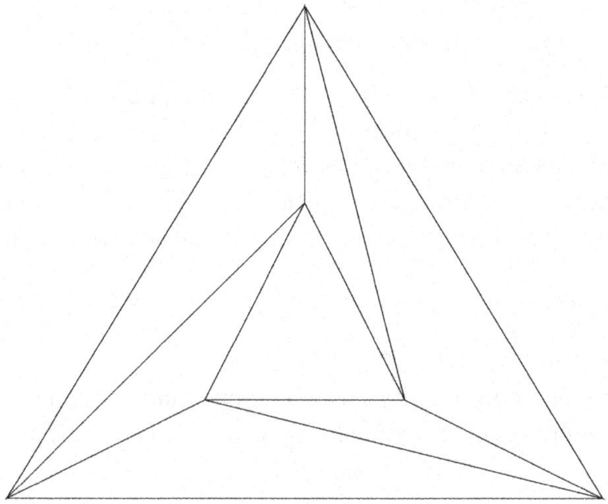

Figure 4.2. The smallest non-regular triangulation.

components of a vector $\mathbf{w}$ that would satisfy the conditions of Definition 4.1.2, and observing that no solution to these inequalities exists.

Simplicial complexes have an intimate connection to commutative algebra via the Stanley-Reisner ideal.

Definition 4.1.4. Let Δ be a simplicial complex of subsets of $[n]$. The *Stanley-Reisner ideal* $I(\Delta)$ is the squarefree monomial ideal

$$I(\Delta) = \left\langle \prod_{i\in\sigma} x_i : \sigma \notin \Delta \right\rangle = \bigcap_{\sigma\in\Delta} \langle x_i : i \notin \sigma \rangle.$$

Example 4.1.5. Let Δ be the first triangulation of Example 4.1.1. Then Δ consists of all subsets of the sets $\{\{1, 2, 4\}, \{2, 4, 5\}, \{2, 3, 5\}\}$, and the Stanley-Reisner ideal $I(\Delta) = \langle x_1x_3, x_1x_5, x_3x_4 \rangle$.

The map from simplicial complexes to squarefree monomial ideals given by taking the Stanley-Reisner ideal is a bijection. It associates a simplicial complex, the Stanley-Reisner complex $\Delta(I)$, to any squarefree monomial ideal I. Specifically, $\Delta(I) = \{\sigma \subseteq [n] : \prod_{i\in\sigma} x_i \notin I\}$.

4.2 The connection with initial ideals of $I_{\mathcal{A}}$

The main result of this lecture is a natural map from initial ideals of the toric ideal $I_{\mathcal{A}}$ to regular triangulations of the configuration $\mathcal{A} = \{\mathbf{a}_1, \ldots, \mathbf{a}_n\}$. We denote by $\mathrm{supp}(\mathbf{u})$ the set $\{i : u_i \neq 0\}$ of a vector $\mathbf{u} \in \mathbb{R}^n$, which we call the *support* of $\mathbf{u}$. Recall the *radical* of an ideal J is the ideal $\mathrm{rad}(J) = \langle f : f^n \in J \text{ for some } n \in \mathbb{N}\rangle$. We say $\mathbf{w}$ is a *generic weight vector* if $\mathrm{in}_{\mathbf{w}}(I_{\mathcal{A}})$ is a monomial ideal.

Theorem 4.2.1. *Let A be a $d \times n$ integer-valued matrix with* $\ker(A) \cap \mathbb{N}^n = \{0\}$, *and let $\mathcal{A} = \{\mathbf{a}_1, \ldots, \mathbf{a}_n\}$ be the columns of A. Let $J =$* $\mathrm{in}_{\mathbf{w}}(I_{\mathcal{A}})$ *for some generic weight vector* $\mathbf{w}$. *Then* $\mathrm{rad}(J)$ *is the Stanley-Reisner ideal of the simplicial complex* $\Delta_{\mathbf{w}}$.

Proof. Let $F \subseteq [n]$. Then F is a face of the simplicial complex associated to $\Delta_{\mathbf{w}}$ if and only if there is a $\mathbf{c} \in \mathbb{R}^d$ with $\mathbf{a}_i \cdot \mathbf{c} \leq \mathbf{w}_i$ for $1 \leq i \leq n$, with equality exactly when $i \in F$. Let P be the polyhedron $\{\mathbf{c} : \mathbf{a}_i \cdot \mathbf{c} \leq \mathbf{w}_i\}$. Candidates for F thus correspond to faces of P. Choose a cost vector $\mathbf{b} \in \mathbb{R}^d$ that is maximized at the face of P corresponding to F. Then we can find an optimal solution $\mathbf{c}$ to the linear program

$$\text{maximize } \mathbf{c} \cdot \mathbf{b} \text{ subject to } \mathbf{c} \in \mathbb{R}^d,\ \mathbf{a}_i \cdot \mathbf{c} \leq \mathbf{w}_i \text{ for } 1 \leq i \leq n,$$

with $\mathbf{a}_i \cdot \mathbf{c} = \mathbf{w}_i$ exactly when $i \in F$.

Now linear programming duality and complementary slackness imply that this $\mathbf{c}$ exists if and only there is an optimal solution $\mathbf{u} \in \mathbb{R}^n$ to the dual program

$$\text{minimize } \mathbf{u} \cdot \mathbf{w} \text{ subject to } \mathbf{u} \in \mathbb{R}^n,\ \mathbf{u} \geq \mathbf{0},\ \sum_{i=1}^{n} u_i \mathbf{a}_i = \mathbf{b}$$

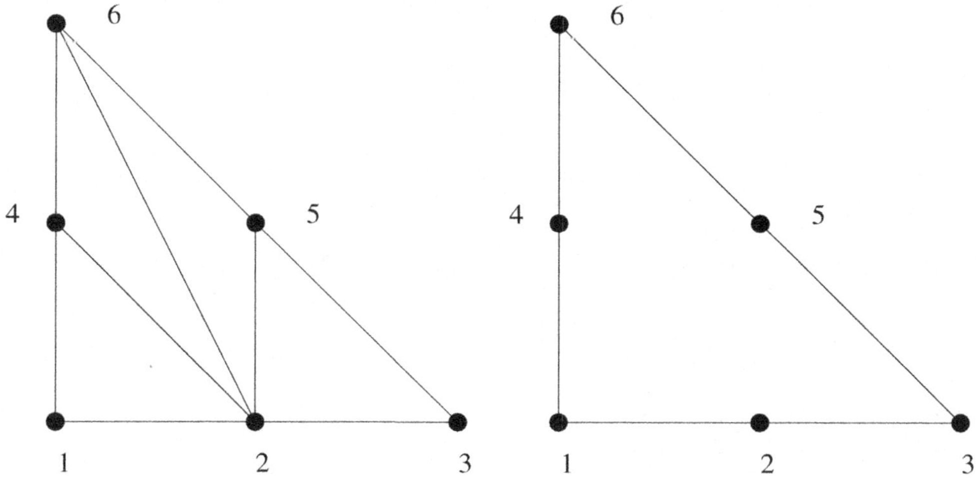

Figure 4.3. The triangulations of Example 4.2.2

with $\text{supp}(\mathbf{u}) = F$. Details and definitions of linear programming duality and complementary slackness may be found in any text on linear programming, such as in Corollary 7.1g and Section 7.9 of [Sch86]. In general the solution $\mathbf{u} \in \mathbb{R}^n$ to the dual program will have rational entries (since the vectors $\mathbf{a}_i$ are integral), but we can replace $\mathbf{b}$ by a suitably large multiple to ensure that $\mathbf{u}$ lies in $\mathbb{N}^n$. But this means that the monomial $\mathbf{x}^{\mathbf{u}}$ has $\text{supp}(\mathbf{u}) = F$, and $\mathbf{x}^{\mathbf{u}} \notin J$, since the solution to this integer program is the unique standard monomial of J of degree $\mathbf{b}$ (see Exercise 3 of Tutorial 3). In fact, no power of $\mathbf{x}^{\mathbf{u}}$ lies in J, since if $\mathbf{u}$ is an integral solution to this dual program then $l\mathbf{u}$ is a solution to the dual program with $\mathbf{b}$ replaced by $l\mathbf{b}$. This condition on $\mathbf{x}^{\mathbf{u}}$ is exactly the condition that $\mathbf{x}^{\mathbf{u}} \notin \text{rad}(J)$. Thus any monomial $\mathbf{x}^{\mathbf{v}}$ in S with $\text{supp}(\mathbf{v}) = F$ does not lie in $\text{rad}(J)$. Each of these implications is reversible, so we conclude that $\text{rad}(J)$ is the Stanley-Reisner ideal of the simplicial complex $\Delta_{\mathbf{w}}$.

Example 4.2.2. Let $\mathcal{A} = \{(1,0,0), (1,1,0), (1,2,0), (1,0,1), (1,1,1), (1,0,2)\}$. Then $I_{\mathcal{A}} = \langle cf - e^2, de - bf, af - d^2, ae - bd, cd - be, ac - b^2 \rangle \subseteq \mathbf{k}[a,b,c,d,e,f]$. For $\mathbf{w} = (10, 1, 10, 5, 1, 1)$, we have $\text{in}_{\mathbf{w}}(I_{\mathcal{A}}) = \langle cf, de, af, ae, cd, ac \rangle$. This is already a radical ideal, and it is the Stanley-Reisner ideal of the triangulation shown in cross-section on the left in Figure 4.3.

When $\mathbf{w} = (1, 10, 1, 10, 10, 1)$, $\text{in}_{\mathbf{w}}(I_{\mathcal{A}}) = \langle e^2, de, be, d^2, bd, b^2 \rangle$. The radical of this ideal is $\langle b, d, e \rangle$. This is the Stanley-Reisner ideal of the second triangulation in Figure 4.3.

We leave it to the reader to check that each of the triangulations in Figure 4.3 is $\Delta_{\mathbf{w}}$ for the corresponding $\mathbf{w}$.

4.3 The secondary fan and the Gröbner fan

In the same way that initial ideals of an ideal correspond to vertices of the state polytope, regular triangulations of a vector configuration correspond to vertices of a

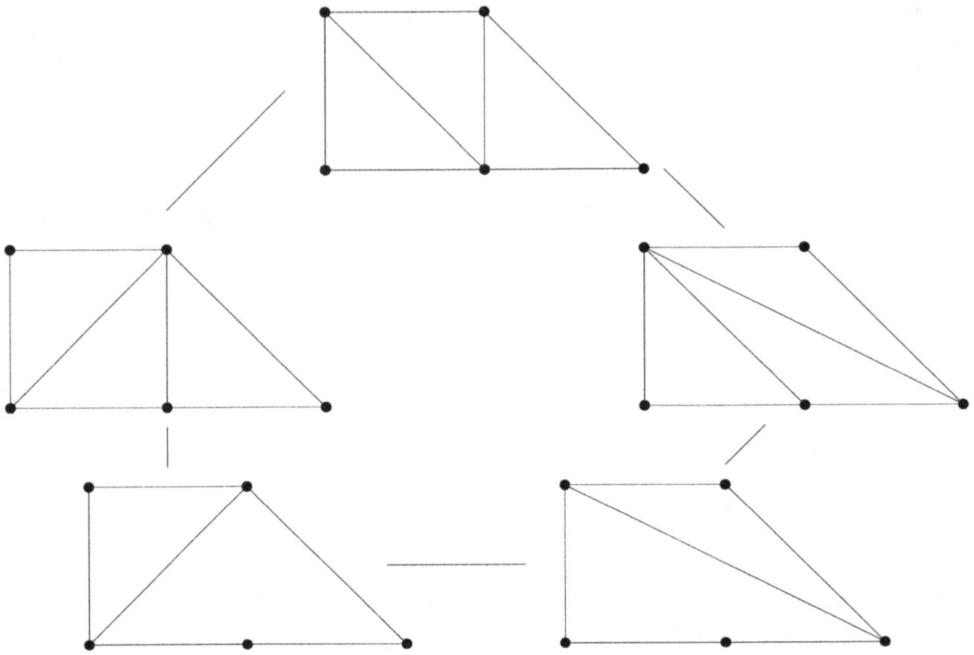

Figure 4.4. The secondary polytope of the configuration of Example 4.1.1.

polytope known as the secondary polytope. Its normal fan is called the secondary fan of $\mathcal{A}$.

Definition 4.3.1. Let $\mathcal{A}$ be a collection of integral vectors in $\mathbb{Z}^d$. The *secondary fan* of $\mathcal{A}$ is the polyhedral fan $S_{\mathcal{A}}$ whose top-dimensional open cones consist of all $\mathbf{w}'$ for which $\Delta_{\mathbf{w}'} = \Delta_{\mathbf{w}}$, as $\Delta_{\mathbf{w}}$ varies over the finitely-many regular triangulations of $\mathcal{A}$.

Remark 4.3.2. In giving Definition 4.3.1 we are omitting a proof that the secondary fan is actually a polyhedral fan. In principle a proof can be given like the one given for the Gröbner fan in Chapter 2, where we define sets of weight vectors for each triangulation, check that their closures are polyhedral cones, and finally check that these cones fit together to form a polyhedral fan.

For example, to check that the closure of these sets are at least convex, note that if $\Delta_{\mathbf{w}} = \Delta_{\mathbf{w}'}$, and $\mathbf{c} \cdot \mathbf{a}_i \leq \mathbf{w}_i$, and $\mathbf{c}' \cdot \mathbf{a}_i \leq \mathbf{w}'_i$ with equality for the same values of i, then $(\lambda\mathbf{c} + (1-\lambda)\mathbf{c}') \cdot \mathbf{a}_i \leq (\lambda\mathbf{w} + (1-\lambda)\mathbf{w}')_i$, for any $0 \leq \lambda \leq 1$, with equality for the same values of i.

Example 4.3.3. For the configuration in Example 4.1.1 the secondary fan has five cones. The corresponding secondary polytope (a pentagon) appears in Figure 4.4, with each vertex replaced by the corresponding triangulation.

Note that although the secondary fan lives in $\mathbb{R}^5$ we have drawn a two-dimensional polytope. This is because the secondary fan has a three dimensional lineality space. In general, if $\mathcal{A}$ is a configuration of n vectors in $\mathbb{Z}^d$ not lying in any hyperplane, then lineality space of the secondary fan is the row space of the corresponding matrix

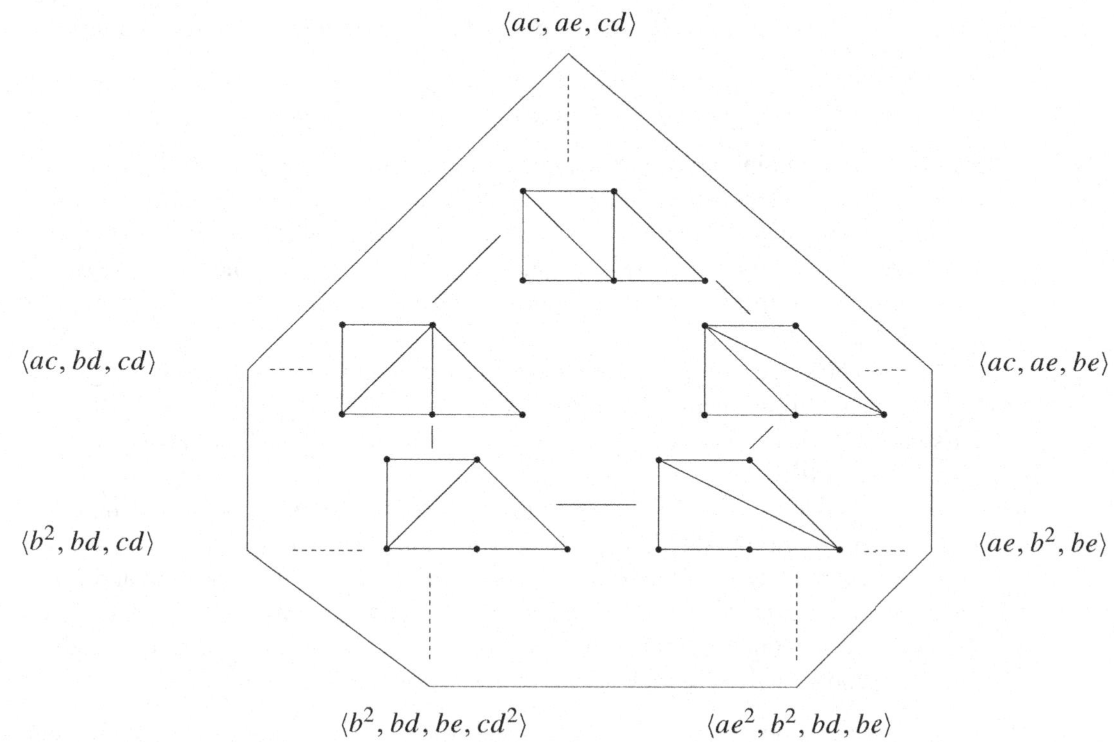

Figure 4.5. The relation between the state and secondary polytope of the configuration of Example 4.1.1.

A with columns the vectors of $\mathcal{A}$, so the secondary polytope of $\mathcal{A}$ will be $(n-d)$-dimensional. Our example consists of five vectors in $\mathbb{R}^3$ (which we draw as five points in $\mathbb{R}^2$), so we have a two-dimensional secondary polytope.

A polyhedral fan $\mathcal{F}'$ *refines* a polyhedral fan $\mathcal{F}$ if every cone of $\mathcal{F}'$ is contained inside some cone of $\mathcal{F}$.

Proposition 4.3.4. *The Gröbner fan of $I_{\mathcal{A}}$ refines the secondary fan of $\mathcal{A}$.*

Proof. If $\text{in}_{\mathbf{w}}(I_{\mathcal{A}}) = \text{in}_{\mathbf{w}'}(I_{\mathcal{A}})$, then Theorem 4.2.1 implies that $\Delta_{\mathbf{w}} = \Delta_{\mathbf{w}'}$.

Example 4.3.5. Let $\mathcal{A}$ be the configuration of Example 4.1.1. The relationship between the state and secondary polytopes of this configuration is shown in Figure 4.5. Here the outer heptagon is the state polytope, with the corresponding initial ideal of $I_{\mathcal{A}}$ written next to the vertex, and the inner pentagon is the secondary polytope. A dotted line connecting an inner vertex to an outer vertex means that the corresponding cone in the secondary fan contains the corresponding cone in the Gröbner fan.

In general this refinement is proper. However there are some special cases where a cone in the secondary fan is not subdivided in the Gröbner cone.

Definition 4.3.6. A triangulation Δ of a collection $\mathcal{A} = \{\mathbf{a}_1, \ldots, \mathbf{a}_n\}$ of integral vectors is *unimodular* if every simplex $\sigma \in \Delta$ satisfies $\mathbb{N}\{\mathbf{a}_i : i \in \sigma\} = \text{pos}(\mathbf{a}_i : i \in \sigma) \cap \mathbb{Z}\{\mathbf{a}_i : 1 \leq i \leq n\}$.

The configuration $\mathcal{A}$ is called *unimodular* if *every* triangulation of $\mathcal{A}$ is unimodular.

Proposition 4.3.7. *If Δ is a regular unimodular triangulation of $\mathcal{A}$, then the cone of the secondary fan corresponding to Δ is not subdivided in the Gröbner fan.*

Proof. Let J be an initial ideal of $I_{\mathcal{A}}$ corresponding to Δ. It suffices to show that J is squarefree, so x_i^2 does not divide any generator for $1 \leq i \leq n$. This suffices because then J is the Stanley-Reisner ideal of a triangulation Δ of $\mathcal{A}$. Since every initial ideal of $I_{\mathcal{A}}$ corresponding to Δ must be contained in the Stanley-Reisner ideal of Δ, and we cannot have proper containment of initial ideals by Corollary 2.2.3, this proves the proposition.

We now show that J is squarefree. Let $\mathbf{x}^{\mathbf{u}}$ be a monomial not in J. Let $\mathbf{b} = \sum_i u_i \mathbf{a}_i$. Let $\sigma \in \Delta$ be a top-dimensional simplex with $\mathbf{b} \in \text{pos}(\mathbf{a}_i : i \in \sigma)$, so $\mathbf{b} = \sum_{i \in \sigma} \lambda_i \mathbf{a}_i$ for some $\lambda_i \geq 0$. The fact that Δ is a unimodular triangulation of $\mathcal{A}$ means that since $\mathbf{b} \in \mathbb{Z}\mathcal{A}$ we have $\lambda_i \in \mathbb{N}$.

By construction we have $\mathbf{x}^{\mathbf{u}} - \mathbf{x}^{\lambda} \in I_{\mathcal{A}}$. We cannot have $\mathbf{x}^{\lambda} \in J$, since $\text{supp}(\lambda) \subseteq \sigma \in \Delta$, and by assumption $\mathbf{x}^{\mathbf{u}} \notin J$. Since J is an initial ideal of $I_{\mathcal{A}}$ the only way this is possible is if $\mathbf{x}^{\mathbf{u}} = \mathbf{x}^{\lambda}$. This means that $\text{supp}(\mathbf{u}) \in \Delta$, so we conclude that $\mathbf{x}^{l\mathbf{u}} \notin J$ for any $l \in \mathbb{N}$. Thus $\mathbf{x}^{\mathbf{u}} \notin \text{rad}(J)$. Since $\mathbf{x}^{\mathbf{u}}$ was an arbitrary monomial not in J, this shows that any monomial not in J is not in $\text{rad}(J)$, so we must have $J = \text{rad}(J)$. Since J is monomial, we conclude that J is squarefree.

The converse, however, is not true because there is a configuration with a nonunimodular triangulation whose secondary cone is not subdivided in the Gröbner fan. In fact, the following is an open problem.

Problem 4.3.8. How does the Gröbner fan refine a given cone in the secondary fan?

When $n - d = 2$, so the state and secondary polytopes are two-dimensional, a two-dimensional cone in the secondary fan is subdivided by adding the rays corresponding to the Hilbert basis of the cone. No such answer is known when $n - d \geq 3$.

Problem 4.3.9. Characterize the top-dimensional open cones in the secondary fan that are not subdivided in the Gröbner fan.

Proposition 4.3.7 says that the cones corresponding to unimodular triangulations are not subdivided, and when $n - d = 2$ these are the only non-subdivided cones. When $n - d \geq 3$ more complicated cones can fail to subdivide.

4.4 Further reading

This lecture covers Chapter 8 of [Stu96].

The Stanley-Reisner ideal was introduced in the work of Stanley and Reisner. A notable early application was Stanley's proof of the Upper Bound Theorem for simplicial spheres, which bounds the number of faces of each dimension a d-dimensional simplicial sphere can have in terms of the number of vertices. Much more information about Stanley-Reisner ideals can be found in [Sta96]. Stanley calls the quotient of the polynomial ring by the Stanley-Reisner ideal the *face ring* of the simplicial complex. Other aspects of squarefree monomial ideals in combinatorics and commutative algebra can be found in [MS05].

The original motivation for regular triangulations can be found in [GKZ94]. A good reference for general information about triangulations will be the forthcoming book [LRS]. The program TOPCOM by Jörg Rambau is the current best method to compute all regular triangulations, and thus the secondary fan. When $\mathcal{A}$ is unimodular, the program CaTS [Jena] computes the Stanley-Reisner ideals of all regular triangulations. CaTS can also be used for arbitrary $\mathcal{A}$ by applying Theorem 4.2.1, though there may be much redundancy in this case. The secondary polytope is a special case of a fiber polytope [BS92].

4.5 Tutorial 4

1. Let $I = \langle x^2y, y^2z, u^3\rangle$ be a monomial ideal in a polynomial ring $k[x, y, z, u]$.
   (a) Find the radical $\sqrt{I}$ of I.
   (b) Find the primary decomposition of $\sqrt{I}$.
   (c) Draw the Stanley-Reisner complex of $\sqrt{I}$.
2. Let $\mathcal{A} = \{(2,0,0), (1,1,0), (0,2,0), (1,0,1), (0,1,1), (0,0,2)\}$. The ideals I_1, I_2 and I_3 described below are three initial ideals of the toric ideal $I_{\mathcal{A}}$ in the polynomial ring $k[a, b, c, d, e, f]$, where the variable a corresponds to the first point $(2, 0, 0)$ and so on. Find the triangulation corresponding to each of these initial ideals.

$$I_1 = \langle e^2, de, d^2, cd, bd, b^2\rangle$$

$$I_2 = \langle b^2, bd, be, ae^2, af, bf, cf\rangle$$

$$I_3 = \langle ac, cd, ae, de, af, cf\rangle$$

Algorithm 4.5.1 ([Stu96, Algorithm 7.2]). Computing the Graver basis of a toric ideal using Lawrence liftings.

Consider a matrix A in $\mathbb{Z}^{d\times n}$ and construct the enlarged matrix called the *Lawrence lifting* of A

$$\Lambda(A) = \begin{pmatrix} A & \mathbf{0} \\ \mathbf{1} & \mathbf{1} \end{pmatrix}$$

where $\mathbf{0}$ and $\mathbf{1}$ are the $d \times n$ zero and $n \times n$ identity matrices, respectively. The matrices A and $\Lambda(A)$ have isomorphic kernels, and so the toric ideal is a homogeneous prime ideal

$$I_{\Lambda(A)} = \langle \mathbf{x}^{\mathbf{u}^+}\mathbf{y}^{\mathbf{u}^-} - \mathbf{x}^{\mathbf{u}^-}\mathbf{y}^{\mathbf{u}^+} \; : \; \mathbf{u} \in \ker(A)\rangle$$

in the polynomial ring $k[x_1, \ldots, x_n, y_1, \ldots, y_n]$.

For a Lawrence type matrix $\Lambda(A)$, the Graver basis and universal Gröbner basis are the same, and they are equal to any reduced Gröbner basis for the ideal $I_{\Lambda(A)}$ [Stu96,

Theorem 7.1]. This fact gives us the following algorithm for computing the Graver basis of a given matrix.

Step 1. Choose any term order on $k[x_1, \ldots, x_n, y_1, \ldots, y_n]$, and compute the reduced Gröbner basis $\mathcal{G}$ of $I_{\Lambda(A)}$.

Step 2. Substitute 1 for $y_1, \ldots, y_n$ in $\mathcal{G}$. The resulting subset of $k[x_1, \ldots, x_n]$ is the Graver basis $Gr_{\mathcal{A}}$.

In the following exercises, let $\mathcal{A}$ be a set of vectors $\{(3,0), (2,1), (1,2), (0,3)\}$ in $\mathbb{Z}^2$. The toric ideal $I_{\mathcal{A}} = \langle xz - y^2, xw - yz, yw - z^2\rangle$ in the polynomial ring $k[x, y, z, w]$ corresponds to the projective twisted cubic.

3. Use Algorithm 4.5.1 and the Hoşten-Sturmfels Algorithm described in Tutorial 3 to compute the Graver basis of $I_{\mathcal{A}}$.
4. Show that the universal Gröbner basis $\mathcal{U}_{\mathcal{A}}$ for $I_{\mathcal{A}}$ is the same as the Graver basis found in Exercise 3. To argue that each element of $Gr_{\mathcal{A}}$ is in $\mathcal{U}_{\mathcal{A}}$ you may use the fact that $\mathcal{C}_{\mathcal{A}} \subseteq \mathcal{U}_{\mathcal{A}} \subseteq Gr_{\mathcal{A}}$ (where $\mathcal{C}_{\mathcal{A}}$ is the set of circuits of $I_{\mathcal{A}}$; see Definition 3.4.8), and in the case of a non-circuit compute some initial ideal to show that the presence of that element in $\mathcal{U}_{\mathcal{A}}$ is necessary. *Hint:* Try initial ideals with respect to some term orders or simple weight vectors.
5. Use the program CaTS [Jena], which was introduced in Chapter 3, to compute all initial ideals of $I_{\mathcal{A}}$.
6. Suppose $\mathcal{A} = \{\mathbf{a}_1, \ldots, \mathbf{a}_n\} \subset \mathbb{N}^d$ and $\pi : \mathbb{N}^n \longrightarrow \mathbb{Z}^d$ as defined in Section 3.2 is a semigroup homomorphism such that $\pi(\mathbf{u}) = u_1\mathbf{a}_1 + \ldots + u_n\mathbf{a}_n$. A vector $\mathbf{b} \in \mathbb{N}\mathcal{A}$ is called a *Gröbner degree* if for some binomial $\mathbf{x}^{\mathbf{u}^+} - \mathbf{x}^{\mathbf{u}^-} \in \mathcal{U}_{\mathcal{A}}$, $\pi(\mathbf{u}^+) = \pi(\mathbf{u}^-) = \mathbf{b}$. If $\mathbf{b}$ is a Gröbner degree, the polytope $\mathrm{conv}(\pi^{-1}(\mathbf{b}))$ is called a *Gröbner fiber*.

   The *inner normal fan* of a polytope P is defined similarly to the outer normal fan, but this time the cones consist of minimizing vectors. In other words, for each face F of P, the *inner normal cone* is the set $\{\mathbf{c} \in \mathbb{R}^n : \mathrm{face}_{-\mathbf{c}} P = F\}$.

   (a) ([Stu96, Theorem 7.15]) For $\mathcal{A} = \{\mathbf{a}_1, \ldots, \mathbf{a}_n\} \subset \mathbb{N}^d$, let P be the Minkowski sum of all Gröbner fibers $\mathrm{conv}(\pi^{-1}(\mathbf{b}))$. Show that the inner normal fan of P is the Gröbner fan of $I_{\mathcal{A}}$.

   Now let $\mathcal{A}$ correspond to the projective twisted cubic as above.

   (b) Find the Gröbner fibers of $\mathcal{A}$.
   (c) Compute the Minkowski sum of the Gröbner fibers that you found and compare your answer with the combinatorial description of the state polytope given by CaTS in Exercise 5.

7. (a) Find the initial complexes of $I_{\mathcal{A}}$ (an *initial complex* of $I_{\mathcal{A}}$ is the Stanley-Reisner complex corresponding to the radical of an initial ideal of $I_{\mathcal{A}}$).
   (b) Find all triangulations of $\mathcal{A}$.
   (c) Show that all triangulations of $\mathcal{A}$ are regular, and find a vector corresponding to each triangulation.

(d) Examine your findings above to see how they fit with Theorem 4.2.1.
(e) Find the secondary polytope of $\mathcal{A}$.

4.6 Solutions to Tutorial 4

1. (a) $\sqrt{I} = \langle xy, yz, u \rangle$. In general, if an ideal in a polynomial ring is generated by monomials $M_1, \ldots, M_n$, then its radical is generated by $M'_1, \ldots, M'_n$, where each M'_i is obtained by replacing every power greater than 1 on the variables in M_i by 1 (prove this!).
   (b) $\sqrt{I} = \langle x, z, u \rangle \cap \langle y, u \rangle$. There are several algorithms for finding primary decomposition of monomial ideals; see [Eis94, Chapter 3] and [Vil01, Corollary 5.1.13]. In Macaulay 2 the relevant code would be

```
i1 :  R = QQ[x,y,z,u];
i2 : I = monomialIdeal(x*y,y*z,u);
o2 : MonomialIdeal of R
i3 : ass I
o3 = {monomialIdeal (y, u), monomialIdeal (x, z, u)}
o3 : List
i4 : primaryDecomposition I
o4 = {monomialIdeal (y, u), monomialIdeal (x, z, u)}
o4 : List
```

   The *associated primes* of I are the radicals of the primary ideals appearing in a primary decomposition of I; these are given by `ass I` in the code above.
   (c)

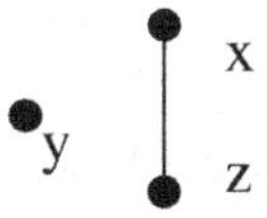

2. The radicals of these ideals are: $\sqrt{I_1} = \langle e, b, d \rangle$, $\sqrt{I_2} = \langle b, ae, af, cf \rangle$ and $\sqrt{I_3} = \langle ac, cd, ae, de, af, cf \rangle$, which are respectively from left to right the Stanley-Reisner ideals of the following three simplicial complexes:

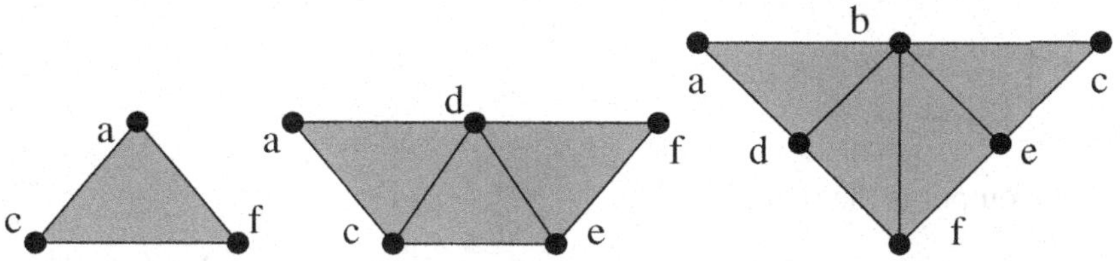

These simplicial complexes correspond, respectively, to the three following triangulations of $\mathcal{A}$:

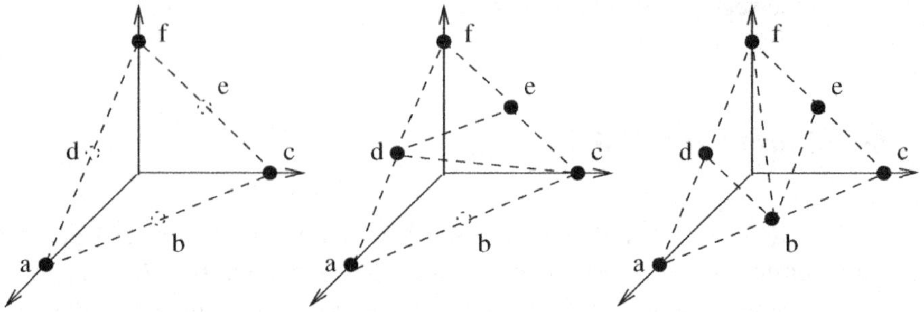

3.

$$\Lambda(A) = \begin{pmatrix} 3 & 2 & 1 & 0 & 0 & 0 & 0 & 0 \\ 0 & 1 & 2 & 3 & 0 & 0 & 0 & 0 \\ 1 & 0 & 0 & 0 & 1 & 0 & 0 & 0 \\ 0 & 1 & 0 & 0 & 0 & 1 & 0 & 0 \\ 0 & 0 & 1 & 0 & 0 & 0 & 1 & 0 \\ 0 & 0 & 0 & 1 & 0 & 0 & 0 & 1 \end{pmatrix}$$

Pick a weight vector **w** and apply HSAlg from Tutorial 3 to the list LA that contains the rows of the Lawrence matrix as follows. In the following code we add a row of 1's to the matrix, as Macaulay 2 will not accept multidegrees that start with 0. This does not affect the toric ideal, since the vector $(1, 1, 1, 1, 1, 1, 1, 1)$ is in the row span of $\Lambda(A)$, and hence the kernel of the matrix will remain the same.

```
i1 : LA = {{1,1,1,1,1,1,1,1}, {3,2,1,0,0,0,0,0},
           {0,1,2,3,0,0,0,0}, {1,0,0,0,1,0,0,0},
           {0,1,0,0,0,1,0,0}, {0,0,1,0,0,0,1,0},
           {0,0,0,1,0,0,0,1}};

i2 : w={1,1,1,1,1,1,1,1};

i3 : HSAlg(LA,w)

o3 = | x_2x_4x_7^2-x_3^2x_6x_8 x_1x_4x_6x_7-x_2x_3x_5
       x_8x_1x_3x_6^2-x_2^2x_5x_7 x_1x_4^2x_7^3-x_3^3
       x_5x_8^2x_1^2x_4x_6^3-x_2^3x_5^2x_8 |

                1        5
o3 : Matrix R  <--- R
```

The output is the set

$$\{x_1x_3x_6^2 - x_2^2x_5x_7,\ x_1x_4x_6x_7 - x_2x_3x_5x_8,\ x_3^2x_6x_8 - x_2x_4x_7^2,\ x_1^2x_4x_6^3 - x_2^3x_5^2x_8,\ x_1x_4^2x_7^3 - x_3^3x_5x_8^2\}.$$

After setting $x_5 = x_6 = x_7 = x_8 = 1$, and $x_1 = x, x_2 = y, x_3 = z, x_4 = w$, we have the Graver basis

$$Gr_{\mathcal{A}} = \{xz - y^2,\ xw - yz,\ z^2 - yw,\ x^2w - y^3,\ xw^2 - z^3\}.$$

4. The four elements $z^2 - yw, xz - y^2, xw^2 - z^3$ and $x^2w - y^3$ are circuits, and so they belong to the universal Gröbner basis $\mathcal{U}_{\mathcal{A}}$.

   To show the remaining element $xw - yz$ is in $\mathcal{U}_{\mathcal{A}}$, we need to find an initial ideal of $I_{\mathcal{A}}$ for which one of the terms of this binomial is a minimal generator. If we use the reverse lexicographic order $x \succ y \succ w \succ z$, we get $\text{in}_{\succ}(I_{\mathcal{A}}) = \langle yw, xw, y^2\rangle$, and so we need the basis element $xw - yz$ to obtain the generator xw for the initial ideal.

   Alternatively, one could use a weight vector, say $\mathbf{w} = (1, 1, 0, 1)$. Then by Lemma 2.4.2, since $Gr_{\mathcal{A}}$ contains a universal Gröbner basis, $\text{in}_{\mathbf{w}}(I_{\mathcal{A}})$ is generated by $\text{in}_{\mathbf{w}}(g)$ where $g \in Gr_{\mathcal{A}}$. So

$$\text{in}_{\mathbf{w}}(I_{\mathcal{A}}) = \langle yw, xw, y^2, xw^2, x^2w - y^3\rangle = \langle yw, xw, y^2\rangle,$$

   so the binomial $xw - yz$ must be in $\mathcal{U}_{\mathcal{A}}$.

5. To use the program CaTS, put $\mathcal{A}$ into a file called "twisted.dat" (so the content of twisted.dat is simply: {(3,0) (2,1) (1,2) (0,3)}). Then run the command:

```
cats -p2 -e -i twisted.dat
```

   CaTS will read the input file twisted.dat, and write the output in a file, which in this case is called "twisted.list". This file will contain the following data (the term `Vtx` refers to a vertex of the state polytope):

```
Vtx: 0 (2 facets/3 binomials/degree 2)
    Initial ideal:{z^2, y*z, y^2}
  Facet Binomials:{# z^2-y*w,# y^2-x*z}
Vtx: 1 (2 facets/4 binomials/degree 3)
    Initial ideal:{y^2, y*z, z^3, y*w}
  Facet Binomials:{# z^3-x*w^2,# y*w-z^2}
Vtx: 2 (2 facets/4 binomials/degree 3)
    Initial ideal:{y^2, y*z, y*w, x*w^2}
  Facet Binomials:{# y*z-x*w,# x*w^2-z^3}
Vtx: 3 (2 facets/3 binomials/degree 2)
    Initial ideal:{y^2, x*w, y*w}
  Facet Binomials:{# y^2-x*z,# x*w-y*z}
Vtx: 4 (2 facets/3 binomials/degree 2)
    Initial ideal:{x*z, x*w, y*w}
  Facet Binomials:{# x*z-y^2,# y*w-z^2}
Vtx: 5 (2 facets/4 binomials/degree 3)
    Initial ideal:{y^3, x*z, y*z, z^2}
  Facet Binomials:{# y^3-x^2*w,# x*z-y^2}
Vtx: 6 (2 facets/4 binomials/degree 3)
```

```
    Initial ideal:{x*z, y*z, z^2, x^2*w}
  Facet Binomials:{# y*z-x*w,# x^2*w-y^3}
Vtx: 7 (2 facets/3 binomials/degree 2)
    Initial ideal:{x*z, z^2, x*w}
  Facet Binomials:{# z^2-y*w,# x*w-y*z}
```

6. (a) It is enough to show that the two fans have the same maximal cones. Suppose $\mathbf{w}$ and $\mathbf{w}'$ are two vectors in the interior of a maximal cone of the inner normal fan of P. We have

$$\text{face}_{-\mathbf{w}}(P) = \text{face}_{-\mathbf{w}'}(P).$$

This equality transfers to each summand of the Minkowski sum, so that we have, equivalently, for each Gröbner degree $\mathbf{b}$

$$\text{face}_{-\mathbf{w}}(\text{conv}(\pi^{-1}(\mathbf{b}))) = \text{face}_{-\mathbf{w}'}(\text{conv}(\pi^{-1}(\mathbf{b}))) = \mathbf{u} \in \text{conv}(\pi^{-1}(\mathbf{b})).$$

In other words

$$\mathbf{w} \cdot \mathbf{u} < \mathbf{w} \cdot \mathbf{v},\ \ \mathbf{w}' \cdot \mathbf{u} < \mathbf{w}' \cdot \mathbf{v} \text{ for all } \mathbf{v} \in \text{conv}(\pi^{-1}(\mathbf{b}))$$

which is equivalent to saying that $\mathbf{w}$ and $\mathbf{w}'$ pick exactly the same (unique) standard monomial $\mathbf{x}^{\mathbf{u}}$ in Gröbner degree $\mathbf{b}$. This means that that $\text{in}_{\mathbf{w}}(I_{\mathcal{A}})$ and $\text{in}_{\mathbf{w}'}(I_{\mathcal{A}})$ have the same generating set in Gröbner degree $\mathbf{b}$. By varying the choice of $\mathbf{b}$ among all Gröbner degrees, we get, equivalently, $\text{in}_{\mathbf{w}}(I_{\mathcal{A}}) = \text{in}_{\mathbf{w}'}(I_{\mathcal{A}})$. This means that $\mathbf{w}$ and $\mathbf{w}'$ belong to the same maximal cone of the Gröbner fan, which proves our claim.

(b) We find the Gröbner degrees:

$$\begin{array}{rll}
yw - z^2 & \mathbf{u}^+ = (0,1,0,1) & \pi(\mathbf{u}^+) = (2,4) \\
xw - yz & \mathbf{u}^+ = (1,0,0,1) & \pi(\mathbf{u}^+) = (3,3) \\
xz - y^2 & \mathbf{u}^- = (0,2,0,0) & \pi(\mathbf{u}^-) = (4,2) \\
xw^2 - z^3 & \mathbf{u}^- = (0,0,3,0) & \pi(\mathbf{u}^-) = (3,6) \\
x^2w - y^3 & \mathbf{u}^- = (0,3,0,0) & \pi(\mathbf{u}^-) = (6,3)
\end{array}$$

We can easily compute each of the fibers:

(1) $\pi^{-1}((2,4)) = \{(0,0,2,0), (0,1,0,1)\}$

(2) $\pi^{-1}((3,3)) = \{(0,1,1,0), (1,0,0,1)\}$

(3) $\pi^{-1}((4,2)) = \{(0,2,0,0), (1,0,1,0)\}$

(4) $\pi^{-1}((3,6)) = \{(0,1,1,1), (0,0,3,0), (1,0,0,2)\}$

(5) $\pi^{-1}((4,2)) = \{(1,1,1,0), (0,3,0,0), (2,0,0,1)\}$

(c) We are looking at the Minkowski sum:

$$\text{conv}(\pi^{-1}((2,4))) + \text{conv}(\pi^{-1}((3,3))) + \text{conv}(\pi^{-1}((4,2))) + \text{conv}(\pi^{-1}((3,6))) + \text{conv}(\pi^{-1}((6,3))).$$

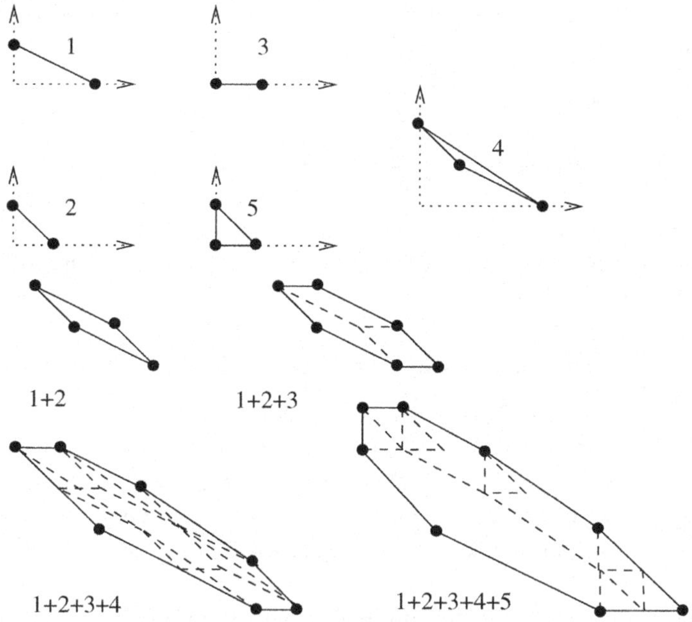

Figure 4.6. The Minkowski sum of the Gröbner fibers

To draw this polytope, first note that it is enough to consider a projection onto $\mathbb{R}^2$. This is because the last two coordinates of the vectors in each fiber determine that particular vector. To see this, suppose $\mathbf{u}$ and $\mathbf{v}$ are two vectors in the fiber $\pi^{-1}(\mathbf{b})$ with $u_3 = v_3$ and $u_4 = v_4$. This means that $A\mathbf{u} = A\mathbf{v}$, where

$$A = \begin{pmatrix} 3 & 2 & 1 & 0 \\ 0 & 1 & 2 & 3 \end{pmatrix}$$

is the matrix of the twisted cubic. It follows that, as the last two coordinates of $\mathbf{u}$ and $\mathbf{v}$ are equal,

$$\begin{pmatrix} 3 & 2 \\ 0 & 1 \end{pmatrix} \begin{bmatrix} u_1 \\ u_2 \end{bmatrix} = \begin{pmatrix} 3 & 2 \\ 0 & 1 \end{pmatrix} \begin{bmatrix} v_1 \\ v_2 \end{bmatrix}$$

and since the matrix $\left(\begin{smallmatrix} 3 & 2 \\ 0 & 1 \end{smallmatrix}\right)$ is nonsingular, we conclude that $u_1 = v_1$ and $u_2 = v_2$, and therefore $\mathbf{u} = \mathbf{v}$.

In Figure 4.6 all the fibers as well as the Minkowski sum are drawn in $\mathbb{R}^2$.

The following explanation clarifies the correspondence between the polytope P and the state polytope. Consider the first initial ideal $\langle z^2, yz, y^2\rangle$ that CaTS gave us in Exercise 5. The Gröbner cone corresponding to this initial ideal, is the closure of the cone

$$C = \{\mathbf{w} \in \mathbb{R}^n_{\geq 0} \ : \ \text{in}_{\mathbf{w}}(I) = \langle z^2, yz, y^2\rangle\}.$$

By Lemma 2.4.2, we can use the Graver basis we found in Exercise 3 to describe C as

$$C = \{\mathbf{w} \in \mathbb{R}^n_{\geq 0} \,:\, 2w_3 > w_2 + w_4,\ 2w_2 > w_1 + w_3,\ 3w_3 > w_1 + 2w_4,$$
$$3w_2 > 2w_1 + w_4,\ w_2 + w_3 > w_1 + w_4\}.$$

Every vertex of P is a sum of vertices of each Gröbner fiber. The inequalities appearing in the description of C tell us which vertex of each Gröbner fiber we should take:

$$\mathbf{v}_1 = (0, 0, 2, 0),\ \mathbf{v}_2 = (0, 2, 0, 0),\ \mathbf{v}_3 = (0, 0, 3, 0),$$
$$\mathbf{v}_4 = (0, 3, 0, 0),\ \mathbf{v}_5 = (0, 1, 1, 0).$$

The vertex $\mathbf{v}_1 + \ldots + \mathbf{v}_5$ of P corresponds to the vertex of the state polytope corresponding to the initial ideal $\langle z^2, y^2, yz\rangle$. Similarly, one can match the remaining 7 initial ideals with the remaining 7 vertices of P.

7. (a) We have already found the initial ideals in Exercise 5. It is also possible to get a list of the initial ideals without any information about the state polytope by using the file "twisted.dat" created earlier and running the command `cats -p2+ < twisted.dat`. The output will be the following.

```
{{z^2, y*z, y^2},
{y^2, y*z, z^3, y*w},
{y^2, y*z, y*w, x*w^2},
{y^2, x*w, y*w},
{x*z, x*w, y*w},
{y^3, x*z, y*z, z^2},
{x*z, y*z, z^2, x^2*w},
{x*z, z^2, x*w}}
```

So the initial ideals, up to radicals, are:

$$\langle y, z\rangle,\ \langle xw, y\rangle,\ \langle xz, xw, yw\rangle,\ \langle z, xw\rangle,$$

which are the Stanley-Reisner ideals of the (initial) complexes (from left to right)

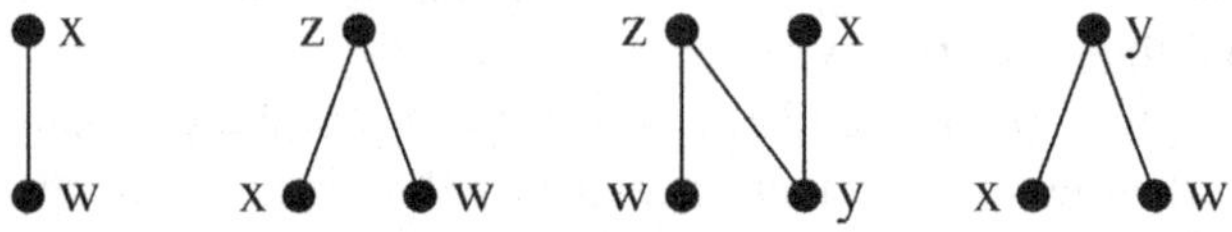

(b) Let the numbers $1, \ldots, 4$ represent the points $\mathbf{a}_1 = (3, 0)$, $\mathbf{a}_2 = (2, 1)$, $\mathbf{a}_3 = (1, 2)$ and $\mathbf{a}_4 = (0, 3)$ of $\mathcal{A}$, respectively. Then the four triangulations are $T_1 = \{\{1, 2\}, \{2, 3\}, \{3, 4\}\}$, $T_2 = \{\{1, 3\}, \{3, 4\}\}$, $T_3 = \{\{1, 2\}, \{2, 4\}\}$ and $T_4 = \{\{1, 4\}\}$. These are shown below respectively, from left to right:

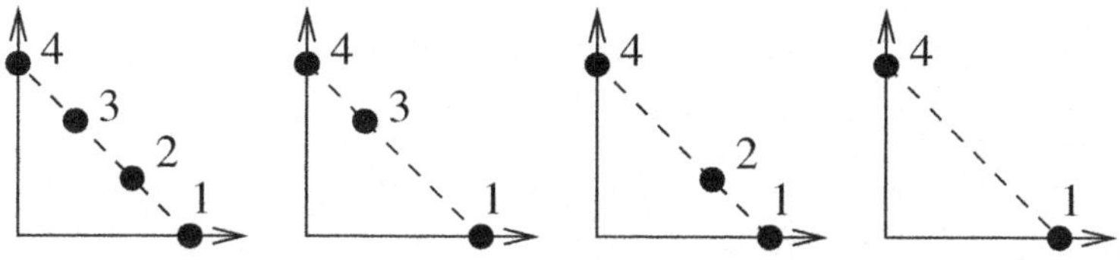

(c) The four triangulations in part (7a) correspond respectively to the following weight vectors: $\mathbf{w}_1 = (1, 0, 0, 1)$, $\mathbf{w}_2 = (1, 1, 0, 1)$, $\mathbf{w}_3 = (1, 0, 1, 1)$ and $\mathbf{w}_4 = (0, 1, 1, 0)$. These choices for the $\mathbf{w}_i$ are not unique; in fact, there are infinitely many options.

We show that, for example, $\mathbf{w}_2 = (1, 1, 0, 1)$ is the appropriate weight vector for $T_2 = \{\{1, 3\}, \{3, 4\}\}$; i.e. $T_2 = \Delta_{\mathbf{w}_2}$. A similar method works for T_1, T_3 and T_4.

Consider the face $\{1, 3\}$. We would like to find $\mathbf{c} = (c_1, c_2) \in \mathbb{R}$ such that

$$\begin{aligned} \mathbf{a}_1 \cdot \mathbf{c} = 3c_1 &= (\mathbf{w}_2)_1 = 1 \\ \mathbf{a}_2 \cdot \mathbf{c} = 2c_1 + c_2 &< (\mathbf{w}_2)_2 = 1 \\ \mathbf{a}_3 \cdot \mathbf{c} = c_1 + 2c_2 &= (\mathbf{w}_2)_3 = 0 \\ \mathbf{a}_4 \cdot \mathbf{c} = 3c_2 &< (\mathbf{w}_2)_4 = 1 \end{aligned}$$

Solving this system, we arrive at $\mathbf{c} = (1/3, -1/6)$.

Now we need to show that $\{1\}$, $\{3\}$, $\{3, 4\}$ and $\{4\}$ are also faces by showing that for each one of them there exists such a $\mathbf{c}$ satisfying a similar set of equations above (the only thing that changes in each case is that the (in)equality signs move around depending which face you are looking at). For these four faces, we see that, respectively, the pairs $\mathbf{c} = (1/3, -1/3)$, $\mathbf{c} = (-1/3, 1/6)$, $\mathbf{c} = (-2/3, 1/3)$ and $\mathbf{c} = (-1, 1/3)$ work.

(d)

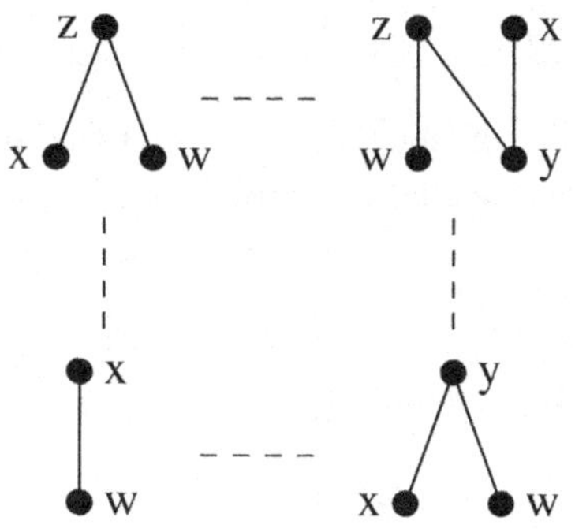

5

Resolutions

5.1 Basics of minimal free resolutions

An important object in commutative algebra is the (minimal) free resolution of a graded module over a polynomial ring. In this chapter we will study minimal free resolutions of toric ideals in $S = \mathbf{k}[x_1, \ldots, x_n]$. For a detailed treatment of the material we touch upon here, see the book by Miller and Sturmfels [MS05]. See Chapters 5 and 6 of [CLO98] for a thorough introduction to free resolutions of polynomial ideals and quotients of polynomial rings, algorithms for their computation, and applications.

Given an ideal I of S, a free resolution $\mathcal{F}$ of the S-module S/I is an exact sequence of the form

$$\mathcal{F} : 0 \longleftarrow S/I \longleftarrow F_0 \xleftarrow{\phi_1} F_1 \xleftarrow{\phi_2} F_2 \xleftarrow{\phi_3} \cdots \xleftarrow{\phi_r} F_r \cdots$$

where each F_i is a free S-module (hence of the form S^{β_i} for some $\beta_i \in \mathbb{N}$). If $F_l \neq 0$ and $0 = F_{l+1} = F_{l+2} = \cdots$, then $\mathcal{F}$ is finite of *length* l:

$$\mathcal{F} : 0 \longleftarrow S/I \longleftarrow F_0 \xleftarrow{\phi_1} F_1 \xleftarrow{\phi_2} F_2 \xleftarrow{\phi_3} \cdots \xleftarrow{\phi_l} F_l \longleftarrow 0.$$

In this chapter we will consider only minimal free resolutions of ideals. Before we define this precisely, let us see an example using Macaulay 2.

```
i1 : S = QQ[x,y];
i2 : I = ideal(x^2-x, y^2-y, x*y)
i3 : F = res coker gens I
      1       3       2
o3 = S  <-- S  <-- S  <-- 0

     0       1       2      3

o3 : ChainComplex
i4 : F.dd
```

```
               1                         3
o4 = 0 : S  <-------------------- S  : 1
              | x2-x xy y2-y |
               3                         2
     1 : S  <-------------------- S  : 2
              {2} | -y   0     |
              {2} | x-1  -y+1  |
              {2} | 0    x     |
               2
     2 : S  <----- 0 : 3
                   0
o4 : ChainComplexMap
```

This computation produces a minimal free resolution of length two of S/I where $I = \langle x^2 - x, y^2 - y, xy\rangle$ with $F_0 = S$, $F_1 = S^3$, $F_2 = S^2$. The rank of F_i in this minimal resolution is called the ith *Betti number* of S/I. We can choose bases for the S-modules F_i allowing the maps ϕ_1 and ϕ_2 to be given by matrices. The entries in ϕ_1 are the three minimal generators of I and thus, image$(\phi_1) = I =$ kernel$(S \longrightarrow S/I)$. Each column in the matrix ϕ_2 is called a *syzygy* (relation) of I. For instance, the first column gives a relation on the generators of I:

$$(-y)(x^2 - x) + (x - 1)(xy) + 0(y^2 - y) = 0.$$

The computation shows that the module of (first) syzygies of I, which is kernel$(\phi_1) =$ image(ϕ_2), is generated by the two columns of ϕ_2. However, kernel$(\phi_2) = 0$ since if

$$\begin{pmatrix} -y \\ x-1 \\ 0 \end{pmatrix} h_1 + \begin{pmatrix} 0 \\ -y+1 \\ x \end{pmatrix} h_2 = 0$$

then $h_1 = h_2 = 0$. Thus the resolution stops after two steps. The degrees of the entries (columns) in the maps ϕ_1 and ϕ_2 are recorded in the *Betti diagram* of $\mathcal{F}$:

```
i5 : betti F

o5 = total: 1 3 2
         0: 1 . .
         1: . 3 2
```

The Betti diagram is to be thought of as a matrix with rows and columns indexed by $0, 1, 2, \ldots$. The column index i denotes the step of the resolution. The top row "`o5 = total: 1 3 2`" indicates the Betti numbers of the steps of the resolution. The row index j is used as follows: the entry in position $\{j, i\}$ of the Betti diagram is the number of entries (columns) of degree $j + i$ in the map ϕ_i. Thus the bottom right entry "2" indicates that ϕ_2 has two degree $(3 = 1 + 2)$ entries (columns). The degree of the syzygy $(-y, x-1, 0)$ is three since in the syzygy relation

$$(-y)(x^2-x)+(x-1)(xy)+0(y^2-y)=0,$$

the highest degree of a term is three. In Macaulay 2 display format, the degrees of the syzygies in ϕ_1 are recorded on the left side of the matrix giving ϕ_2.

In the 1890s Hilbert proved that every ideal in the polynomial ring has a finite free resolution which is known as Hilbert's Syzygy Theorem.

Theorem 5.1.1 (Hilbert's Syzygy Theorem). *Let $S = \mathbf{k}[x_1, \ldots, x_n]$. Then every finitely generated S-module has a free resolution of length at most n.*

This theorem implies that every ideal I in S has a resolution with at most $n+1$ free S-modules in it. The length of a shortest finite free resolution of S/I is called the *projective dimension* of S/I. This is an important invariant of S/I and can be computed in Macaulay 2 using the command `pdim`.

```
i6 : pdim coker gens I
o6 = 2
```

In this chapter, we are interested in graded minimal free resolutions of ideals. If the polynomial ring S is graded by an abelian group A, then we can write $S = \oplus_{a\in A} S_a$ where S_a is the $\mathbf{k}$-vector space of all homogeneous polynomials of degree a. In the usual total degree grading where $\deg(x_i) = 1$ for $i = 1, \ldots, n$, the group $A = \mathbb{Z}$ and $S = \oplus_{t\in\mathbb{Z}} S_t$ where $S_t = 0$ for all $t < 0$ and $S_0 = \mathbf{k}$. An S-module M is graded by A if there exists subgroups $M_a, a \in A$ such that (i) $M = \oplus_{a\in A} M_a$ and (ii) $S_a M_b \subset M_{a+b}$. When $A = \mathbb{Z}$, graded ideals are just homogeneous ideals under the usual total degree grading. In this chapter we only consider gradings of S by abelian groups of the form $\mathbb{Z}^n/\mathcal{L}$ where $\mathcal{L}$ is a sublattice of $\mathbb{Z}^n$ such that $\mathcal{L} \cap \mathbb{N}^n = \{\mathbf{0}\}$. See Chapter 8 in [MS05] for a discussion of valid multigradings in the context of resolutions.

The module M *twisted* by $d \in A$ is defined to be

$$M(d) = \oplus_{a\in A} M(d)_a \text{ where } M(d)_a := M_{d+a}.$$

Example 5.1.2. For a simple example, consider $S(-2) = \oplus_{t\in\mathbb{Z}} S_{-2+t}$. This twisted ring is generated (as an S-module) by 1 which lies in $S_0 = S_{-2+2}$ and hence has degree two in $S(-2)$.

The twisted free modules we will see are of the form $S(-d_1) \oplus S(-d_2) \oplus \cdots \oplus S(-d_m)$ generated by the standard basis elements $e_1, \ldots, e_m$ of degrees $d_1, \ldots, d_m$ respectively.

Definition 5.1.3. Let M and N be S-modules graded by A. A homomorphism $\phi : M \to N$ is a *graded homomorphism* of degree d if $\phi(M_a) \subset N_{a+d}$.

Example 5.1.4. If M is a graded S-module generated by homogeneous elements $f_1, \ldots, f_m$ of degrees $d_1, \ldots, d_m$ respectively, then we get the graded homomorphism $\phi : S(-d_1) \oplus S(-d_2) \oplus \cdots \oplus S(-d_m) \longrightarrow M$ of degree zero which sends $e_i \mapsto f_i$.

We saw earlier that homomorphisms in free resolutions are given by matrices. A graded homomorphism of degree zero

$$S(-d_1) \oplus \cdots \oplus S(-d_p) \longrightarrow S(-c_1) \oplus \cdots \oplus S(-c_m)$$

is defined by an $m \times p$ matrix U whose ijth entry u_{ij} is a homogeneous element of S of degree $d_j - c_i$ for all i, j. We call such a matrix a *graded* matrix over S.

Definition 5.1.5. If M is a graded S-module, then a *graded resolution* of M is a resolution of the form

$$0 \longleftarrow M \longleftarrow F_0 \xleftarrow{\phi_1} F_1 \xleftarrow{\phi_2} F_2 \cdots \longleftarrow$$

where each F_i is a twisted free graded module $S(-d_1) \oplus \cdots \oplus S(-d_{p_i})$ and each map ϕ_i is a graded homomorphism of degree zero (given by graded matrices).

Remark 5.1.6. In the Macaulay 2 display format, maps in resolutions always go from right to left as in Definition 5.1.5. However, it is more usual in the literature to have maps between modules go from left to right. We use the Macaulay 2 format in this chapter when talking about a full resolution to match the Macaulay 2 outputs on examples.

The Hilbert Syzygy Theorem holds also for graded resolutions which guarantees finite graded free resolutions of length at most n for every S-module. Among these the most useful are the minimal graded resolutions defined as follows.

Definition 5.1.7. Let $\mathcal{F}$ be a graded resolution

$$0 \longleftarrow M \longleftarrow F_0 \xleftarrow{\phi_1} F_1 \xleftarrow{\phi_2} F_2 \longleftarrow \cdots$$

of a graded S-module M.

1. The resolution $\mathcal{F}$ is *minimal* if the non-zero entries in the graded matrices ϕ_l have positive degree for every $l \geq 1$.
2. If $\mathcal{F}$ is a minimal graded resolution of M, then the number of copies of $S(-d)$ in F_i is the graded Betti number $\beta_{i,d}$.

Theorem 3.13 in [CLO98, Chapter 6] proves that any two minimal resolutions of M are isomorphic. The above examples and explanations are taken from Chapter 6 of [CLO98]. We refer the reader to that chapter for more details.

The ideal $I_{\{5,10,25,50\}}$ from Chapter 3 is homogeneous under the grading $\deg(a) = 5, \deg(b) = 10, \deg(c) = 25$, and $\deg(d) = 50$. Computing the minimal graded free resolution of this homogeneous ideal using Macaulay 2, we get the following.

```
i1 : S = QQ[a..d, Degrees => {5,10,25,50}, MonomialSize
              => 16]
i2 : I = ideal(c^2-d, a*b^2-c, a^2-b)
i3 : G = res coker gens I
i4 : G.dd
```

```
             1                                3
o4 = 0 : S  <------------------------ S   : 1
                 | a2-b ab2-c c2-d |

             3                                    3
     1 : S  <-------------------------------- S   : 2
                 {10} | -ab2+c -c2+d 0      |
                 {25} | a2-b    0     -c2+d |
                 {50} | 0       a2-b  ab2-c |

             3                         1
     2 : S  <------------------- S   : 3
                 {35} | c2-d   |
                 {60} | -ab2+c |
                 {75} | a2-b   |

             1
     3 : S  <----- 0 : 4
                 0

i5 : betti G
o5 = total: 1 3 3 1
         0: 1 . . .
         9: . 1 . .
        24: . 1 . .
        33: . . 1 .
        49: . 1 . .
        58: . . 1 .
        73: . . 1 .
        82: . . . 1
```

Note that in line `i1` we prescribe the grading on S. Here $F_0 = S$, $F_1 = S(-10) \oplus S(-25) \oplus S(-50)$, $F_2 = S(-35) \oplus S(-60) \oplus S(-75)$ and $F_3 = S(-85)$. Check that each of the matrices displayed above is a graded matrix with respect to the above degrees. The Betti diagram has been truncated to remove all rows with no non-zero entries. The projective dimension of S/I is three and *regularity* (which is the index of the last row in the Betti diagram) is 82. The values of all the non-zero Betti numbers are one. The above is a graded resolution where S has been graded by the semigroup $\mathbb{N}\{5, 10, 25, 50\}$ or alternately by the group $\mathbb{Z}^4/\mathcal{L}$ where $\mathcal{L} = \ker_{\mathbb{Z}}([5\ 10\ 25\ 50])$.

5.2 Free resolutions of toric ideals

Our goal in the rest of this chapter is to see what can be said about minimal free graded resolutions of toric ideals using the rich polyhedral and combinatorial information they carry. If $I_{\mathcal{A}}$ is a toric ideal, we assume that the lattice $\mathcal{L} = \ker_{\mathbb{Z}}(A)$ has the property that $\mathcal{L} \cap \mathbb{N}^n = \{\mathbf{0}\}$. The abelian group $\mathbb{Z}^n/\mathcal{L}$ grades both the polynomial ring S and the S-module $S/I_{\mathcal{A}}$ via $\deg(x_i) := \mathbf{e}_i + \mathcal{L}$ where $\mathbf{e}_i$ is the ith unit vector in $\mathbb{Z}^n$. Note that this implies that the set of degrees of polynomials in S is the monoid $\mathbb{N}^n/\mathcal{L}$. This monoid is isomorphic to $\mathbb{N}\mathcal{A}$ via the map

$$\pi : \mathbb{Z}^n/\mathcal{L} \to \mathbb{Z}\mathcal{A}, \quad \mathbf{u} + \mathcal{L} \mapsto A\mathbf{u}.$$

Thus $\deg(x_i) = \mathbf{a}_i$, and $I_{\mathcal{A}}$ and $S/I_{\mathcal{A}}$ are graded by $\mathbb{N}\mathcal{A}$. A monomial $\mathbf{x}^{\mathbf{u}}$ is said to be of $\mathcal{A}$-degree $A\mathbf{u}$.

We first show that the graded Betti numbers of $I_{\mathcal{A}}$ can be calculated from homology groups of certain simplicial complexes that come from polyhedra associated to $\mathcal{L}$. For each $\mathbf{b} \in \mathbb{N}\mathcal{A}$ ($\cong \mathbb{N}^n/\mathcal{L}$) define a simplicial complex $\Delta_{\mathbf{b}}$ as follows:

$$\begin{aligned}\Delta_{\mathbf{b}} &:= \{\sigma \subseteq [n] : \sigma \subseteq \operatorname{supp}(\mathbf{u}) \text{ for some } \mathbf{u} \in \pi^{-1}(\mathbf{b})\} \\ &= \left\{\sigma \subseteq [n] : \mathbf{b} - \sum_{i\in\sigma} \mathbf{a}_i \in \mathbb{N}\mathcal{A}\right\}.\end{aligned}$$

Consider the minimal multigraded free resolution of $S/I_{\mathcal{A}}$ where $\deg(x_i) = \mathbf{a}_i$ for $i = 1, \ldots, n$. Then the Betti numbers are of the form $\beta_{i,\mathbf{b}}$ which denotes the size of a basis for the ith syzygy module of degree $\mathbf{b}$ or equivalently, the rank of $S(-\mathbf{b})$ in F_i.

Theorem 5.2.1 ([Stu96, Theorem 12.12]). *The multigraded Betti number $\beta_{i,\mathbf{b}}$ equals the rank of the $(i-1)$th reduced homology group $\widetilde{H}_{i-1}(\Delta_{\mathbf{b}}, \mathbf{k})$ of the simplicial complex $\Delta_{\mathbf{b}}$.*

This theorem was proved originally in [CM91,CP93].

Corollary 5.2.2. The toric ideal $I_{\mathcal{A}}$ has a minimal generator of degree $\mathbf{b}$ if and only if $\beta_{1,\mathbf{b}} = rank(\widetilde{H}_0(\Delta_{\mathbf{b}}, \mathbf{k})) > 0$. The latter condition is equivalent to the simplicial complex $\Delta_{\mathbf{b}}$ being disconnected.

Example 5.2.3. Consider the ideal $I_{\mathcal{A}}$ where $\mathcal{A} = \{5, 10, 25, 50\}$ graded by $\mathbb{N}\mathcal{A}$. We saw that $a^2 - b$, $ab^2 - c$, $c^2 - d$ are minimal generators of $I_{\mathcal{A}}$. The monomials of the same $\mathcal{A}$-degree as the generators can be computed either via CaTS or via Macaulay 2 when the $\mathcal{A}$-degree is small.

```
i1 : S = QQ[a..d, Degrees => {5, 10,25,50},
                    MonomialSize => 16]
i2 : basis({10},S)
o2 = | a2 b |

i3 : basis({25},S)
o3 = | a5 a3b ab2 c |

i4 : basis({50},S)
o4 = | a10 a8b a6b2 a5c a4b3 a3bc a2b4 ab2c b5 c2 d |
```

Listing just the maximal cells in the simplicial complexes, we get: $\Delta_{10} = \{\{1\}, \{2\}\}$, $\Delta_{25} = \{\{1, 2\}, \{3\}\}$ and $\Delta_{50} = \{\{1, 2, 3\}, \{4\}\}$, all of which are disconnected.

From the Betti diagram we see that there is a syzygy of degree 85 in the third step of the resolution. Thus we expect that $\text{rank}(\widetilde{H}_2(\Delta_{85}, \mathbf{k})) > 0$.

```
i5 : basis({85},S)

o5 = | a17 a15b a13b2 a12c a11b3 a10bc a9b4 a8b2c a7b5
       a7c2 a7d a6b3c a5b6 a5bc2 a5bd a4b4c a3b7 a3b2c2
       a3b2d a2b5c a2c3 a2cd ab8 ab3c2 ab3d b6c bc3 bcd |
```

Then $\Delta_{85} = \{\{1, 2, 3\}, \{1, 2, 4\}, \{1, 3, 4\}, \{2, 3, 4\}\}$ which is an empty tetrahedron and hence $\widetilde{H}_2(\Delta_{85}, \mathbf{k}) \neq 0$.

Peeva and Sturmfels extend Theorem 5.2.1 to lattice ideals graded by $\mathbb{N}^n/\mathcal{L}$. Toric ideals are special examples of lattice ideals. As a corollary they obtain the following theorem.

Theorem 5.2.4 ([PS98b, Theorem 2.3]). *The projective dimension of $S/I_{\mathcal{A}}$ as an S-module is at most $2^{n-d} - 1$.*

Note that $n - d = \text{codim}(I_{\mathcal{A}}) = \text{rank}(\mathcal{L})$. In order to prove the above theorem and for the rest of our discussion, it is worthwhile to look at the fiber $P_{\mathbf{b}} := \pi^{-1}(\mathbf{b}) = \{\mathbf{x} \in \mathbb{N}^n : A\mathbf{x} = \mathbf{b}\}$. Choose a matrix $B \in \mathbb{Z}^{n\times(n-d)}$ such that $\mathcal{L} = \{B\mathbf{z} : \mathbf{z} \in \mathbb{Z}^{n-d}\}$. Let $\mathbf{u}_0 \in P_{\mathbf{b}}$. Then

$$\begin{aligned} P_{\mathbf{b}} &= \{\mathbf{u} \in \mathbb{N}^n : \mathbf{u} + \mathcal{L} = \mathbf{u}_0 + \mathcal{L}\} = \{\mathbf{u} \in \mathbb{N}^n : \mathbf{u}_0 - \mathbf{u} \in \mathcal{L}\} \\ &= \{\mathbf{u} \in \mathbb{N}^n : \mathbf{u}_0 - \mathbf{u} = B\mathbf{z} \text{ for some } \mathbf{z} \in \mathbb{Z}^{n-d}\} \\ &\cong \{\mathbf{z} \in \mathbb{Z}^{n-d} : \mathbf{u} = \mathbf{u}_0 - B\mathbf{z} \geq 0\} \\ &= \{\mathbf{z} \in \mathbb{Z}^{n-d} : B\mathbf{z} \leq \mathbf{u}_0\} =: P'_{\mathbf{b}} \end{aligned}$$

Thus we have created $P'_{\mathbf{b}} \subset \mathbb{Z}^{n-d}$, a bijective copy of $P_{\mathbf{b}}$ in $\mathbb{N}^n$, via the affine linear transformation that takes $\mathbf{u} \in P_{\mathbf{b}} \mapsto \mathbf{z} \in \mathbb{Z}^{n-d}$ such that $\mathbf{u} = \mathbf{u}_0 - B\mathbf{z}$. It can be checked that a different choice of $\mathbf{u}_0 \in P_{\mathbf{b}}$ will lead to the same isomorphic copy $P'_{\mathbf{b}}$ up to translation. So we can pick any $\mathbf{u}_0$ in this construction.

The set $P'_{\mathbf{b}} = Q_{\mathbf{b}} \cap \mathbb{Z}^{n-d}$ where $Q_{\mathbf{b}}$ is the polyhedron

$$Q_{\mathbf{b}} := \{\mathbf{z} \in \mathbb{R}^{n-d} : B\mathbf{z} \leq \mathbf{u}_0\}.$$

This implies that

$$\Delta_{\mathbf{b}} = \{\sigma \subseteq [n] : \sigma \subseteq \text{supp}(\mathbf{u}_0 - B\mathbf{z}),\ \mathbf{z} \in P'_{\mathbf{b}}\}.$$

Thus $\mathbf{z} \in P'_{\mathbf{b}}$ contributes $\sigma = \{j \in [n] : B_j \cdot \mathbf{z} < (\mathbf{u}_0)_j\}$ which is the index set of all inequalities in $B\mathbf{z} \leq \mathbf{u}_0$ that do not hold at equality at $\mathbf{z}$.

Sketch of proof of Theorem 5.2.4. The projective dimension of $I_{\mathcal{A}}$ is the maximum integer l such that $\beta_{l,\mathbf{b}} \neq 0$ for some $\mathbf{b} \in \mathbb{N}\mathcal{A}$. By Theorem 5.2.1, $\beta_{l,\mathbf{b}} = \text{rank}(\widetilde{H}_{l-1}(\Delta_{\mathbf{b}}, \mathbf{k}))$. If $Q_{\mathbf{b}}$ has a lattice point $\mathbf{z}$ in its interior, then $[n] \in \Delta_{\mathbf{b}}$ which implies that $\Delta_{\mathbf{b}}$ has the same homology groups as the n-ball and $\widetilde{H}_{l-1}(\Delta_{\mathbf{b}}, \mathbf{k}) = 0$ for all l. Thus if $\beta_{l,\mathbf{b}} \neq 0$ then all lattice points in $P'_{\mathbf{b}}$ lie on the boundary of $Q_{\mathbf{b}}$. Such polyhedra are said to be *empty* or *lattice point free*.

Let $F_1, \ldots, F_s$ be the facets of $\Delta_{\mathbf{b}}$. Then by definition there exists $\mathbf{z}_1, \ldots, \mathbf{z}_s$ in $P'_{\mathbf{b}}$ such that $F_i = \text{supp}(\mathbf{u}_0 - B\mathbf{z}_i)$. Suppose $s > 2^{n-d}$. Then there exists two vectors, say $\mathbf{z}_1$ and $\mathbf{z}_2$, that have the same odd-even parity for their components. This implies that $\mathbf{z} = \frac{1}{2}(\mathbf{z}_1 + \mathbf{z}_2) \in \mathbb{Z}^{n-d}$. Further, $B\mathbf{z}_1 \leq \mathbf{u}_0$, $B\mathbf{z}_2 \leq \mathbf{u}_0$ implies that $B\mathbf{z} = B\left(\frac{1}{2}(\mathbf{z}_1 + \mathbf{z}_2)\right) \leq \mathbf{u}_0$. Thus $\mathbf{z} \in P'_{\mathbf{b}}$ which corresponds to $\frac{1}{2}(\mathbf{u}_1 + \mathbf{u}_2) \in P_{\mathbf{b}}$ where $\mathbf{u}_1 = \mathbf{u}_0 - B\mathbf{z}_1$ and $\mathbf{u}_2 = \mathbf{u}_0 - B\mathbf{z}_2$. Therefore, $\text{supp}\left(\frac{1}{2}(\mathbf{u}_1 + \mathbf{u}_2)\right) = \text{supp}(\mathbf{u}_1) \cup \text{supp}(\mathbf{u}_2) \in \Delta_{\mathbf{b}}$ which contradicts that $F_1 = \text{supp}(\mathbf{u}_1)$, $F_2 = \text{supp}(\mathbf{u}_2)$ are facets in $\Delta_{\mathbf{b}}$. Thus $s \leq 2^{n-d}$. Now computing homology of $\Delta_{\mathbf{b}}$ we see that homology of $\Delta_{\mathbf{b}}$ vanishes above dimension 2^{n-d}. □

The above discussions show that lattice point free polytopes of the form $Q_{\mathbf{b}}$ play a fundamental role in the resolution of the toric ideal $I_{\mathcal{A}}$. Such polytopes have been studied extensively in discrete geometry and optimization. They play a central role in the work on integer programming due to the mathematical economist Scarf at Yale University [BSS95,Sca81,Sca86]. Using Scarf's work, Bayer, Peeva and Sturmfels gave explicit resolutions of generic monomial and lattice ideals in [BPS98,BS98,PS98a,PS98b]. The complete material in these papers is beyond the scope of this chapter. However, we will conclude with the main results in [PS98a] which deals with minimal free resolutions of generic lattice ideals. A lattice ideal is *generic* if it is generated by binomials of full support.

Example 5.2.5 ([PS98a, Example 4.5]). The toric ideal $I_{\{20,24,25,31\}}$ is generic. It is minimally generated by $\{a^4 - bcd,\ a^3c^2 - b^2d^2,\ a^2b^3 - c^2d^2,\ ab^2c - d^3,$
$b^4 - a^2cd,\ b^3c^2 - a^3d^2,\ c^3 - abd\}$.

Definition 5.2.6. For a $\mathbf{b} \in \mathbb{N}\mathcal{A}$, let gcd($\mathbf{b}$) denote the greatest common divisor of $\{\mathbf{x}^{\mathbf{u}},\ \mathbf{u} \in P_{\mathbf{b}}\}$. Then $P_{\mathbf{b}}$ is called *basic* if $\gcd(\mathbf{b}) = 1$ and $\gcd(\{\mathbf{x}^{\mathbf{u}} : \mathbf{u} \in P_{\mathbf{b}}\} \setminus \{\mathbf{x}^{\mathbf{a}}\}) \neq 1$ for any $\mathbf{x}^{\mathbf{a}}$ such that $\mathbf{a} \in P_{\mathbf{b}}$.

Example 5.2.7. Resolving $S/I_{\{20,24,25,31\}}$ when S is graded by $\mathbb{N}\{20, 24, 25, 31\}$, we get the following.

```
i7 : F = res coker gens I
        1       7       12       6
o7 = S   <-- S   <-- S    <-- S   <-- 0

        0       1       2        3      4
      1                                                             7
0:S <---------------------------------------------------------- S
     |c3-abd a4-bcd ab2c-d3 b4-a2cd a3c2-b2d2 a2b3-c2d2 b3c2-a3d2|
```

```
       7                                                                12
1:S <-------------------------------------------------------------- S
    {75} | -bd -a3 0   d2  0   0   -ab2 -a2d -b3 0   0    0    |
    {80} | -c2 -bd -cd 0   0   d2  0    0    0   -b3 -b2c 0    |
    {93} | 0   0   0   -ab -b2 -bc c2   0    0   0   a3   -a2c |
    {96} | 0   0   -a2 0   ac  0   0    -c2  -ad -cd 0    d2   |
    {110}| a   c   0   0   d   0   0    0    0   0   0    b2   |
    {112}| 0   0   b   c   0   0   -d   0    0   a2  0    0    |
    {122}| 0   0   0   0   0   a   0    b    c   0   -d   0    |

       12                                   6
2 : S    <---------------------------------- S  : 3
           {130} | d  0  0  0   -b2 0   |
           {135} | 0  -d 0  0   0   -b2 |
           {136} | -c 0  -d 0   0   0   |
           {137} | b  0  0  -a2 0   0   |
           {141} | -a c  0  0   0   0   |
           {142} | 0  -b -c 0   0   0   |
           {143} | 0  0  -b 0   0   a2  |
           {146} | 0  a  0  -d  0   0   |
           {147} | 0  0  a  0   d   0   |
           {152} | 0  0  0  c   0   d   |
           {153} | 0  0  0  -b  c   0   |
           {158} | 0  0  0  0   a   c   |
       6
3 : S    <----- 0 : 4
           0
```

The truncated Betti diagram of this graded minimal free resolution is:

```
i9 : betti F
total: 1 7 12 6
    0: 1 .  . .
   74: . 1  . .
   79: . 1  . .
   92: . 1  . .
   95: . 1  . .
  109: . 1  . .
  111: . 1  . .
  121: . 1  . .
  128: . .  1 .
  133: . .  1 .
  134: . .  1 .
  135: . .  1 .
  139: . .  1 .
  140: . .  1 .
  141: . .  1 .
  144: . .  1 .
  145: . .  1 .
```

```
150: . . 1 .
151: . . 1 .
156: . . 1 .
158: . . . 1
163: . . . 1
164: . . . 1
174: . . . 1
175: . . . 1
180: . . . 1
```

From the Betti diagram we see that $\beta_{3,178} = 1$. The monomials of degree 178 are $a^4b^2c^2, a^3cd^3, ab^4d^2, b^3c^3d$. Check that P_{178} is basic and

$$\Delta_{178} = \{\{1,2,3\},\{1,3,4\},\{1,2,4\},\{2,3,4\}\}$$

by computing the monomials of $\mathcal{A}$-degree 178 in the multigraded ring `S`. See Example 5.2.9. This simplicial complex has spherical homology in degree two.

We will see that all homology fibers of generic toric ideals are basic.

Lemma 5.2.8 ([PS98a, Lemma 2.4]). *Let* $P_{\mathbf{b}}$ *be basic and* $\mathbf{a} \in P_{\mathbf{b}}$. *Then* $P_{\mathbf{b}}\backslash\{\mathbf{a}\}$ *is again a basic fiber after* $\min\{\mathbf{v} \in P_{\mathbf{b}}\backslash\{\mathbf{a}\}\}$ *has been subtracted from all the elements in* $P_{\mathbf{b}}\backslash\{\mathbf{a}\}$.

Example 5.2.9. By abuse of notation we let $P_{178} = \{a^4b^2c^2, a^3cd^3, ab^4d^2, b^3c^3d\}$.
$P_{178}\backslash\{a^4b^2c^2\}$ gives the basic fiber $P_{147} = \{a^3cd^2, ab^4d, b^3c^3\}$,
$P_{178}\backslash\{a^3cd^3\}$ gives the basic fiber $P_{130} = \{a^4c^2, ab^2d^2, bc^3d\}$,
$P_{178}\backslash\{ab^4d^2\}$ gives the basic fiber $P_{153} = \{a^4b^2c, a^3d^3, b^3c^2d\}$,
$P_{178}\backslash\{b^3c^3d\}$ gives the basic fiber $P_{158} = \{a^3b^2c^2, a^2cd^3, b^4d^2\}$.
Note that the fifth column in the map ϕ_3 corresponds to the second syzygy of degree 178 and it is a combination of first syzygies of degrees 130, 147, 153, 158. Hence the basic fibers gotten from P_{178} are again homology fibers.

This observation is, in fact, the main theorem of [PS98a].

Definition 5.2.10 ([PS98a, Definition 3.1]). The *algebraic Scarf complex* of $I_{\mathcal{A}}$ is the complex of free S-modules

$$\mathcal{F}_{\mathcal{L}} = \bigoplus_{P_{\mathbf{b}} \text{ basic}} SE_{\mathbf{b}}$$

where $E_{\mathbf{b}}$ is a basis vector in homological degree $|\mathbf{b}|-1$. The basis degrees considered are modulo the action of $\mathcal{L}$ by translation.

The differential is

$$\delta(E_{\mathbf{b}}) = \sum_{m \in P_{\mathbf{b}}} \operatorname{sign}(m, P_{\mathbf{b}}) \gcd(P_{\mathbf{b}}\backslash\{m\}) E_{P_{\mathbf{b}}\backslash\{m\}}.$$

Here we have abused notation by identifying monomials with their exponent vectors. The number $\operatorname{sign}(m, P_{\mathbf{b}})$ is $(-1)^{l+1}$ if m is in the lth position in the lexicographic ordering of monomials in $P_{\mathbf{b}}$.

Theorem 5.2.11 ([PS98a, Theorem 4.2]). *If the toric ideal $I_{\mathcal{A}}$ is generic then the algebraic Scarf complex $\mathcal{F}_{\mathcal{L}}$ is the minimal free resolution of $S/I_{\mathcal{A}}$.*

Example 5.2.12. Continuing our example of the generic toric ideal $I_{\{20,24,25,31\}}$, $\delta(E_{178}) = (-1)^{1+1}dE_{147} + (-1)^{2+1}b^2E_{130} + (-1)^{3+1}cE_{153} + (-1)^{4+1}aE_{158} = -b^2E_{130} + dE_{147} + cE_{153} - aE_{158}$ which is the syzygy in the fifth column of ϕ_3.

If the toric ideal is not generic, then the Scarf complex is contained in a minimal free resolution of the ideal [PS98a, Theorem 3.2]. The results by Peeva and Sturmfels hold for all lattice ideals although we have stated them here only for toric ideals.

5.3 Tutorial 5

As we saw in Lecture 5 Macaulay 2 can calculate minimal free resolutions of S/I using the command `F = res I` (or equivalently `F = res coker gens I`). You can also display the maps using `F.dd` or the Betti diagram using `betti F`. Sometimes Macaulay 2 does not keep track of the entire calculation when you run the command `res I`. If you are having trouble viewing the maps, try finding the resolution using the command `complete res I` which will force the Macaulay 2 to keep track of all relevant information required in order to display the maps. Also you might want the command `S = ring I` which sets the name of the ring in which I is defined to be S. It is useful whenever your ring was defined inside a function and hence does not have a short name.

1. Using Macaulay 2 find a minimal free resolution and Betti diagram of $S/I_{\mathcal{A}}$ for each of the following matrices. You may use either of the algorithms of Tutorial 3 to compute $I_{\mathcal{A}}$.

   (a) $A = \begin{pmatrix} 1 & 2 & 3 \end{pmatrix}$

   (b) $A = \begin{pmatrix} 1 & 1 & 1 & 1 \\ 0 & 1 & 3 & 4 \end{pmatrix}$

   (c) $A = \begin{pmatrix} 1 & 2 & 2 & 3 \\ 0 & 1 & 3 & 4 \end{pmatrix}$

2. Let $R = \mathbf{k}[x]/\langle x^2 \rangle$ and let $I = \langle x \rangle \subset R$. What should a minimal free resolution of R/I look like? Now use Macaulay 2 to calculate a minimal resolution of R/I. Is the answer what you expected?
3. Let $S = \mathbf{k}[x_1, x_2, \ldots, x_n]$ and let $I \subset S$ be an ideal. Suppose $\mathcal{G}_{\succ} = \{g_1, g_2, \ldots, g_s\}$ is a Gröbner basis of I with respect to any monomial order $\succ$. Define

$$s_{ij} = \frac{\mathrm{lcm}(\mathrm{in}_{\succ}(g_i), \mathrm{in}_{\succ}(g_j))}{\mathrm{in}_{\succ}(g_i)}\mathbf{e}_i - \frac{\mathrm{lcm}(\mathrm{in}_{\succ}(g_i), \mathrm{in}_{\succ}(g_j))}{\mathrm{in}_{\succ}(g_j)}\mathbf{e}_j - \mathbf{a}_{ij}$$

   where $\mathbf{a}_{ij}$ is the vector of coefficients of the g_k's in the expansion of the S-polynomial S-pair(g_i, g_j) in terms of $\mathcal{G}_{\succ}$ and $\mathbf{e}_i$ is the standard basis vector in $\mathbb{R}^s$.

   For example, consider the ideal I generated by $\mathcal{G} = \{x^2 - x, xz - x, y - z\} \subset \mathbb{Q}[x, y, z]$. Call the polynomials g_1, g_2, and g_3 in the order listed. The set $\mathcal{G}$ is a

Gröbner basis for I with respect to the lexicographic ordering. The S-pair of g_1 and g_2 is

$$z\, g_1 - x\, g_2 = z(x^2 - x) - x(xz - x) = x^2 - xz$$

In order to find $\mathbf{a}_{12}$, we rewrite $x^2 - zx$ in terms of the Gröbner basis elements. We find that $x^2 - zx = g_1 - g_2$, so $\mathbf{a}_{12} = \mathbf{e}_1 - \mathbf{e}_2$. Therefore s_{12} is the vector

$$\frac{\operatorname{lcm}(x^2, xz)}{x^2}\mathbf{e}_1 - \frac{\operatorname{lcm}(x^2, xz)}{xz}\mathbf{e}_2 - (\mathbf{e}_1 - \mathbf{e}_2) = (z-1)\mathbf{e}_1 + (-x+1)\mathbf{e}_2.$$

These vectors are very useful because of the following theorem.

Schreyer's Theorem [CLO98, Theorem 3.3]. *With S, I, and $\mathcal{G}_{\succ}$ as above, the set $\{s_{ij} \,:\, 1 \le i, k \le s\}$ generates $\mathrm{Syz}(I)$, the syzygies on I, as an S-module.*

[CLO98, p. 238] Consider the following matrices:

$$M = \begin{pmatrix} x^2 - x & xy & y^2 - y \end{pmatrix} \quad N = \begin{pmatrix} y & 0 \\ -x+1 & y-1 \\ 0 & -x \end{pmatrix}$$

Let I be the ideal generated by the entries of M.

(a) Verify that the matrix product MN equals the 1×2 matrix of all zeros. Explain why this shows that the module generated by the columns of the matrix N is contained in $Syz(I)$.

To show that $Syz(I)$ is, in fact, generated by the columns of N, we can use Schreyer's Theorem.

(b) Check that the generators for I form a Gröbner basis for I with respect to the lexicographic order.
(c) Compute the syzygies s_{12}, s_{13}, s_{23} obtained from the S-polynomials on the generators of I. By Schreyer's Theorem, these three syzygies generate $Syz(I)$.
(d) How are the the columns of N related to the generators s_{12}, s_{13}, s_{23} of S? Why does N only have two columns?

4. The ideal of the twisted cubic is the toric ideal of $A = \left(\begin{smallmatrix} 1 & 1 & 1 & 1 \\ 0 & 1 & 2 & 3 \end{smallmatrix}\right)$.

(a) Compute the minimal graded free resolution of $S/I_{\mathcal{A}}$ using the weight vector $\{1, 1, 1, 1\}$.
(b) For each degree in which there is a nonzero Betti number, calculate the fiber and then the simplicial complex $\Delta_{\mathbf{b}}$. A helpful Macaulay 2 command is `basis(b,S)` which gives the monomials of S corresponding to the vectors in the fiber $\pi^{-1}(\mathbf{b})$. Do these complexes have the homology of a simplicial sphere?

Corollary 5.2.2 says that the toric ideal $I_{\mathcal{A}}$ has a minimal generator in degree $\mathbf{b}$ if and only if $\Delta_{\mathbf{b}}$ is disconnected. Verify this statement for the three generators of the twisted cubic.
(c) Check that all the fibers from part b are basic. Using these basic fibers, write out the algebraic Scarf complex. How is it related to the minimal free resolution?

5.4 Solutions to Tutorial 5

1. Using the Hoşten-Sturmfels algorithm function from Tutorial 3 and a weight vector of all ones, we get the following resolutions and Betti diagrams. The Conti-Traverso algorithm may also be used to get the same result by changing "HSAlg" to "CTAlg" and using the same input. The remainder of the problem is identical.

   (a) We display the differentials in the resolution below.

```
i2 : Ia = ideal HSAlg({{1,2,3}},{1,1,1});
i3 : S = ring Ia;
i4 : (complete res Ia).dd

             1                                        2
o4 = 0 : S  <--------------------------------- S   : 1
                 | x_1^2-x_2 x_1x_2-x_3 |

             2                              1
     1 : S  <-------------------------- S   : 2
                 {2} | -x_1x_2+x_3 |
                 {3} | x_1^2-x_2   |

             1
     2 : S  <----- 0 : 3
                 0

o4 : ChainComplexMap

i5 : betti res Ia

o5 = total: 1 2 1
         0: 1 . .
         1: . 1 .
         2: . 1 .
         3: . . 1
```

   (b) In this example the differential matrices are large, and the entire resolution does not fit easily on the page when they are included, so only the modules and arrows are displayed (to achieve this, we used `res Ib` here rather than `(complete res Ib).dd`).

```
i6 : Ib = ideal HSAlg({{1,1,1,1},{0,1,3,4}},
            {1,1,1,1});
i7 : S = ring Ib;
i8 : res Ib
```

```
         1      4      4      1
o8 = S  <-- S  <-- S  <-- S  <-- 0

     0      1      2      3      4

o8 : ChainComplex
i9 : betti res Ib
o9 = total: 1 4 4 1
         0: 1 . . .
         1: . 1 . .
         2: . 3 4 1
```

(c) Again the differentials have not been displayed.

```
i10 : Ic=ideal HSAlg({{1,2,2,3},{0,1,3,4}},
          {1,1,1,1});
i11 : S = ring Ic;
i12 : res Ic

          1      5      6      2
o12 = S  <-- S  <-- S  <-- S  <-- 0

      0      1      2      3      4

o12 : ChainComplex
i13 : betti res Ic
o13 = total: 1 5 6 2
          0: 1 . . .
          3: . 1 . .
          5: . 1 . .
          6: . 1 1 .
          7: . 1 2 .
          8: . 1 2 1
          9: . . 1 1
```

2. In a polynomial ring an ideal with one generator has no syzygies, however, over $R = \mathbf{k}[x]/\langle x^2\rangle$, multiplication by x sends the element x to x^2 which is zero in R. Now for the syzygies on the syzygies, we again want syzygies on x, so again we get x. Repeating this process we get the infinite repeating resolution:

$$0 \longleftarrow R/I \longleftarrow R \overset{(x)}{\longleftarrow} R \overset{(x)}{\longleftarrow} R \overset{(x)}{\longleftarrow} R \longleftarrow \cdots .$$

Putting this ring and ideal into Macaulay 2, we find that the resolution stops after two steps. The computer cannot handle the infinite resolution.

```
i14 : R=(QQ[x])/ideal(x^2);
```

```
i15 : res ideal(x)

       1       1       1
o15 = R  <-- R  <-- R

      0       1       2

o15 : ChainComplex
```

3. (a) Since the generators of I are the three entries of the matrix M, any column vector $\mathbf{v}$ such that $M\mathbf{v} = 0$ is a syzygy on the generators of I. The matrix MN contains only zeros, so each column of N is a syzygy on the generators of I, and hence the module the columns generate must be contained in the syzygy module $\mathrm{Syz}(I)$.

   (b) To check that these polynomials form a Gröbner basis, we calculate the S-pairs. Let $g_1 = x^2 - x$, $g_2 = xy$ and $g_3 = y^2 - y$ and $G = \{g_1, g_2, g_3\}$. We will use the notation $h \equiv 0 \mod G$ to mean that the remainder obtained by dividing the polynomial h by the ordered list of polynomials of G is 0.

$$\begin{aligned}\text{S-pair}(g_1, g_2) &= \frac{x^2y}{x^2}(x^2 - x) - \frac{x^2y}{xy}(xy) = -xy = -g_2 \\ &\equiv 0 \mod G\end{aligned}$$

$$\begin{aligned}\text{S-pair}(g_1, g_3) &= \frac{x^2y^2}{x^2}(x^2 - x) - \frac{x^2y^2}{y^2}(y^2 - y) = -xy^2 + x^2y \\ &= (xy)(x - y) = g_2(x - y) \equiv 0 \mod G\end{aligned}$$

$$\begin{aligned}\text{S-pair}(g_2, g_3) &= \frac{xy^2}{xy}(xy) - \frac{xy^2}{y^2}(y^2 - y) = xy = g_2 \\ &\equiv 0 \mod G\end{aligned}$$

   Since all three S-pairs reduce to 0, G is a Gröbner basis for I.

   (c) From our calculations above, we see

$$s_{12} = \frac{x^2y}{x^2}\mathbf{e}_1 - \frac{x^2y}{xy}\mathbf{e}_2 - (-\mathbf{e}_2) = y\mathbf{e}_1 + (-x + 1)\mathbf{e}_2$$

$$s_{13} = \frac{x^2y^2}{x^2}\mathbf{e}_1 - \frac{x^2y^2}{y^2}\mathbf{e}_3 - (x - y)\mathbf{e}_2 = y^2\mathbf{e}_1 - (x - y)\mathbf{e}_2 - x^2\mathbf{e}_3$$

$$s_{23} = \frac{xy^2}{xy}\mathbf{e}_2 - \frac{xy^2}{y^2}\mathbf{e}_3 - \mathbf{e}_2 = (y - 1)\mathbf{e}_2 - x\mathbf{e}_3$$

   (d) The syzygies s_{12} and s_{23} are the columns of N. The remaining syzygy $s_{13} = (y^2, -(x - y), -x^2) = ys_{12} + xs_{23}$ so it is in the module generated by s_{12} and s_{23}. Hence it is unnecessary to include it in a minimal generating set for $\mathrm{Syz}(I)$.

4. (a) Use Macaulay 2 and the Hoşten-Sturmfels algorithm to find the ideal I and the resolution of S/I.

```
i16 : I = ideal HSAlg({{1,1,1,1},{0,1,2,3}},{1,1,1,1});
i18 : (complete res I).dd

              1
o18 = 0 : S  <---
                                                         3
      ------------------------------------------------ S  : 1
      | x_2^2-x_1x_3 x_2x_3-x_1x_4 x_3^2-x_2x_4 |

              3                        2
      1 : S  <----------------------- S  : 2
                  {2,2} | -x_3 x_4  |
                  {2,3} | x_2  -x_3 |
                  {2,4} | -x_1 x_2  |

              2
      2 : S  <----- 0 : 3
                  0

o18 : ChainComplexMap
```

(b) Looking at the resolution, we see that there are generators in multidegrees $(2,2)$, $(2,3)$, and $(2,4)$. We add the multidegree of any monomial from the first column of the differential map from S^2 to S^3 to the multidegree of its row, and we find $(3,4)$ is the multidegree of one of the first syzygies. Similarly, using the second column, we find the other first syzygy has multidegree $(3,5)$.

The fiber in multidegree $\mathbf{b}$ can be calculated in Macaulay 2 by finding a basis for $S_{\mathbf{b}}$ using the command `basis`. For example if $\mathbf{b} = (2,2)$:

```
i19 : basis({2,2},S)

o19 = | x_1x_3 x_2^2 |

                    1        2
o19 : Matrix S   <--- S
```

So the maximal faces of the simplicial complex $\Delta_{(2,2)}$ are $\{1,3\}$ and $\{2\}$, which means $\Delta_{(2,2)}$ consists of three vertices and one edge connecting two of the vertices. In other words, $\Delta_{(2,2)}$ is contractible to a 0-sphere (see Figure 5.1), and hence has the same homology as a simplicial sphere.

For $\mathbf{b} = (2,3)$, the fiber is $\{x_1x_4, x_2x_3\}$, and for $\mathbf{b} = (2,4)$, the fiber is $\{x_2x_4, x_3^2\}$. The simplicial complexes $\Delta_{(2,3)}$ and $\Delta_{(2,4)}$ are both contractible to 0-spheres (see Figure 5.1) which we expected because they are generators of the ideal.

Now we check the statement for the first syzygies. For $\mathbf{b} = (3,4)$, the fiber is $\{x_1x_2x_4, x_1x_3^2, x_2^2x_3\}$. The maximal faces of the simplicial complex are

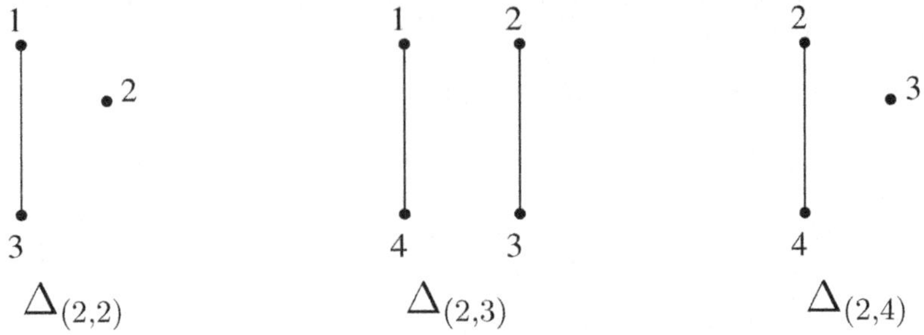

Figure 5.1. Some simplicial complexes

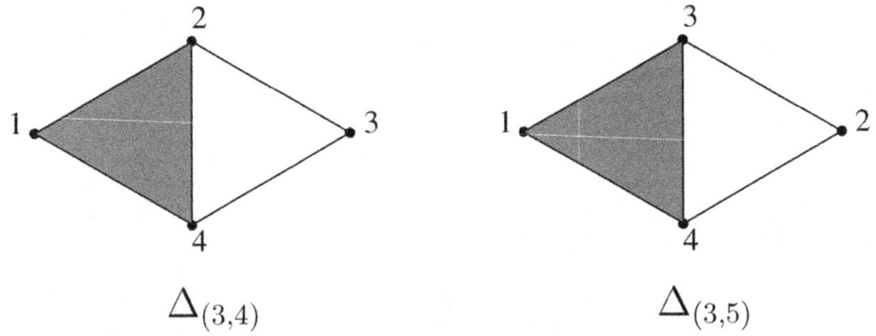

Figure 5.2. Some simplicial complexes

$\{1, 2, 4\}$, $\{1, 3\}$, and $\{2, 3\}$ which contracts to a simplicial 1-sphere (see Figure 5.2).

Similarly for $\mathbf{b} = (3, 5)$, the fiber is $\{x_1x_3x_4, x_2^2x_4, x_2x_3^2\}$. Again we see that the associated simplicial complex contracts to a simplicial 1-sphere (see Figure 5.2).

Looking back at the simplicial complexes associated to the minimal generators, it is clear that each is indeed disconnected, while the simplicial complexes associated to higher syzygies are connected. This verifies Corollary 5.2.2.

(c) The fiber for $\mathbf{b} = (2, 2)$ is basic since the greatest common divisor of x_1x_3 and x_2^2 is 1. Similarly for $\mathbf{b} = (2, 3)$ and $\mathbf{b} = (2, 4)$ it is clear that the fiber is basic since the gcd of each pair of generators is 1.

For $\mathbf{b} = (3, 4)$ we can see that $\gcd(x_1x_2x_4, x_1x_3^2, x_2^2x_3) = 1$. Removing one element at a time, we see that $\gcd(x_1x_3^2, x_2^2x_3) = x_3$, $\gcd(x_1x_2x_4, x_2^2x_3) = x_2$, and $\gcd(x_1x_2x_4, x_1x_3^2) = x_1$. Hence this fiber is basic. Similarly for $\mathbf{b} = (3, 5)$ calculations show the fiber is basic.

All five fibers are basic so the Scarf complex will be exactly the same as the minimal free resolution. Notice that even though for generic toric ideals all fibers are basic, the reverse is not true since the twisted cubic is not generic.

6

Connections to Algebraic Geometry

6.1 Introduction

In this chapter we will explain the connections between the theory we have covered in the first five lectures and toric varieties as they occur in algebraic geometry. We also make additional connections between commutative algebra and toric geometry. More familiarity with the basics of modern algebraic geometry is assumed here. For simplicity we will always work over the field $\mathbb{C}$.

6.2 A brief introduction to toric varieties

To an algebraic geometer, a toric variety is a variety defined by the combinatorial data of a polyhedral fan. The name "toric" comes from the fact that a toric variety is precisely a d-dimensional normal variety X that contains a dense copy of the algebraic torus $(\mathbb{C}^*)^d$ and with an action of $(\mathbb{C}^*)^d$ on X extending the action of $(\mathbb{C}^*)^d$ on itself by multiplication. For example, $\mathbb{C}^d$ is a toric variety, with $(t \cdot v)_i = t_i v_i$ for $t \in (\mathbb{C}^*)^d, v \in \mathbb{C}^d$.

We begin by defining an affine toric variety. Let $C \subseteq \mathbb{R}^d$ be a rational polyhedral cone. The word *rational* here means that every one-dimensional ray of C contains a lattice point (a point with all integer coordinates). Equivalently, it means that all the facet inequalities for C can be written using rational numbers. We will also assume that C is *pointed*, which means that there is a vector $\mathbf{v}$ with $\mathbf{v} \cdot \mathbf{x} \geq 0$ for all $\mathbf{x} \in C$. Let $D = C \cap \mathbb{Z}^d$ be the semigroup of lattice points inside C, and let $R = \mathbb{C}[D]$ be the corresponding semigroup algebra. The algebra R has as a vector-space basis the elements $t^{\mathbf{v}}$ for all $\mathbf{v} \in D$, with multiplication given by $\mathbf{t}^{\mathbf{v}}\mathbf{t}^{\mathbf{v}'} = \mathbf{t}^{\mathbf{v}+\mathbf{v}'}$. The *affine toric variety* X_C is defined to be $\mathrm{Spec}(R)$.

Example 6.2.1. Let C be the cone in $\mathbb{R}^2$ generated by $(1, 0)$ and $(2, 3)$. This is illustrated in Figure 6.1. The semigroup D is generated by $(1, 0)$, $(1, 1)$, and $(2, 3)$, so $R = \mathbb{C}[x, xy, x^2y^3]$.

Recall from Definition 3.4.4 of Chapter 3 that a *Hilbert basis* of the cone C is a finite generating set for $C \cap \mathbb{Z}^d$. Since C is pointed, its Hilbert basis is unique.

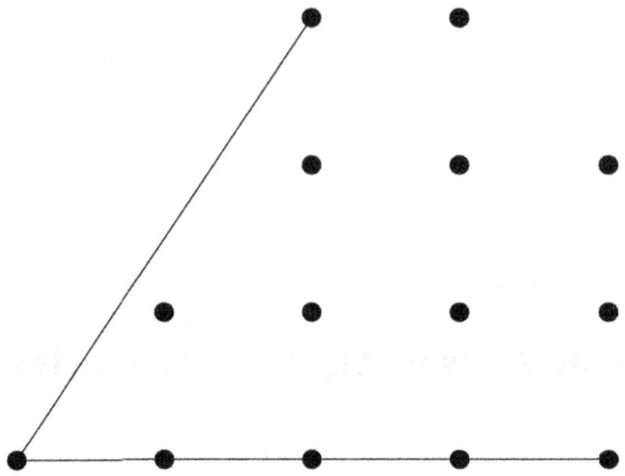

Figure 6.1.

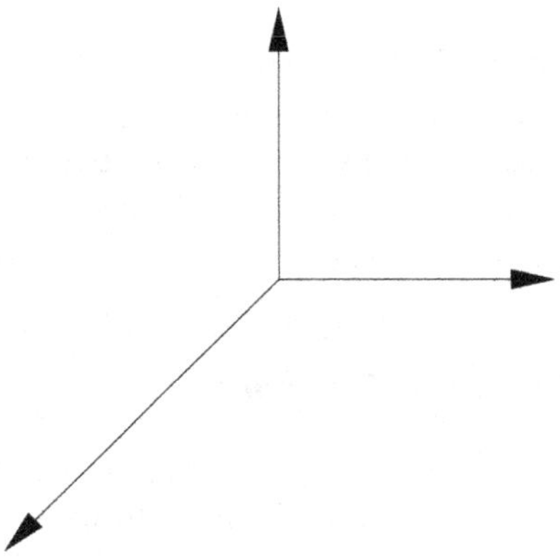

Figure 6.2.

Let $\mathcal{A} = \{\mathbf{a}_1, \ldots, \mathbf{a}_n\}$ be the Hilbert basis for C, and let $S = \mathbb{C}[x_1, \ldots, x_n]$. Then $R = S/I_{\mathcal{A}}$, where $I_{\mathcal{A}}$ is a toric ideal in the sense of Lecture 3.

Example 6.2.2. Let C be the cone of Example 6.2.1. Then $R = \mathbb{C}[x, y, z]/I_{\mathcal{A}}$, where $\mathcal{A} = \{(1, 0,), (1, 1), (2, 3)\}$.

A general toric variety is defined by a polyhedral fan Σ. For each cone $\sigma \in \Sigma$, we assign the affine chart $A_\sigma = X_{\sigma^\vee}$, where $\sigma^\vee = \{\mathbf{x} \in \mathbb{R}^n : \mathbf{x} \cdot \mathbf{y} \geq 0 \text{ for all } \mathbf{y} \in \sigma\}$ is the polar cone of σ. If σ and σ' share a face $\tau \in \Sigma$, then A_τ is an open subvariety of A_σ and of $A_{\sigma'}$. We then glue together A_σ and $A_{\sigma'}$ along the subvariety A_τ. The resulting variety X_Σ is the toric variety corresponding to the fan Δ.

Example 6.2.3. Let Σ be the fan in Figure 6.2, and let $\sigma_1 = \text{pos}((1, 0), (0, 1))$, $\sigma_2 = \text{pos}((0, 1), (-1, -1))$, and $\sigma_3 = \text{pos}((-1, -1), (1, 0))$ be its three maximal cones. The dual cones are $\sigma_1^\vee = \text{pos}((1, 0), (0, 1))$, $\sigma_2^\vee = \text{pos}((-1, 0), (-1, 1))$, and $\sigma_3^\vee = \text{pos}((0, -1), (1, -1))$. These are shown in Figure 6.3.

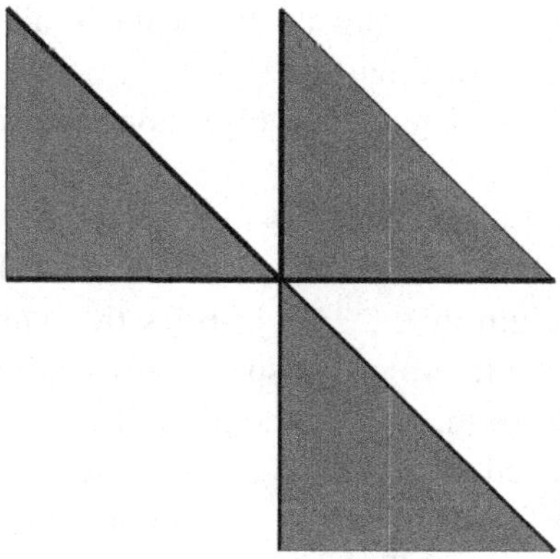

Figure 6.3.

Our three coordinate patches are now $A_i = \operatorname{Spec}(R_i)$, $i = 1, 2, 3$, where $R_1 = \mathbb{C}[x, y]$, $R_2 = \mathbb{C}[x^{-1}, y/x]$, $R_3 = \mathbb{C}[x/y, y^{-1}]$. These correspond, in order, to the three coordinate patches $z \neq 0$, $x \neq 0$, and $y \neq 0$ of $\mathbb{P}^2 = \operatorname{Proj}(\mathbb{C}[x, y, z])$. Since the overlaps are also consistent, we conclude that $X_\Sigma = \mathbb{P}^2$.

The construction of $\mathbb{P}^n$ as a toric variety is similar, with the rays of the fan Σ in $\mathbb{R}^n$ generated by $\mathbf{e}_1, \ldots, \mathbf{e}_n$ and $-\sum_{i=1}^n \mathbf{e}_i$. The n-dimensional cones in Σ are the positive hull of all but one of the above rays. There are thus $n + 1$ of these cones. Each cone gives one of the standard affine patches for $\mathbb{P}^n$.

If our fan Σ is the normal fan of an integral polytope P, then X_Σ is a *projective* toric variety. A map of X_Σ into some projective space is obtained by placing P at height one in $\mathbb{R}^{d+1}$ and letting $\mathcal{A}$ be the Hilbert basis for $\operatorname{pos}(P \times \{1\})$. If P is sufficiently large we then have $\operatorname{Proj}(S/I_\mathcal{A}) \cong X_\Sigma$. There are many different choices of a polytope P with normal fan Σ. Each choice corresponds to the choice of an ample divisor on X_Σ. Demanding that P be sufficiently large is ensuring that this ample divisor is very ample.

6.3 The Cox homogeneous coordinate ring

One of the most important themes in algebraic geometry and commutative algebra is that algebro-geometric questions about projective space $\mathbb{P}^n$ can be translated into commutative algebra questions about the graded polynomial ring in $n + 1$ variables. This section introduces an analogous object for toric varieties, the Cox homogeneous coordinate ring. We shall see that much of what is "nice" about projective space carries over to this setting.

6.3.1 Projective space

We will start by reviewing well-known facts about projective space and graded polynomial rings.

1. The homogeneous coordinate ring of $\mathbb{P}^n$ is the graded polynomial ring $S = \mathbb{C}[x_0, \ldots, x_n]$. There is a distinguished ideal in S, $\mathfrak{m} = \langle x_0, \ldots, x_n \rangle$, which is called the *irrelevant ideal*. The *degree* of a monomial in S is the total degree in the variables x_i.
2. The variety $\mathbb{P}^n$ is the quotient of $\mathbb{C}^{n+1} - (0, \ldots, 0)$ by the diagonal action of $\mathbb{C}^*$ on the coordinates. This construction is just the identification of $\mathbb{P}^n$ with the set of lines through the origin in $\mathbb{C}^{n+1}$. It justifies the names "coordinate ring" and "irrelevant ideal"; a point in projective space has coordinates $(x_0 : \cdots : x_n)$ except that the vanishing locus of $\mathfrak{m}$, namely the origin $(0, \ldots, 0)$, is disallowed.
3. (Nullstellensatz) A graded ideal $I \subset S$ gives rise to a subvariety $V(I)$ of $\mathbb{P}^n$. The variety $V(I)$ is empty if and only if $\mathfrak{m}^k \subset I$ for some integer k.
4. (Ideal variety correspondence) The map $I \to V(I)$ is a 1-1 correspondence from graded radical ideals I contained in $\mathfrak{m}$ to subvarieties of $\mathbb{P}^n$.

   More generally in algebraic geometry, subschemes (possibly not reduced) are given by ideal sheaves. The correspondence above generalizes when we consider general saturated ideals I.
5. A finitely generated graded module M over S gives rise to a coherent sheaf $\widetilde{M}$. Moreover, every coherent sheaf comes this way from some M.

6.3.2 Construction of homogeneous coordinate ring

Our goal will be to generalize all of the properties of the last section to general toric varieties. Recall that a toric variety X_Σ was defined as a variety arising from a rational polyhedral fan $\Sigma \subset \mathbb{R}^n$. Let $\Sigma(1)$ denote the set of one-dimensional cones (rays) of Σ. Given a ray $\rho \in \Sigma(1)$, let $\mathbf{n}_\rho$ be the first lattice point on ρ.

The various ρ correspond exactly to the $(n-1)$-dimensional orbits of the torus action on X_Σ. Their closures D_ρ are the codimension-one irreducible subvarieties of X_Σ. We will consider the group, isomorphic to $\mathbb{Z}^{\Sigma(1)}$, of formal integer linear combinations of the D_ρ. This is the group of torus invariant (Weil) divisors. Following this terminology we will call a linear combination $D = \sum a_\rho D_\rho$ a divisor, where the sum is over all $\rho \in \Sigma(1)$.

There is a natural map $\mathbb{Z}^n \to \mathbb{Z}^{|\Sigma(1)|}$ sending $\mathbf{m} \in \mathbb{Z}^n$ to $\sum(\mathbf{m} \cdot \mathbf{n}_\rho) D_\rho$. Two divisors are said to be *linearly equivalent* if their difference is in the image of this map. The cokernel of this map is a group $\mathrm{Cl}(X)$, called the *divisor class group* of X. Given a divisor D, let $[D]$ denote its image in $\mathrm{Cl}(X)$.

With all of this terminology straight we can finally define the homogeneous coordinate ring S_X of a toric variety X.

Definition 6.3.1. Let $S_X = \mathbb{C}[x_\rho \ : \ \rho \in \Sigma(1)]$. A monomial $\mathbf{x}^D := \prod x_\rho^{a_\rho}$ corresponds to a divisor $D = \sum a_\rho D_\rho$. The *degree* of $\mathbf{x}^D$ is defined to be $[D] \in \mathrm{Cl}(X)$.

Thus S_X is multigraded by $\mathrm{Cl}(X)$, which can be any abelian group. We shall see that this S_X satisfies many of the properties of the coordinate ring for projective space. The next step is to find the appropriate notion of irrelevant ideal. Up until now we have only used the one-dimensional cones. For each cone $\sigma \in \Sigma$ let $\mathbf{x}^{\widehat{\sigma}} = \prod_{\rho \notin \sigma(1)} x_\rho$. Here $\sigma(1)$ is the set of one-dimensional generators of σ.

Definition 6.3.2. The *irrelevant ideal* $B \subset S_X$ is

$$B = \langle \mathbf{x}^{\widehat{\sigma}} \ : \sigma \in \Sigma \rangle.$$

It is easy to see that B is generated by the $\mathbf{x}^{\widehat{\sigma}}$ for the maximal cones $\sigma \in \Sigma$.

Example 6.3.3 (Affine space). Affine space $\mathbb{C}^n$ is a toric variety generated by one maximal cone spanned by the standard basis vectors $\mathbf{e}_i = (0, \dots, 1, \dots, 0)$ with a 1 in the ith coordinate. The coordinate ring is now $\mathbb{C}[x_1, \dots, x_n]$. In this case the map $\mathbb{Z}^n \to \mathbb{Z}^{|\Sigma(1)|}$ is an isomorphism and $\mathrm{Cl}(X) = \{0\}$. Hence the ring is entirely in degree 0 and can be considered ungraded.

Example 6.3.4 (Projective space). Projective space $\mathbb{P}^n$ is a toric variety with $\Sigma(1) = \{\mathbf{e}_1, \dots, \mathbf{e}_n, -\mathbf{e}_1 - \cdots - \mathbf{e}_n\}$. Here we recover the homogeneous coordinate ring in $n+1$ variables with the usual grading, since $\mathrm{Cl}(\mathbb{P}^n) = \mathbb{Z}$. Every set of n rays span a maximal cone, hence the irrelevant ideal is $\langle x_0, \dots, x_n \rangle$ as expected.

Example 6.3.5 (Weighted projective space). Consider the complete fan in $\mathbb{R}^2$ with one-dimensional generators the set $\{(1, 0), (0, 1), (-a, -b)\}$. The coordinate ring is $\mathbb{C}[x_0, x_1, x_2]$, is graded by $\mathrm{Cl}(X) = \mathbb{Z}$, and the irrelevant ideal is $\langle x_0, x_1, x_2 \rangle$ just as in the case of $\mathbb{P}^2$. However, the variables x_0 and x_1 have degrees a and b respectively while x_2 has degree 1. This is called the *weighted projective space* of type $(a, b, 1)$.

Example 6.3.6 ($\mathbb{P}^r \times \mathbb{P}^s$). One can realize the product of two projective spaces as a toric variety by taking the product of the associated complete fans. There are now $r+s+2$ one-dimensional cones. The coordinate ring is $\mathbb{C}[x_0, \dots, x_r, y_0, \dots, y_s]$. The irrelevant ideal is generated by the monomials $x_i y_j$, and the grading is by $\mathrm{Cl}(X) = \mathbb{Z}^2$ where the x variables have degree $(1, 0)$ while the y variables have degree $(0, 1)$.

6.3.3 Quotient construction

We now emulate the quotient construction of projective space. Consider

$$Z = V(B) = \left\{ t \in \mathbb{C}^{\Sigma(1)} \ : \ \prod_{\rho \notin \sigma} t_\rho = 0 \ \text{ for all } \sigma \in \Sigma \right\}$$

We will construct X as a quotient of $\mathbb{C}^{|\Sigma(1)|} - Z$. For any toric variety X of dimension n, the group $\mathrm{Cl}(X)$ is finitely generated of rank $s - n$ where $s = |\Sigma(1)|$. The corresponding complex multiplicative group $G = \mathrm{Hom}(\mathrm{Cl}(X), \mathbb{C}^*)$ is the product of $(\mathbb{C}^*)^{s-n}$ and a finite group. G acts on $\mathbb{C}^{|\Sigma(1)|}$ by $g \cdot t = (g([D_\rho])t_\rho)$. Recall from Definition 2.3.9 that the n-dimensional fan Σ is simplicial if every maximal cone of Σ has n generators.

Theorem 6.3.7. *Let X be the toric variety determined by the fan Σ, and let $Z = V(B) \subset \mathbb{C}^{|\Sigma(1)|}$ be as above.*

(i) *The set $\mathbb{C}^{|\Sigma(1)|} - Z$ is invariant under the action of the group G.*
(ii) *X is naturally isomorphic to the categorical quotient of $\mathbb{C}^{|\Sigma(1)|} - Z$ by G.*

(iii) *If Σ is simplicial then X is called a simplicial toric variety. If X is simplicial then X is the actual geometric quotient.*

This theorem implies that a point in X is always an equivalence class of points in $\mathbb{C}^{|\Sigma(1)|}$. If X is simplicial then the only disallowed points are those in the irrelevant set Z. In the nonsimplicial case we have to disallow more points, namely those for which the G orbits are not closed.

Proof. The basic idea behind the proof is to notice that if we set S_σ to be the localization of S at $\mathbf{x}^{\widehat{\sigma}}$, then $\mathbb{C}^{|\Sigma(1)|} - Z$ is covered by the affine open sets $U_\sigma = \mathrm{Spec}(S_\sigma)$. The connection with the affine cover of the toric variety is that $\mathbb{C}[\sigma^\vee \cap \mathbb{Z}^n] = (S_\sigma)_0 = S_\sigma^G$. A standard result in invariant theory tells us that the quotient U_σ/G is equal to $\mathrm{Spec}(S_\sigma^G)$. It remains to check the gluing. Showing that this is actually a geometric quotient in the simplicial case is harder. □

6.3.4 Homogeneous coordinate ring and algebraic geometry

Our next result generalizes Conditions 3 and 4 of Section 6.3.1 when X is a simplicial toric variety. For any graded ideal $I \subset S$ the set $V(I) - Z$ determines a closed subset of X, $V_X(I)$ via the above quotient. By abuse of notation we can write

$$V_X(I) = \{t \in X \ : \ f(t) = 0 \text{ for all } f \in I\},$$

where t is taken in homogeneous coordinates on X.

Theorem 6.3.8. *Let X be a simplicial toric variety and $B = \langle x^{\widehat{\sigma}} \ : \ \sigma \in \Sigma\rangle$ the irrelevant ideal.*

(i) *(The Toric Nullstellensatz) For any graded ideal $I \subset S$, $V_X(I) = 0$ if and only if $B^k \subset I$ for some integer k.*
(ii) *(The Toric Ideal-Variety Correspondence) The map $I \to V_X(I)$ induces a one to one correspondence between radical graded ideals of S contained in B and subvarieties of X.*

Finally we can consider more general graded modules over S. A module M is said to be $\mathrm{Cl}(X)$-*graded* if there is a direct sum decomposition:

$$M = \sum_{\alpha \in \mathrm{Cl}(X)} M_\alpha$$

which respects the multiplication, so $S_\alpha M_\beta \subseteq M_{\alpha+\beta}$.

As in the projective space case, we can construct a sheaf $\widetilde{M}$ on X from a graded module M.

Theorem 6.3.9. *For any toric variety X, every (quasi)-coherent sheaf on X is of the form $\widetilde{M}$ for some graded S-module M.*

6.4 Open questions on normality

The toric ideals for which $\mathcal{A}$ is the Hilbert basis of a rational cone are an important subclass of all toric ideals. We saw in this chapter that geometrically they are the defining ideals of affine toric varieties. Algebraically, they are the set of normal toric ideals, which are the toric ideals $I_{\mathcal{A}}$ such that $S/I_{\mathcal{A}}$ is integrally closed. If $\mathcal{A} = \{\mathbf{a}_1, \ldots, \mathbf{a}_n\} \subseteq \mathbb{Z}^d$ then $\mathcal{A}$ is called *normal* if $I_{\mathcal{A}}$ is normal. This is the case if the semigroup $\mathbb{N}\mathcal{A} = \{\sum_i \lambda_i \mathbf{a}_i : \lambda_i \in \mathbb{N}\}$ is normal, in the sense that if $\mathbf{b} \in \mathbb{Z}^d$ with $l\mathbf{b} \in \mathbb{N}\mathcal{A}$ for some $l \in \mathbb{N}$, $l > 0$, then $\mathbf{b} \in \mathbb{N}\mathcal{A}$. We list here some open questions about normal toric ideals.

A configuration $\mathcal{A}$ is called *graded* if there is a $\mathbf{c} \in \mathbb{Z}^d$ with $\mathbf{a}_i \cdot \mathbf{c} = 1$ for all $\mathbf{a}_i \in \mathcal{A}$.

6.4.1 Degrees of Gröbner bases

It is well-known (Sturmfels refers to it as a "folklore result" [Stu96, Theorem 13.14]) that if $\mathcal{A} \subseteq \mathbb{Z}^d$ is graded, and $\mathbb{N}\mathcal{A}$ is normal, then $I_{\mathcal{A}}$ is generated in degree at most d. The following question, however, is still open.

Question 6.4.1. Let $\mathcal{A} \subseteq \mathbb{Z}^d$ be a graded configuration for which $\mathbb{N}\mathcal{A}$ is normal. Is there a Gröbner basis for $I_{\mathcal{A}}$ consisting of binomials of degree at most d?

There are some results on this topic due to Hoşten, O'Shea, and Thomas [HT03], [OT05]. In [OT05] it is shown that such a Gröbner basis exists for the Δ-*normal* configurations defined in [HT03].

Definition 6.4.2. A configuration $\mathcal{A} \subseteq \mathbb{Z}^d$ is Δ-*normal* if there is a triangulation Δ of $\mathcal{A}$ for which if σ is a maximal cone of Δ then the Hilbert basis of $C_\sigma = \mathrm{pos}(\mathbf{a}_i : i \in \sigma)$ is $\mathcal{A} \cap C_\sigma$.

We note that the term order inducing such a Gröbner basis cannot always be chosen to be a lexicographic or reverse-lexicographic order.

6.4.2 Projective normality

In the next two sections we consider the case where X_σ is a smooth projective variety.

Let X_Σ be the projective toric variety determined by a polytope $P \subseteq \mathbb{R}^d$. The variety X_σ is smooth if every cone in Σ is simplicial and *unimodular* (the first lattice points on the extreme rays generate the semigroup of lattice points in the cone). In this case we also call the corresponding polytope P smooth.

For a general P if we take the embedding of X_Σ into projective space as $\mathrm{Proj}(S/I_{\mathcal{A}})$ where $\mathcal{A}$ is the configuration of lattice points inside the polytope, the resulting ideal $I_{\mathcal{A}}$ will not be normal. However there is some evidence that this is never the case when X_Σ is smooth.

Question 6.4.3. If X_Σ is a smooth projective toric variety, defined by a polytope P, and $\mathcal{A}$ is the configuration of lattice points inside the polytope $P \times \{1\} \subseteq \mathbb{R}^{d+1}$, is $\mathbb{N}\mathcal{A}$ a normal semigroup?

This question is asking whether the corresponding embedding into projective space is projectively normal. When P is two-dimensional, the answer to the question is yes for any polytope (smooth or not). This follows from the fact that every polygon in the plane has a unimodular triangulation.

This is a special case of a question raised by Oda [Oda97]. In polytope language Oda's question is the following.

Question 6.4.4. Let P be a smooth lattice polytope in $\mathbb{R}^d$, and let Q be a lattice polytope whose normal fan is a coarsening of that of P. Is every lattice point in $P+Q$ the sum of a lattice point in P and a lattice point in Q?

Question 6.4.3 asks this question for $Q = kP$. In algebro-geometric language Question 6.4.4 asks whether the map

$$H^0(X_\Sigma, P) \times H^0(X_\Sigma, Q) \to H^0(X_\Sigma, P+Q)$$

is surjective, where P, Q, $P+Q$ in this statement denote the corresponding ample or nef divisors. Question 6.4.4 has been answered affirmatively by Fakhruddin [Fak02] when P and Q are two-dimensional.

6.4.3 Quadratic generation

Another question one can ask about smooth projective toric varieties is about the degrees of generators of their defining ideals.

Question 6.4.5. Let X_Σ be a smooth toric variety, defined by a polytope P, and let $\mathcal{A}$ be the configuration of lattice points inside the polytope $P \times \{1\} \subseteq \mathbb{Z}^{d+1}$. Is $I_\mathcal{A}$ generated by quadrics? In other words, is the maximal degree of a minimal generator of $I_\mathcal{A}$ equal to two?

This question is still open, and suspected to have a negative answer, even in the case where $\mathcal{A}$ defines a normal semigroup.

6.4.4 Higher syzygies

Questions about projective normality and quadratic generation are special cases of questions about whether the toric variety has the property N_p.

Definition 6.4.6. Let P be a lattice polytope in $\mathbb{Z}^d$, corresponding to the toric variety X_Σ. Let $\mathcal{A}$ be the set of lattice points in $P \times \{1\}$, let $C = \text{pos}(\mathcal{A})$, and let D be the semigroup $C \cap \mathbb{Z}^{d+1}$. As before, we set S to be the polynomial ring with one generator for every lattice point in $P \times \{1\}$, and R to be the S-module $\mathbb{C}[D]$, with $x_i \cdot \mathbf{t}^\mathbf{v} = \mathbf{t}^{\mathbf{a}_i+\mathbf{v}}$.

Let

$$0 \longrightarrow F_k \longrightarrow \dots \longrightarrow F_1 \longrightarrow F_0 \longrightarrow R \longrightarrow 0$$

be the minimal free resolution of R as an S-module, so each F_i is the direct sum of shifts of S.

We say the line bundle corresponding to P has property N_p if $F_0 = S$ and $F_i = \oplus S[-i-1]$ for $1 \leq i \leq p$.

Note that the condition N_0 means that the embedding into projective space defined by P is projectively normal ($\mathbb{N}\mathcal{A}$ is a normal semigroup, as all Hilbert basis elements of C are at height one). This means that $R = S/I_{\mathcal{A}}$. The condition N_1 means that $I_{\mathcal{A}}$ is quadratically generated.

Hering, Schenck, and Smith [HSS05] have recently proved the following theorem about when a toric variety has property N_p.

Theorem 6.4.7. *Let P be a lattice polytope in $\mathbb{Z}^d$, corresponding to the toric variety X_Σ. Then the line bundle corresponding to $(d - 1 + p)P$ has property N_p.*

Their proof, however, is not at all combinatorial. This leads to the following question.

Question 6.4.8. Give a combinatorial proof of Theorem 6.4.7.

This question has been answered in the case $p = 0, 1$. The result that the semigroup generated by the lattice points in $(d - 1)P$ is normal (N_0) is a result of Ewald and Wessels [EW91]. Another proof is given in [LTZ93]. The fact that the when $\mathcal{A}$ is the set of lattice points in dP the ideal $I_{\mathcal{A}}$ is quadratically generated appears in [BGT97]. These proofs are all combinatorial, and inspired the work in [HSS05].

6.5 Further reading

One of the canonical introductions to toric varieties in their algebro-geometric formulation is Fulton's book [Ful93]. Another good reference, which does not assume any algebraic geometry background is Ewald [Ewa96]. The Cox homogeneous coordinate ring was introduced for simplicial toric varieties by Cox in [Cox95]. It was generalized to all toric varieties by Mustaţa [Mus02]. More about the connection between toric ideals and the toric varieties discussed here can be found in Chapter 13 of [Stu96], and [Stu97]. The latter contains some other (still) open questions about toric ideals.

Index

References

[AF92] D. Avis and K. Fukuda, *A pivoting algorithm for convex hulls and vertex enumeration of arrangements and polyhedra*, Discrete Comput. Geom. **8** (1992), no. 3, 295–313, ACM Symposium on Computational Geometry (North Conway, NH, 1991). MR MR1174359 (93h:68137)

[AL94] W. W. Adams and P. Loustaunau, *An introduction to gröbner bases*, American Mathematical Society, 1994, Graduate Studies in Mathematics **III**.

[Bay82] D. Bayer, *The division algorithm and the Hilbert scheme*, Ph.D. thesis, Harvard University, 1982.

[BGT97] W. Bruns, J. Gubeladze, and N. V. Trung, *Normal polytopes, triangulations, and Koszul algebras*, J. Reine Angew. Math. **485** (1997), 123–160. MR 99c:52016

[BM88] D. Bayer and I. Morrison, *Standard bases and geometric invariant theory. I. Initial ideals and state polytopes*, J. Symbolic Comput. **6** (1988), no. 2-3, 209–217, Computational aspects of commutative algebra. MR 90e:13001

[BPS98] D. Bayer, I. Peeva, and B. Sturmfels, *Monomial resolutions*, Mathematical Research Letters **5** (1998), 31–46.

[BS92] L. J. Billera and B. Sturmfels, *Fiber polytopes*, Annals of Mathematics **135** (1992), 527–549.

[BS98] D. Bayer and B. Sturmfels, *Cellular resolutions of monomial modules*, Journal für die Reine und Angewandte Mathematik **502** (1998), 123–140.

[BSS95] I. Barany, H. E. Scarf, and D. Shallcross, *The topological structure of maximal lattice point free convex bodies: the general case*, Integer Programming and Combinatorial Optimization (Berlin) (E. Balas and J. Clausen, eds.), Lecture Notes in Computer Science 920, Springer-Verlag, 1995, pp. 244–251.

[Buc65] B. Buchberger, *On finding a vector space basis of the residue class ring modulo a zero dimensional polynomial ideal*, Ph.D. thesis, University of Innsbruck, Austria, 1965, (German).

[CKM97] S. Collart, M. Kalkbrener, and D. Mall, *Converting bases with Gröbner walks*, Journal of Symbolic Computation **24** (1997), 465–469.

[CLO97] D. Cox, J. Little, and D. O'Shea, *Ideals, varieties, and algorithms*, second ed., Undergraduate Texts in Mathematics, Springer-Verlag, New York, 1997, An introduction to computational algebraic geometry and commutative algebra. MR MR1417938 (97h:13024)

[CLO98] D. Cox, J. Little, and D. O'Shea, *Using algebraic geometry*, Springer-Verlag, New York, 1998, Second edition.

[CM91] A. Campillo and C. Marijuan, *Higher relations for a numerical semigroup*, Sém. Théorie Nombres Bordeaux **3** (1991), 249–260.

[COC] *CoCoA 4.1*, Available from ftp://cocoa.dima.unige.it/cocoa.

[Cox95] D. A. Cox, *The homogeneous coordinate ring of a toric variety*, Journal of Algebraic Geometry **4** (1995), no. 1, 17–50.

[CP93] A. Campillo and P. Pison, *L'idéal d'un semi-group de type fini*, Comptes Rendues Acad. Sci. Paris, Série I **316** (1993), 1303–1306.

[CT91] P. Conti and C. Traverso, *Buchberger algorithm and integer programming*, Applied Algebra, Algebraic Algorithms and Error-Correcting Codes (H. F. Mattson, T. Mora, and T. R. N. Rao, eds.), Lecture Notes in Computer Science 539, Springer-Verlag, 1991, pp. 130–139.

[Eis94] D. Eisenbud, *Commutative algebra with a view towards algebraic geometry*, Springer Graduate Texts in Mathematics, 1994.

[EW91] G. Ewald and U. Wessels, *On the ampleness of invertible sheaves in complete projective toric varieties*, Results Math. **19** (1991), no. 3-4, 275–278. MR 92b:14028

[Ewa96] G. Ewald, *Combinatorial convexity and algebraic geometry*, Graduate Texts in Mathematics, vol. 168, Springer-Verlag, New York, 1996. MR 97i:52012

[Fak02] N. Fakhruddin, *Multiplication maps of linear systems on smooth projective toric surfaces*, arXiv:math.AG/0208178, 2002.

[Ful93] W. Fulton, *Introduction to toric varieties*, Annals of Mathematics Studies 131, Princeton University Press, Princeton, NJ, 1993, The William H. Roever Lectures in Geometry.

[GKZ94] I. M. Gel'fand, M. M. Kapranov, and A. V. Zelevinsky, *Discriminants, resultants, and multidimensional determinants*, Mathematics: Theory & Applications, Birkhäuser Boston Inc., Boston, MA, 1994. MR 95e:14045

[GP02] G-M. Greuel and G. Pfister, *A Singular introduction to commutative algebra*, Springer, 2002.

[GPS05] G.-M. Greuel, G. Pfister, and H. Schönemann, SINGULAR *3.0*, A Computer Algebra System for Polynomial Computations, Centre for Computer Algebra, University of Kaiserslautern, 2005, http://www.singular.uni-kl.de.

[GS] D. R. Grayson and M. E. Stillman, *Macaulay 2, a software system for research in algebraic geometry*, Available at http://www.math.uiuc.edu/Macaulay2/.

[Hem] R. Hemmecke, *4ti2*, Available at http://www.4ti2.de.

[HS95] S. Hoşten and B. Sturmfels, *GRIN: an implementation of Gröbner bases for integer programming*, Integer programming and combinatorial optimization (Copenhagen, 1995), Lecture Notes in Comput. Sci., vol. 920, Springer, Berlin, 1995, pp. 267–276. MR MR1367987

[HSS05] M. Hering, H. Schenck, and G. G. Smith, *Syzygies, multigraded regularity and toric varieties*, arXiv:math.AG/0502240, 2005.

[HT00] B. Huber and R.R. Thomas, *Computing Gröbner fans of toric ideals*, Experimental Mathematics **9** (2000), 321–331.

[HT03] S. Hoşten and R. R. Thomas, *Gomory integer programs*, Math. Program. **96** (2003), no. 2, Ser. B, 271–292, Algebraic and geometric methods in discrete optimization. MR 1 993 049

[Jena] A. N. Jensen, *CaTS, a software package for computing state polytopes of toric ideals*, Available at http://www.soopadoopa.dk/ anders/cats/cats.html.

[Jenb] A. N. Jensen, *Gfan, a software system for Gröbner fans*, Available at http://home.imf.au.dk/ajensen/software/gfan/gfan.html.

[Jen05] A. N. Jensen, *A non-regular Groebner fan*, arXiv:math.CO/0501352, 2005.

[KR00] M. Kreuzer and L. Robbiano, *Computational commutative algebra I*, Springer, Berlin, 200.

[LRS] J. De Loera, J. Rambau, and F. Santos, *Triangulations*, Forthcoming book.

[LTZ93] J. Y. Liu, L. E. Trotter, Jr., and G. M. Ziegler, *On the height of the minimal Hilbert basis*, Results Math. **23** (1993), no. 3-4, 374–376. MR 94f:52029

[Mac01] D. Maclagan, *Antichains of monomial ideals are finite*, Proc. Amer. Math. Soc. **129** (2001), no. 6, 1609–1615 (electronic). MR 2002f:13045

[MM82] E. W. Mayr and A. R. Meyer, *The complexity of the word problems for commutative semigroups and polynomial ideals*, Adv. in Math. **46** (1982), no. 3, 305–329. MR MR683204 (84g:20099)

[MR88] T. Mora and L. Robbiano, *The Gröbner fan of an ideal*, J. Symbolic Comput. **6** (1988), no. 2-3, 183–208, Computational aspects of commutative algebra. MR 90d:13004

[MS05] E. Miller and B. Sturmfels, *Combinatorial commutative algebra*, Graduate Texts in Mathematics, vol. 227, Springer-Verlag, New York, 2005. MR MR2110098

[Mus02] M. Mustaţa, *Vanishing theorems on toric varieties*, The Tohoku Mathematical Journal. Second Series **54** (2002), no. 3, 451–470.

[Oda97] T. Oda, *Problems on Minkowski sums of convex lattice polytopes*, Unpublished, 1997.

[OT05] E. O'Shea and R.R. Thomas, *Toric initial ideals of δ-normal configurations: Cohen-Macaulayness and degree bounds*, Journal of Algebraic Combinatorics **21** (2005), 247–268.

[PS98a] I. Peeva and B. Sturmfels, *Generic lattice ideals*, Journal of the American Mathematical Society **11** (1998), 363–373.

[PS98b] I. Peeva and B. Sturmfels, *Syzygies of codimension 2 lattice ideals*, Mathematische Zeitschrift **229** (1998), 163–194.

[Rob86] L. Robbiano, *On the theory of graded structures*, J. Symbolic Comput. **2** (1986), no. 2, 139–170. MR 87j:13010

[Sca81] H. E. Scarf, *Production sets with indivisibilities. Part I. Generalities*, Econometrica **49** (1981), 1–32.

[Sca86] H. E. Scarf, *Neighborhood systems for production sets with indivisibilities*, Econometrica **54** (1986), 507–532.

[Sch86] A. Schrijver, *Theory of linear and integer programming*, Wiley-Interscience Series in Discrete Mathematics and Optimization, New York, 1986.

[Sta96] R. P. Stanley, *Combinatorics and commutative algebra*, second ed., Progress in Mathematics, vol. 41, Birkhäuser Boston Inc., Boston, MA, 1996. MR 98h:05001

[Stu96] B. Sturmfels, *Gröbner bases and convex polytopes*, University Lecture Series, vol. 8, American Mathematical Society, Providence, RI, 1996. MR 97b:13034

[Stu97] B. Sturmfels, *Equations defining toric varieties*, Algebraic geometry—Santa Cruz 1995, Proc. Sympos. Pure Math., vol. 62, Amer. Math. Soc., Providence, RI, 1997, pp. 437–449.

[Stu02] B. Sturmfels, *Solving systems of polynomial equations*, vol. 97, American Mathematical Society, Providence, RI, 2002.

[Vil01] R. Villarreal, *Monomial algebras*, vol. 238, Marcel Dekker, Inc., New York, 2001.

[Zie95] G. M. Ziegler, *Lectures on polytopes*, Graduate Texts in Mathematics, vol. 152, Springer-Verlag, New York, 1995. MR 96a:52011

Part II

Topics in Commutative Algebra and Combinatorics

based on

International Conference on Commutative Algebra and Combinatorics

organised by

Bhaskaracharya Pratishthana, Pune
Harish-Chandra Research Institute, Allahabad

Lecture Notes

In Memory of *Vishwa Chander Dumir* (1943–2006)

Winfried Bruns
Editor

Preface

On the occasion of the first International Meeting of the American Mathematical Society in India held by the Indian Academy of Sciences and Indian National Science Academy at Indian Institute of Sciences, Bangalore from December 17–21 the Bhaskaracharya Pratishthana, Pune and the Harish-Chandra Research Institute, Allahabad organised an International Conference on Commutative Algebra and Combinatorics at Allahabad, 08–13 December, 2003.

This volume contains the notes of the four series of lectures given by W. Bruns, S. Hosţen, I. Swanson and N. V. Trung at the conference. T. Hibi was unable to attend the conference, but has kindly supplied notes for this volume in cooperation with H. Ohsugi.

The conference was a tribute to Vishwa Chander Dumir on the occasion of his 60th birthday. His work with H. Anand and H. Gupta on a combinatorial enumeration problem, called *magic squares* by R. Stanley, was one of the starting points of combinmatorial commutative algebra. The Anand–Dumir–Gupta conjectures were proved by Stanley using methods of commutative algebra. This development led to a renaissance of the theory of Hilbert functions and made combinatorially defined algebras a focal point of commutative algebra.

The article by Bruns gives an introduction to the Anand–Dumir–Gupta conjectures and their proof. The notes by Hoşten surveys a large class of combinatorially defined ideals in polynomial rings, and the contribution by Ohsugi and Hibi is devoted to algebraic ststistics, the perhaps most recent facet of combinatorial commutative algebra. Swanson treats primary decomposition, a classical subject going back to Lasker and Noether, for which computational methods have now become available. Castelnuovo–Mumford regularity, surveyed by Trung, is a very useful measure for the complexity of ideals and modules and of great interest in recent research.

While these notes were prepared, the sad news of Dumir's untimely death on February 3, 2006 reached us. We are grateful to Professor R. J. Hans-Gill for writing an obituary. The great mathematician Dumir and his excellent work will be kept in lively remembrance by all of us.

The conference was partially supported by the following institutions:

National Board for Higher Mathematics, India

Department of Science and Technology, Government of India

International Mathematical Union

National Science Foundation, U.S.A.

We are grateful to them for their generous support.

The editor and contributors of this volume are very grateful to the Bhaskaracharya Pratishthana for organizing an excellent conference, Our warm thanks go to Professor Shashikant A. Katre and Ms. Manjusha Joshi for their excellent handling of all the technical and organizational work that arises at such a large event.

We thank the Harish-Chandra Research Institute for providing a very pleasant and comfortable environment. We are deeply indebted to Professor Ravi S. Kulkarni, former Director of the Harish-Chandra Research Institute, for his encouragement during all the stages of organization and also for taking keen interest in publishing these notes in the lecture note series of the Ramanujan Mathematical Society.

Last, but not least, the contributors and the editor cordially thank Professor Jugal K. Verma. His constant efforts were a decisive factor for the success of the conference and the publication of these notes. He has been a driving force in promoting Commutative Algebra in the Indian mathematical community.

March 2006 **Winfried Bruns**

International Organizing Committee

Chairman	*W. Bruns*, University of Osnabrück
Members	*A. Conca*, University of Genoa, Italy
	I. Swanson, New Mexico State University, Las Cruces, U.S.A.
	T. Hibi, Osaka University, Japan
	R. V. Gurjar, Tata Institute of Fundamental Research, Mumbai, India
	S. Adhikari, Harish-Chandra Research Institute, Allahabad, India
	R. Thomas, University of Washington at Seattle, U.S.A.
	N. V. Trung, Institute of Mathematics, Hanoi, Vietnam
	S. A. Katre, University of Pune, India
	R. A. Rao, Tata Institute of Fundamental Research, Mumbai, India

Local Organizing Committee

Chairman	*R. V. Gurjar*, Tata Institute of Fundamental Research, Mumbai, India & Trustee, Bhaskaracharya Pratishthana
Secretary & Treasurer	*S. A. Katre*, University of Pune, & Trustee, Bhaskaracharya Pratishthana
Local Secretary	*S. D. Adhikari*, Harish-Chandra Research Institute, Allahabad, India
Co-ordinator	*R. A. Rao*, Tata Institute of Fundamental Research, Mumbai, India
Co-ordinator	*J. K. Verma*, Indian Institute of Technology, Mumbai, India
Member	*R. L. Karandikar*, Indian Statistical Institute, Delhi, India

List of Participants

Speakers of Series of Expository Talks

Winfried Bruns
FB Mathematik/Informatik,
University of Osnabrück,
49069 Osnabrück,
Germany.
wbruns@uos.de

Serkan Hoşten
San Francisco State University,
Department of Mathematics,
San Francisco, CA 94132,
U.S.A.
serkan@fener.sfsu.edu

Irena Swanson
Reed College,
3203 SE Woodstock Blvd.,
Portland, OR 97202,
U.S.A.
iswanson@reed.edu

Ngô Việt Trung
Vien Toan Hoc,
Institute of Mathematics,
P.O. Box 631 Boho,
Hanoi, Vietnam.
nvtrung@math.ac.vn

Speakers of Conference Talks

Shreeram S. Abhyankar
Department of Mathematics,
Purdue University,
West Lafayette, IN 47907,
U.S.A.
ram@cs.purdue.edu

Marc Chardin
Institut de Mathématiques de Jussieu,
4, place Jussieu,
F-75252 Paris CEDEX 05,
France.
chardin@math.jussieu.fr

Jaydeep V. Chipalkatti
Department of Mathematics,
University of British Columbia,
Vancouver B.C. V6T 1Z2,
Canada.
chernclass@hotmail.com

Clare D'Cruz
Chennai Mathematical Institute,
Plot D19/D20 SIPCOT IT Park,
Padur PO, Siruseri 603103,
India.
clare@cmi.ac.in

Nguyen Tu Cuong
Vien Toan Hoc,
Institute of Mathematics,
P.O. Box 631 Boho,
Hanoi, Vietnam.
cuongnt@hn.vnn.vn

Harm Derksen
Department of Mathematics,
University of Michigan,
530 Church Street,
Ann Arbor, MI 48109-1043,
U.S.A.
hderksen@umich.edu

Kamran Divaani-Aazar
Institute for Studies in Theoretical,
Physics and Mathematics,
P.O. Box 19395-5746,
Tehran, Iran.
kdivaani@ipm.ir

Sara Faridi
Department of Mathematics & Statistics,
Dalhousie University,
Halifax, NS,
Canada B3H 3J5.
faridi@mathstat.dal.ca

Sudhir R. Ghorpade
Department of Mathematics,
Indian Institute of Technology Bombay,
Powai, Mumbai 400 076, India.
srg@math.iitb.ac.in

Rajendra V. Gurjar
TIFR, Mumbai, India.
gurjar@math.tifr.res.in

Lê Tuân Hoa
Vien Toan Hoc,
Institute of Mathematics,
P.O. Box 631 Boho,
Hanoi, Vietnam.
lthoa@math.ac.vn

A. V. Jayanthan
Indian Institute of Technology Madras,
Chennai India.
jayanav@iitm.ac.in

Ming-chang Kang
Department of Mathematics,
National Taiwan University,
Taipei, Taiwan.
kang@math.ntu.edu.tw

Gregor Kemper
Technische Universitaet Muenchen,
Zentrum Mathematik - M11,
Boltzmannstr. 3,
85748 Garching,
Germany.
kemper@ma.tum.de

Graham Leuschke
Department of Mathematics,
405 Snow Hall,
University of Kansas,
Lawrence, KS 66045, U.S.A.
gleuschke@math.ukans.edu

Diana Maclagan
Department of Mathematics,
Hill Center - Busch Campus,
Rutgers,
The State University of New Jersey,
Picataway, NJ 08854-8019,
U.S.A.
maclagan@math.rutgers.edu

Tony J. Puthenpurakal
Indian Institute of Technology Bombay,
Powai, Mumbai,
India.
tputhen@math.iitb.ac.in

Purusottam Rath
Harisch-Chandra Research Institute,
(H.R.I.)
Chhatnag Road, Jhusi,
Allahabad 211 019,
India.
rath@mri.ernet.in

K. N. Raghavan
Institute of Mathematical Sciences,
C.I.T. Campus,
Taramani, Chennai 600 113,
India.
knr@imsc.res.in

Jugal K. Verma
Indian Institute of Technology Bombay,
Powai, Mumbai,
India.
jkv@math.iitb.ac.in

Santiago Zarzuela
Departament d'Algebra i Geometria,
Universitat de Barcelona,
E-08007 Barcelona,
Spain.
szarz@math.ist.utl.pt

Other Conference Participants

Sukumar Das Adhikari
H.R.I., Allahabad, India.
adhikari@mri.ernet.in

Jitendra Kumar Bajpai
H.R.I., Allahabad, India.
jitendra@mri.ernet.in

Punita Batra
H.R.I., Allahabad, India.
batra@mri.ernet.in

Kalyan Chakraborty
H.R.I., Allahabad, India.
kalyan@mri.ernet.in

Chandan Singh Dalawat
H.R.I., Allahabad, India.
dalawat@mri.ernet.in

Satya Deo
H.R.I., Allahabad, India.
sdeo@mri.ernet.in

Vishwa Chander Dumir (1943–2006)
Panjab University,
Chandigarh, India.

Krishnendu Gongopadhyay
H.R.I., Allahabad, India.
krishnendu@mri.ernet.in

Leah Gold
Department of Mathematics,
Cleveland State University,
2121 Euclid Avenue, RT 1515,
Cleveland, OH 44122-2214,
U.S.A.
lgold@math.tamu.edu

Shashikant A. Katre
Dept. of Mathematics,
University of Pune, India.
sakatre@math.unipune.ernet.in

Amit Khetan
Dept. of Mathematics,
University of Massachusetts, Amherst,
Amherst, MA 01003,
U.S.A.
khetan@math.umass.edu

Inder Bir Singh Passi
Centre for Advanced Study in
Mathematics,
Panjab University,
Chandigarh 160 014,
India.
passi@mri.ernet.in

N. Raghavendra
H.R.I., Allahabad, India.
raghu@mri.ernet.in

B. Ramakrishnan
H.R.I., Allahabad, India.
ramki@mri.ernet.in

Maneesh Thakur
H.R.I., Allahabad, India.
maneesh@mri.ernet.in

R. Thangadurai
H.R.I., Allahabad, India.
thanga@mri.ernet.in

Rekha R. Thomas
Department of Mathematics,
University of Washington,
Seattle, WA 98195-4350,
U.S.A.
thomas@math.washington.edu

Postal address of the Harish-Chandra Research Institute (H.R.I.):

Chhatnag Road, Jhusi, Allahabad 211 019, India.

Publications of Vishwa Chander Dumir

1. (with Rajinder Jeet Hans) *An interesting identity.* Research Bulletin, Panjab University **15** (1964), 353–356.
2. (with Rajinder Jeet Hans) *An inhomogeneous minimum of a class of functions.* Monatsh. Math. **69** (1965), 193–207.
3. (with H. Gupta and H. Anand) *A combinatorial distribution problem.* Duke Math. J. **33** (1966) 757–769.
4. *Lattice Double Covering by Spheres.* Proc. Natl. Inst. Sci. India, Part A **33** (1967), 259–263.
5. *Asymmetric inequalities for non-homogeneous ternary quadratic forms.* Proc. Camb. Philos. Soc. **63** (1967), 291–303.
6. *Inhomogeneous minimum of indefinite quaternary quadratic forms.* Proc. Camb. Philos. Soc. **63** (1967), 277–290.
7. *Positive values of inhomogeneous quaternary quadratic forms, I.* J. Aust. Math. Soc. **8** (1968), 87–101.
8. *Positive values of inhomogeneous quaternary quadratic forms,II.* J. Aust. Math. Soc. **8** (1968), 287–303.
9. *Asymmetric inequality for non-homogeneous ternary quadratic forms.* J. Number Theory **1** (1969), 326–345.
10. *On positive values of indefinite binary quadratic forms.* Math. Student **38** (1970), 177–182.
11. (with R. J. Hans-Gill) *A counter-example in geometry of numbers.* Indian J. pure appl. Math. **3** (1972), 460–465.
12. (with R. J. Hans-Gill) *Lattice double coverings in the plane.* Indian J. pure appl. Math. **3** (1972) 466–480.
13. (with R. J. Hans Gill) *Lattice double packings in the Plane.* Indian J. pure appl. Math. **3** (1972), 481–487.
14. (with D. Singh) *A conjecture of Fejes Toth on saturated systems of circles.* Proc. Camb. Philos. Soc. **74** (1973), 453–460.
15. (with D. Singh) *Saturated systems of symmetric convex domains; results of Egglestion, Bambah and Woods.* Proc. Camb. Philos. Soc. **74** (1973), 107–116.
16. (with R. P. Bambah and R. J. Hans-Gill) *Covering by Star domains.* Indian J. Pure Appl. Math. **8** (1977), 344–350.
17. (with R. J. Hans-Gill) *On a conjecture of Mahler.* Bull. Aust. Math. Soc. **16** (1977), 125–129.
18. (with R. J. Hans-Gill) *Positive values of non-homogeneous quaternary quadratic forms.* Indian J. Pure Appl. Math. **12** (1981), 814–825.
19. (with R. P. Bambah and R. J. Hans-Gill) *On a conjecture of Jackson on non-homogeneous quadratic forms.* J. Number Theory **16** (1983), 403–419.

20. (with R. P. Bambah and R. J. Hans-Gill) *Positive values of non-homogeneous quadratic forms of signature* 1, 2 *and* 3. Topics in classical number theory, Colloq. Budapest 1981, Vol. I, Colloq. Math. Soc. János Bolyai **34** (1984), 111–170.
21. (with R. P. Bambah and R. J. Hans Gill) *Positive values of non-homogeneous indefinite quadratic forms II.* J. Number Theory **18** (1984), 313–341.
22. (with R. P. Bambah and R. J. Hans-Gill) *On an analogue of a problem of Mordell.* Stud. Sci. Math. Hung. **21** (1986), 135–142.
23. (with R. J. Hans-Gill) *View-obstruction problem for* 3*-dimensional spheres.* Monatsh. Math. **101** (1986), 279–289.
24. (with V. K. Grover) *Some asymmetric inequalities for non-homogeneous indefinite binary quadratic forms.* J. Indian Math. Soc. **50** (1986) 21–28.
25. (with R. J. Hans-Gill) *The view obstruction problems for boxes.* J. Indian Math. Soc., New Ser. **57** (1991), 117–122.
26. (with R. J. Hans-Gill) *Markoff type chain for the view obstruction problem for three dimensional cubes.* J. Indian Math. Soc., New Ser. **57** (1991), 123–141.
27. (with R. J. Hans-Gill) *Mathematical Contributions of Professor Hansraj Gupta.* J. Indian Math. Soc., New Ser. **57** (1991), 11–16.
28. (with Ranjeet Sehmi) *Positive values of non-homogeneous indefinite quadratic forms of type* (3, 2). Indian J. Pure Appl. Math. **23** (1992), 827–853.
29. (with Ranjeet Sehmi) *Positive values of non-homogeneous indefinite quadratic forms of signature* 1. Indian J. Pure Appl. Math. **23** (1992), 855–864.
30. (with R. J. Hans-Gill and J. B. Wilker) *Contributions to a general theory of view-obstruction problems.* Can. J. Math. **45** (1993), 517–536.
31. (with R. J. Hans-Gill and J. B. Wilker) *The* ℓ_p*-family of View-Obstruction problems.* Nieuw Arch. Wiskd., IV. Ser. **11** (1993), 225–239.
32. (with R. J. Hans-Gill and J. B. Wilker) *The view-obstruction constants for* ℓ_p*-balls in* $\mathbb{R}_3$., Expo. Math. **11** (1993), 407–417.
33. (with Ranjeet Sehmi) *Positive values of non-homogeneous Minima of Indefinite quadratic forms of type* (2, 5). J. Number Theory **48** (1994), 1–35.
34. (with R. J. Hans-Gill and A. C. Woods) *Values of non-homogeneous indefinite quadratic forms.* J. Number Theory **47** (1994), 190–197.
35. (with R. J. Hans-Gill and J. B. Wilker) *A Markoff-type chain for the View obstruction problem for spheres in* $\mathbb{R}_4$. Monatsh Math. **118** (1994), 205–217.
36. (with Ranjeet Sehmi) *Positive values of non-homogeneous indefinite quadratic forms of type* (1, 4). Proc. Indian Acad. Sci., Math. Sci. **104** (1994), 557–579.
37. (with Ranjeet Sehmi) *Isolated minima of non-homogeneous indefinite quadratic forms of type* (3, 2) *or* (2, 3) J. Indian Math. Soc., New Ser. **61** (1995), 197–212.
38. (with R. J. Hans-Gill, Ranjeet Sehmi) *Positive values of non-homogeneous quadratic forms of type* (2, 4). J.Number Theory **55** (1995), 261–284.
39. (with R. J. Hans-Gill and J. B. Wilker) *Contributions to a general theory of view-obstruction problems, II.* J. Number Theory **59** (1996), 352–373.
40. (with R. J. Hans-Gill and J. B. Wilker) *View-obstruction and a conjecture of Schoenberg.* Indian J. Pure Appl. Math. **27** (1996), 323–342.

41. (with R. J. Hans-Gill and J. B.Wilker) *View-obstruction problem for spheres in* $\mathbb{R}_5$. Monatsh. Math. **122** (1996), 21–34.
42. (with R. J. Hans-Gill) *The second minimum from positive values of non-homogeneous ternary quadratic forms of type* $(1, 2)$. AV Prasad Memorial Volume Ranchi University Mathematical Journal **28** (1997) 65–75.
43. (with R. P. Bambah and R. J. Hans-Gill) *Diophantine inequalities.* Proc. Natl. Acad. Sci. India, Sect. A **68** (1998), 101–114.
44. (with R. P. Bambah and R. J. Hans-Gill) *Non-homogeneous Problems: Conjectures of Minkowski and Watson.* In: R. P. Bambah, V.C. Dumir and R. J. Hans-Gill (eds.), Number theory. Basel: Birkhäuser. Trends in Mathematics (2000), 15–41.
45. *Number Theory, Indian National Science Academy.* Edited by R. P. Bambah, V.C. Dumir and R. J. Hans-Gill, Hindustan Book Agency 1999. Also published as: Number theory. Trends in Mathematics. Basel: Birkhäuser. vii, 527 p.
46. (with R. J. Hans-Gill) *View-obstruction through trajectories of co-dimension three.* In: A. Bezdek (ed.), Discrete geometry. In honor of W. Kuperberg's 60th birthday. New York, NY: Marcel Dekker. Pure Appl. Math., Marcel Dekker **253** (2003), 229–237.
47. (with Anuradha Sharma, G. K. Bakshi and Madhu Raka) *Cyclotomic Numbers and Primitive Idempotents in the Ring* $GF(q)[x]/(x^{p^n} - 1)$. Finite Fields Appl. **10** (2004), 653–673.
48. (with Anuradha Sharma, G. K. Bakshi and Madhu Raka) *Irreducible cyclic codes of length* 2^n. Ars Comb., (2006), to appear.

Table of Contents

Proc. Int. Conf. – Commutative Algebra and Combinatorics No. 4, 2007, pp. 1–38.

Commutative Algebra Arising from the Anand–Dumir–Gupta Conjectures

Winfried Bruns

Universität Osnabrück, FB Mathematik/Informatik, 49069 Osnabrück, Germany
e-mail: wbruns@uos.de

Abstract. In 1966 the Indian mathematicians H. Anand, V. C. Dumir, and H. Gupta investigated a combinatorial distribution problem and formulated some conjectures on the number of solutions. These conjectures were solved and extended by R. Stanley. His solution, based on methods of commutative algebra, was one of the starting points of combinatorial commutative algebra. In this article we describe the conjectures and their proofs and introduce the notions of commutative algebra on which the proofs are based.

Meinem Freund und Lehrer Udo Vetter zum 65. Geburtstag

Introduction

In 1966 the Indian mathematicians H. Anand, V. C. Dumir, and H. Gupta [2] investigated a combinatorial distribution problem that they had learnt from a paper of K. Manjo: *suppose that n distinct objects, each available in r identical copies, are distributed among n persons in such a way that each person receives exactly r objects. What can be said about the number $H(n, r)$ of such distributions?* Based on their investigation of $H(3, r)$, for which they gave an explicit formula, Anand, Dumir, and Gupta formulated several conjectures:

(ADG-1) there exists a polynomial $P_n(r)$ of degree $(n-1)^2$ such that $H(n, r) = P_n(r)$ for all $r \gg 0$;
(ADG-2) $H(n, r) = P_r(n)$ for all $r > -n$; in particular $P_n(-r) = 0, r = 1, \dots, n-1$;
(ADG-3) $P_n(-r) = (-1)^{(n-1)^2} P_n(r-n)$ for all $r \in \mathbb{Z}$.

In the original formulation (ADG-1) is not present. We have split the original conjecture into two parts since (ADG-1) is substantially easier to prove than (ADG-2).

The combinatorial problem can be recast as follows. If we let a_{ij} denote the number of copies of object i that person j receives, then $A = (a_{ij}) \in \mathbb{Z}_+^{n\times n}$ is an $n \times n$ matrix with non-negative integral entries such that

$$\sum_{k=1}^{n} a_{ik} = \sum_{l=1}^{n} a_{lj} = r, \qquad i, j = 1, \dots, n. \tag{1}$$

8	1	6
3	5	7
4	9	2

16	3	2	13
5	10	11	8
9	6	7	12
4	15	14	1

Figure 1. Two famous magic squares

Clearly $H(n,r)$ is the number of such matrices A. The system of equations (1) is part of the definition of *magic squares*. For such squares one usually requires that the sums over their diagonal and anti-diagonal also have the same value r, and true magic $n \times n$ squares, like the ones in Figure 1, must have the entries $1,\dots,n^2$. The 3×3 square has been found in ancient Chinese sources and the 4×4 appears in Albrecht Dürer's engraving *Melancholia* (1514). It has remarkable symmetries and shows the year of its creation.

The system of equations (1) is satisfied by, from a magician's view point, rather trivial objects, like the zero matrix, the identity matrix, and the matrix **1** with all entries equal to 1. Although it is highly doubtful that such matrices possess any witchcraft, mathematicians have called them "magic squares", and we will stick to this terminology.

Generating functions

A very strong tool of enumerative combinatorics, endowed with real magic power, is the concept of generating function. We have to introduce it in order to describe Stanley's extension of the ADG-conjectures. Consider the formal power series

$$H_n(t) = \sum_{r=0}^{\infty} H(n,r)t^r.$$

Then, as we will see, (ADG-1) is equivalent to the fact that $H_n(t)$ is the power series expansion at 0 of a rational function, denoted by the same name,

$$H_n(t) = \frac{h_0 + h_1 t + \cdots + h_u t^u}{(1-t)^{(n-1)^2+1}}, \qquad h_0 = 1, h_1, \dots, h_u \in \mathbb{Z}.$$

(ADG-2) and (ADG-3) can be converted into properties of this function:

$$\text{(ADG-2)} \iff \deg H_n(t) = u - ((n-1)^2+1) = -n,$$

and once (ADG-2) has been proved,

$$\text{(ADG-3)} \iff h_i = h_{u-i}, \quad i = 0,\dots,u.$$

Stanley extended the conjectures by

(ADG-4) $h_0, \dots, h_u \geq 0$;
(ADG-5) $h_0 \leq h_1 \leq \cdots \leq h_{\lceil u/2 \rceil}$.

In [22], [23] he proved (ADG-1)–(ADG-4) via a translation of the problem into commutative algebra. See Stanley's book [27] for more information on the pre-ADG history of the problem and on its solution. The major steps of our notes follow the development in [27].

For results and references on explicit formulas for $H(n, r)$ and a computer approach to them we refer the reader to Ahmed, De Loera and Hemmecke [1].

The commutative algebra way

The basic step is to understand $H(n, r)$ as the Hilbert function of a graded K-algebra, where K is an arbitrary field.

Let $\mathcal{M}_{nr}$ denote the set of solutions to (1) in $\mathbb{Z}_+^{n\times n}$ and set

$$\mathcal{M}_n = \bigcup_{r=0}^{\infty} \mathcal{M}_{nr}$$

Since $A+B \in \mathcal{M}_{n,r+s}$ for $A \in \mathcal{M}_{nr}$, $B \in \mathcal{M}_{ns}$, the set $\mathcal{M}_n$ has an algebraic structure: it is a submonoid of $\mathbb{Z}_+^{n\times n}$ (under addition). By the theorem of Birkhoff-von Neumann the monoid $\mathcal{M}_n$ is generated by $\mathcal{M}_{n1}$, the set of $n \times n$ permutation matrices.

We form the monoid algebra $K[\mathcal{M}_n]$. For this algebra the magic sum r serves as the degree. As just stated, $K[\mathcal{M}_n]$ is finitely generated by the degree 1 elements representing the permutation matrices. Moreover it has Krull dimension $(n-1)^2 + 1$. After these observations (ADG-1) follows immediately from the theory of Hilbert functions. The equation $H(n, r) = P_n(r)$ holds for all $r > \deg H_n(t)$.

The crucial observation for the proof of (ADG-2)–(ADG-4), if approached from the commutative algebra view point, is the normality of the monoid $\mathcal{M}_n$. Then (ADG-4) follows from Hochster's theorem that normal affine monoid algebras are Cohen–Macaulay rings.

For (ADG-2) and (ADG-3) one has to compute the canonical module ω of $K[\mathcal{M}_n]$. By a theorem of Danilov and Stanley it is the ideal I in $K[\mathcal{M}_n]$ generated by all elements corresponding to strictly positive magic squares. This interior ideal I is easily seen to be a principal ideal, generated by the monomial representing the matrix $\mathbf{1}$; it has magic sum n. Since $-\deg H_r(t)$ is the lowest degree of a non-zero element in ω, it follows that $\deg H_r(t) = -n$. This proves (ADG-2).

The Hilbert series of a graded Cohen–Macaulay ring R and its canonical module ω are related by the functional equation $H_\omega(t) = (-1)^d H_R(t^{-1})$, with $d = \dim R$. Since $\omega \cong K[\mathcal{M}_n]$, more precisely $\omega \cong K[\mathcal{M}_n](-n)$ as graded modules, $K[\mathcal{M}_n]$ is a Gorenstein ring whose canonical module is generated by an element of positive degree, and in this case the functional equation appears in the strong version (ADG-3).

We will develop, or at least discuss, all these results, trying to restrict the prerequisites to a level covered by introductions to commutative algebra, for example Atiyah and Macdonald [4] or Sharp [21]. Our main tool will be Noether normalizations and homogeneous systems of parameters, but for the proof of (ADG-2) and (ADG-3) more advanced techniques will be needed. Our general reference is Bruns and Herzog [8].

In 2003 conjecture (ADG-5) was proved by Athanasiadis [3] via a reduction to Stanley's g-theorem for simplicial polytopes. His result has now been extended significantly by Römer and the author [10]. Finding a commutative algebra proof of the g-theorem is still a real challenge.

Other developments

The ADG-conjectures were a strong motivation for Stanley's work, as is quite apparent from [27]. Nevertheless, and needless to say, the title of these notes is an exaggeration. At least three other closely related developments must be mentioned:

(i) Hochster [17] proved his theorem on the Cohen–Macaulay property of normal monoid algebras as a significant step on his way towards the Hochster–Roberts theorem by which invariant rings of linearly reductive groups acting on polynomial rings are Cohen–Macaulay. Normal affine monoid rings arise as invariant rings of torus actions.

(ii) A combinatorial problem generalizing the ADG-conjectures had been studied by E. Ehrhart [12]. Let P be an integral or, more generally, a rational polytope, and let $E(P, r)$ denote the number of integral points in the multiple rP, $r \in \mathbb{Z}_+$. If P is the polytope spanned by the $n \times n$-permutation matrices, then $E(P, r) = H(n, r)$. It is not difficult to interpret $E(P, r)$ for arbitrary P as the Hilbert function of a normal monoid algebra. We will discuss it at the end of Section 6.

(iii) Normal affine monoid rings are the coordinate rings of affine and projective toric varieties, and therefore extremely interesting objects for this branch of algebraic geometry. See Danilov [11], Kempf, Knudsen, Mumford, and Saint-Donat [19] or Fulton [14].

1. Affine monoids

Finite generation

Let us first prove that $\mathcal{M}_n$ is indeed generated by the permutation matrices.

Theorem 1.1 (Birkhoff-von Neumann). *The monoid $\mathcal{M}_n$ is generated by the permutation matrices $A_\pi \in \mathcal{M}_1$, $\pi \in S_n$.*

Proof. Let $B = (b_{ij}) \in \mathcal{M}_n$, $B \neq 0$. We must show that there exists a permutation $\pi \in S_n$ such that $B - A_\pi \in \mathcal{M}_n$. For each row $i = 1, \dots, n$ set

$$C_i = \{j : b_{ij} \neq 0\}.$$

Consider $I \subset \{1, \dots, n\}$. Then $\# \bigcup_{i \in I} C_i \geq \# I$, as the reader may verify. Therefore the marriage theorem implies that we can find a permutation $\pi \in S_n$ such that $b_{i\pi(i)} \neq 0$ for $i = 1, \dots, n$. Clearly $B - A_\pi \in \mathcal{M}_n$. □

Without an explicit reference to the magic sum r we could have defined $\mathcal{M}_n$ as the set of all matrices $A = (a_{ij}) \in \mathbb{Z}_+^{n\times n}$ satisfying the system

$$\sum_{k=1}^{n} a_{ik} = \sum_{l=1}^{n} a_{lj}, \qquad i, j = 1, \dots, n, \tag{2}$$

of linear homogeneous equations. More generally, if M is the set of non-negative solutions of a homogeneous diophantine linear system of equations in m variables, then M is certainly a submonoid of $\mathbb{Z}_+^m$, but it is not immediately clear that M, like $\mathcal{M}_n$, has a finite system of generators.

Definition 1.2. A submonoid $M \subset \mathbb{Z}^m$ is called *affine* if it is finitely generated. It is *normal* if it contains all elements $x \in \mathbb{Z}^m$ such that $cx \in M$ for some $c \in \mathbb{Z}$, $c > 0$.

Normal affine monoids are the discrete analogues of polyhedral cones, and some polyhedral geometry is necessary, or at least very useful, for their exploration.

Theorem 1.3. *Let $C \neq \emptyset$ be a subset of $\mathbb{R}^m$.*

(a) *Then the following are equivalent:*

  (i) *there exist finitely many elements $y_1, \dots, y_n \in \mathbb{R}^m$ such that $C = \mathbb{R}_+y_1 + \dots + \mathbb{R}_+y_n$;*

  (ii) *there exist finitely many linear forms $\lambda_1, \dots, \lambda_s$ such that C is the intersection of the half-spaces $H_i^+ = \{x : \lambda_i(x) \geq 0\}$.*

(b) *If C generates $\mathbb{R}^m$ as a vector space and the representation $C = H_1^+ \cap \dots \cap H_s^+$ is irredundant, then the halfspaces in* (ii) *above are uniquely determined (up to enumeration). Equivalently, the linear forms λ_i are unique up to positive scalar factors.*

(c) *The generating elements $y_1, \dots, y_n$ can be chosen in $\mathbb{Q}^m$ (or $\mathbb{Z}^m$) if and only if the λ_i can be chosen as linear forms with rational (or integral) coefficients.*

(d) *If $Y = \{y_1, \dots, y_n\} \subset \mathbb{Q}^m$, then $\mathbb{Q}^m \cap \mathbb{R}_+Y = \mathbb{Q}_+Y$.*

The reader is referred to Ziegler [29] or [6] for a proof. We use the implication (a) (ii)$\Rightarrow$(i) in order to prove *Gordan's lemma*. It shows that the finite generation of $\mathcal{M}_n$ is not an accident.

Theorem 1.4. *Let $\lambda_1, \dots, \lambda_p$ and $\mu_1, \dots, \mu_q$ be $\mathbb{Z}$-linear forms on $\mathbb{Z}^m$, and set*

$$M = \{x : \lambda_i(x) = 0,\ i = 1, \dots, p\} \cap \{x : \mu_j(x) \geq 0,\ j = 1, \dots, q\}.$$

Then M is a normal affine monoid.

Proof. We can replace every equation $\lambda_i(x) = 0$ by two inequalities so that we may assume $p = 0$. Obviously M is a normal monoid.

In $\mathbb{R}^m$ we consider the cone

$$C = \{x \in \mathbb{R}^m : \mu_j(x) \geq 0,\ j = 1, \dots, q\}$$

after the extension of the μ_j to $\mathbb{R}^m$. Then $M = C \cap \mathbb{Z}^m$. By Theorem 1.3 there exist finitely many elements $y_1, \dots, y_n \in M = C \cap \mathbb{Z}^m$ such that every element $x \in C$ can

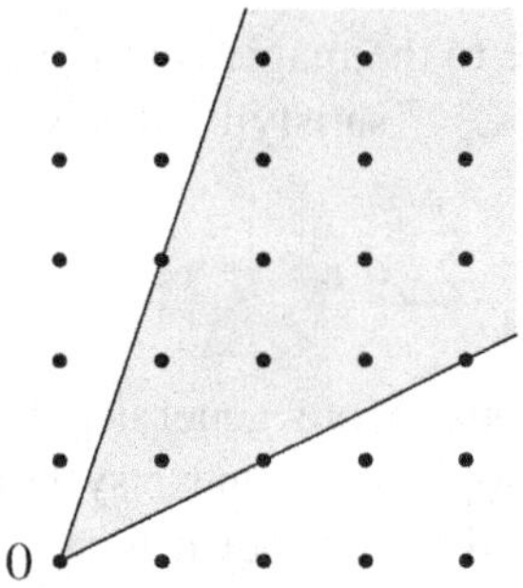

Figure 2. A normal monoid

be written as a linear combination $x = a_1 y_1 + \cdots + a_n y_n$ with non-negative coefficients $a_i \in \mathbb{R}_+$. Then

$$x = x' + x'', \qquad x' = \lfloor a_1 \rfloor y_1 + \cdots + \lfloor a_n \rfloor y_n.$$

Both x', x'' belong to M, and $x'' = b_1 y_1 + \cdots + b_n y_n$ with $b_j \in [0, 1)$, $j = 1, \dots, n$. Thus x'' belongs to the intersection of $\mathbb{Z}^m$ with a bounded set B, and $y_1, \dots, y_n$ together with the elements of the finite set $M \cap B$ generate M as a monoid. □

Normality and purity

The set M of non-negative solutions of a homogeneous linear diophantine system of equations in m variables is a special type of submonoid of $\mathbb{Z}^m_+$: let $G = \operatorname{gp}(M)$ be the subgroup of $\mathbb{Z}^m$ generated by M; then $M = G \cap \mathbb{Z}^m_+$. One says that M is a *pure* submonoid of $\mathbb{Z}^m_+$.

We want to show that every normal affine monoid M can be realized as a pure submonoid of $\mathbb{Z}^s_+$ for suitable s, provided M is *positive*: $x, -x \in M \Rightarrow x = 0$. Positivity is an evident necessary condition for the embeddability into $\mathbb{Z}^s_+$.

For brevity we let "affine" include "positive" in the following since we are mainly interested in normal affine monoids M, and every such monoid decomposes into the direct sum

$$M = U(M) \oplus M' \tag{3}$$

where $U(M) = \{x \in M : -x \in M\}$ and M' is normal, affine and positive. In fact, the normality of M implies that $\mathbb{Z}^m / U(M)$ is torsionfree. Hence $U(M)$ is a free direct summand of $\mathbb{Z}^m$, and we can choose M' as the image of M under the natural projection $\mathbb{Z}^m \mapsto \mathbb{Z}^m / U(M)$. In our context, direct summands that are groups are almost irrelevant.

Proposition 1.5. *Let $M \subset \mathbb{Z}^m$ be a affine monoid (positive, as understood). Then the following are equivalent:*

(a) *M is normal;*
(b) *$M = \mathbb{R}_+ M \cap \operatorname{gp}(M)$;*
(c) *M is isomorphic to a pure submonoid of $\mathbb{Z}^s_+$ for suitable s.*

Proof. Suppose that $M \subset \mathbb{Z}^m$ is normal, and let $x \in \mathbb{R}_+ M \cap \operatorname{gp}(M)$. The cone $\mathbb{R}_+ M$ is generated by finitely many elements $y_1, \dots, y_n \in M$, and $x = a_1 y_1 + \dots + a_n y_n$ with $a_1, \dots, a_n \in \mathbb{Q}_+$ by Theorem 1.3. Let $a \in \mathbb{Z}$, $a > 0$, be a common denominator for $a_1, \dots, a_n$. Then $ax \in M$, and $x \in M$ by hypothesis. This shows (a) $\Rightarrow$ (b).

For (b) $\Rightarrow$ (c) we identify the subgroup $\operatorname{gp}(M)$ of $\mathbb{Z}^m$ with $\mathbb{Z}^r$, $r = \operatorname{rank} \operatorname{gp}(M)$. The number r is called the *rank* of M. We can replace $\mathbb{Z}^m$ by $\mathbb{Z}^r$. Let $C = \mathbb{R}_+ M$ be the cone generated by M in $\mathbb{R}^r$. By Theorem 1.4 there exist finitely many linear forms $\sigma_1, \dots, \sigma_s$ on $\mathbb{R}^r$ with integral coefficients such that:

(i) C is the intersection of the half-spaces $H_i^+ = \{x \in \mathbb{R}^r : \sigma_i(x) \geq 0\}$, $i = 1, \dots, s$;
(ii) the representation $C = H_1^+ \cap \dots \cap H_s^+$ is irredundant;
(iii) $\sigma_i(\mathbb{Z}^r) = \mathbb{Z}$, $i = 1, \dots, s$.

Up to enumeration, the forms $\sigma_1, \dots, \sigma_s$ are uniquely determined. (Property (iii) can be achieved by extracting the greatest common divisor of the coefficients.) We call them the *support forms*.

Consider the $\mathbb{Z}$-linear map $\sigma : \mathbb{Z}^r \to \mathbb{Z}^s$, $\sigma(x) = (\sigma_1(x), \dots, \sigma_s(x))$. Since M is positive, σ is injective. It maps M isomorphically on a submonoid M' of $\mathbb{Z}_+^s$. Every element x of $\operatorname{gp}(M')$ belongs to $\sigma(\operatorname{gp}(M))$, and $x \in M'$ if and only if $\sigma(x) \in \mathbb{Z}_+^s$.

The remaining implication (c) $\Rightarrow$ (a) is trivial. □

Monoid algebras

Let K be a field and M a commutative monoid. Then the monoid algebra $K[M]$ is a vector space with a basis indexed by the elements of M for which the addition in M serves as the multiplication table. If $M \subset \mathbb{Z}^m$ is an affine monoid, then $K[M]$ is a subalgebra of $K[\mathbb{Z}^m]$ since the monoid embedding $M \hookrightarrow \mathbb{Z}^m$ extends to an algebra embedding $K[M] \hookrightarrow K[\mathbb{Z}^m]$. The monoid algebra $K[\mathbb{Z}^m]$ is isomorphic to the Laurent polynomial ring $K[X_1^{\pm 1}, \dots, X_m^{\pm 1}]$ via the assignment

$$a = (a_1, \dots, a_m) \mapsto X^a = X^{a_1} \cdots X^{a_m}.$$

It follows that the monoid algebras $K[M]$, M affine, can be identified with subalgebras of the Laurent polynomial rings $K[X_1^{\pm 1}, \dots, X_m^{\pm 1}]$ that are generated by finitely many monomials. By Hilbert's basis theorem such algebras are Noetherian, and so we can apply the rich structure theory of Noetherian rings to the study of affine monoids.

The terms "normal" and "pure" have been chosen judiciously:

Theorem 1.6. *Let M be an affine monoid.*

(a) *Then M is normal if and only if $K[M]$ is a normal ring, i. e. integrally closed in its field of fractions.*
(b) *$M \subset \mathbb{Z}_+^s$ is a pure submonoid if and only if $K[M]$ is a pure subalgebra of $K[\mathbb{Z}_+^s] = K[Y_1, \dots, Y_s]$, i. e. $K[\mathbb{Z}_+^s]$ splits into the direct sum $K[M] \oplus T$ of $K[M]$ and a $K[M]$-submodule T of $K[Y_1, \dots, Y_s]$.*

Proof.

(a) Let $x \in \mathrm{gp}(M)$ such that $mx \in M$ for some $m \in \mathbb{Z}$, $m > 0$. For the algebra $K[M]$ and the monomial $\tilde{x} \in K[\mathrm{gp}(M)] \subset \mathrm{QF}(K[M])$ with exponent vector x this implies $\tilde{x}^m \in K[M]$. If $K[M]$ is normal, then $\tilde{x} \in K[M]$ or, equivalently, $x \in M$.

Conversely, let M be normal, and set $C = \mathbb{R}_+ M$. We identify $\mathrm{gp}(M)$ with $\mathbb{Z}^r$. Then $M = \mathbb{Z}^r \cap C$ and $C = H_1^+ \cap \cdots \cap H_s^+$ where each H_i^+ is a closed halfspace defined by a linear form with integral coefficients. Thus

$$M = N_1 \cap \cdots \cap N_s, \qquad N_i = \mathbb{Z}^r \cap H_i^+.$$

Consequently

$$K[M] = K[N_1] \cap \cdots \cap K[N_s].$$

Since an intersection of normal domains is normal it is sufficient that $K[N]$ is normal for the intersection N of $\mathbb{Z}^r$ with a half-space H^+ defined by an integral linear form. Now

$$N \cong \mathbb{Z}^{r-1} \oplus \mathbb{Z}_+,$$

as follows from the splitting (3) of N: for $x \in \mathbb{Z}^r$ one has $x, -x \in N$ if and only if x lies in the hyperplane bounding the halfspace H^+. Thus $K[N]$ is isomorphic to a partial Laurent polynomial ring $K[Y_1^{\pm 1}, \dots, Y_{r-1}^{\pm 1}, Z]$. It is even a factorial ring, which is certainly normal.

(b) Let $R = K[Y_1, \dots, Y_s]$ and $S = K[M]$. Set $W = \mathbb{Z}_+^s \setminus M$ and let U be the vector subspace of $R = K[\mathbb{Z}_+^s]$ generated by the monomials with exponent vector in W. The purity of M translates into the condition $M + W \subset W$ or, equivalently, $SU \subset U$. Since $R = S \oplus U$ as a K-vector space, it is of course sufficient for the splitting of R that U is an S-submodule of $K[Y_1, \dots, Y_S]$.

Conversely, if $R = S \oplus T$ for some S-submodule T, then one certainly has the equation

$$I = IR \cap S$$

for all ideals I of S since the ideal IR of R decomposes into the direct sum $I \oplus IT$ (the reader should verify this claim). Take I as the principal ideal generated by the monomial corresponding to $x \in M$. If $x + W$ contains an element $y \in M$, then IR contains the monomial corresponding to y, an element of S which, on the other hand, is not in I. □

In a more general context, one uses a weaker property to define the purity of a ring extension; see [8, p. 293].

In order to avoid cumbersome formulations we will sometimes consider the monoids as subsets of their algebras, identifying each monoid element with the monomial of which it is the exponent vector.

The class group

We have used the support forms associated with the facets of the cone $\mathbb{R}_+M$ in order to realize $R = K[M]$ as a pure subalgebra of a polynomial ring. Let us observe that they also define prime ideals in R that determine the ideal theory of R to a great extent. Let σ_i be a support form and set

$$\mathfrak{p}_i = \sum_{x\in M,\ \sigma_i(x)\geq 1} Kx.$$

Then $\mathfrak{p}_i$ is an ideal of R, and even prime:

$$K[M]/\mathfrak{p}_i \cong K[M \cap F_i]$$

where F_i is the facet of $\mathbb{R}_+M$ belonging to σ_i. The *symbolic powers* of $\mathfrak{p}_i$ are defined as follows:

$$\mathfrak{p}_i^{(c)} = \sum_{x\in M,\ \sigma_i(x)\geq c} Kx.$$

One ideal in $K[M]$ will receive our special attention, namely

$$\mathfrak{p}_1 \cap \cdots \cap \mathfrak{p}_s.$$

It is the "interior" ideal generated by all monomials in the interior of the cone $\mathbb{R}_+M$.

More generally, let R be a Noetherian normal domain (or just a Krull domain). A subset J of the field $\mathrm{QF}(R)$ of fractions of R is a *fractional ideal* if there exists $x \in R$, $x \neq 0$, such that xJ is an ideal of R. A fractional ideal is divisorial if $(J^{-1})^{-1} = J$; here $J^{-1} = \{x \in K : xJ \subset R\}$ can be identified with $\mathrm{Hom}_R(J, R)$. From the view point of linear algebra, the divisorial fractional ideals are exactly the reflexive ones: $(J^{-1})^{-1} = J$ if and only if the natural homomorphism $J \to \mathrm{Hom}_R(\mathrm{Hom}(J, R), R)$ is an isomorphism.

The isomorphism classes of divisorial ideals are parameterized by a classical invariant, the *class group*; see Fossum [13]. (In the number-theoretic situation R is a Dedekind domain, and over such domain every fractional ideal is divisorial.) For a normal affine monoid ring $R = K[M]$ the class group can be calculated as follows: let x be a monomial in the interior of $\mathbb{R}_+M$. Then $R[x^{-1}]$ contains all monomials corresponding to elements in $\mathbb{Z}^d$, and therefore is the Laurent polynomial ring $L = K[\mathbb{Z}^d]$, a factorial ring. Since x is a monomial, the ideal Rx has a vector space of monomials, and by the normality of M the quotient y/x belongs to M if and only if $\sigma_i(y) \geq \sigma_i(x)$; therefore

$$Rx = \mathfrak{p}_1^{(v_1)} \cap \cdots \cap \mathfrak{p}_s^{(v_s)}, \qquad v_i = \sigma_i(x).$$

Nagata's theorem [13] implies that the class group is generated by the classes of the divisorial prime ideals containing x. These are exactly the monomial prime ideals $\mathfrak{p}_1, \ldots, \mathfrak{p}_s$ corresponding to the facets, as defined above. This implies: for each divisorial ideal J there exist integers $c_1, \ldots, c_s \in \mathbb{Z}$ such that

$$J \cong \mathfrak{p}_1^{(c_1)} \cap \cdots \cap \mathfrak{p}_s^{(c_s)}. \tag{4}$$

Figure 3 illustrates this construction. See Bruns and Gubeladze [7] for a detailed study of the class group of a normal monoid algebra.

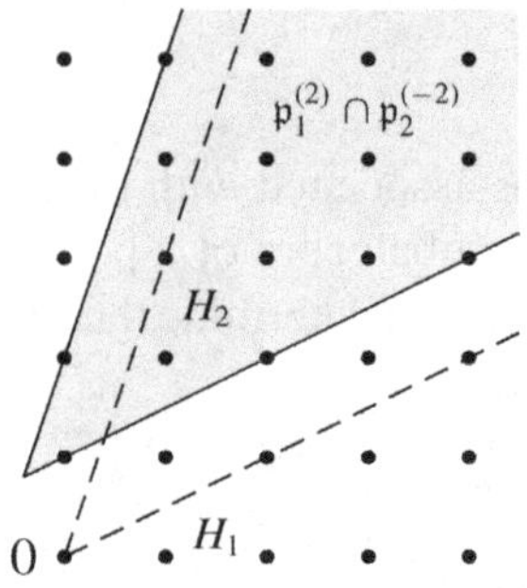

Figure 3. A divisorial fractional ideal

2. Graded rings and modules

The monoid algebra $R = K[\mathcal{M}_n]$ has a natural decomposition as a vector space: since $\mathcal{M}_n = \bigcup_{r=0}^{\infty} \mathcal{M}_{nr}$ we have $R = \bigoplus_{r=0}^{\infty} R_r$ where R_r is the vector subspace spanned by the monomials corresponding to the $n \times n$ magic squares of magic sum r. Since $\mathcal{M}_{nr} + \mathcal{M}_{ns} \subset \mathcal{M}_{n,r+s}$, this decomposition is compatible with the multiplicative structure: $R_r R_s \subset R_{r+s}$. Thus R is a graded ring:

Definition 2.1. A *grading* of a ring R is a decomposition $R = \bigoplus_{i \in \mathbb{Z}} R_i$ of R as an abelian group such that $R_i R_j \subset R_{i+j}$ for all i, j. A *graded ring* is a ring together with a grading.

A *grading* of a module M over a graded ring R is a decomposition $M = \bigoplus_{i \in \mathbb{Z}} M_i$ of R as an abelian group such that $R_i M_j \subset M_{i+j}$ for all i, j. A *graded R-module* is an R-module together with a grading.

The elements of M_i are *homogeneous* of degree i.

The main example of a graded ring is the polynomial ring $S = K[X_1, \dots, X_n]$ with the grading in which the graded component of degree i is the vector space spanned by the monomials of total degree i. More generally we can define a grading on S by assigning degrees $g_i \in \mathbb{Z}$ to the indeterminates X_i and setting

$$\deg X_1^{a_1} \cdots X_n^{a_n} = \sum_{j=1}^{n} a_j g_j.$$

The degree i component is again spanned by the monomials of degree i.

Let M be an affine monoid. Then a $\mathbb{Z}$-linear map $\gamma : \mathrm{gp}(M) \to \mathbb{Z}$ is called a grading on M. It induces a grading on $K[M]$ in which the degree i component is spanned by the elements $x \in M$ with $\gamma(x) = i$. We say that γ is *positive* if $\gamma(x) > 0$ for all $x \in M$, $x \neq 0$. It follows from Proposition 1.5 that every (normal) affine monoid admits a positive grading, as does every submonoid of $\mathbb{Z}_+^s$.

A submodule U of M is graded if $U = \bigoplus_{i \in \mathbb{Z}} U \cap M_i$. In this case the residue class module M/U is naturally graded, too, since $M/U = \bigoplus_{i \in \mathbb{Z}} M_i/(U \cap M_i)$ by a natural isomorphism.

The classical example of a graded submodule is an ideal I in the polynomial ring S as above that is generated by homogeneous polynomials. Then S/I is the homogeneous coordinate ring of the projective variety defined by I.

A *homogeneous* R-linear map $\varphi : M \to N$ of graded R-modules preserves the graded structure: one has $\varphi(M_i) \subset N_i$. The kernel and the cokernel of a graded homomorphism are graded R-modules.

In order to make certain natural linear maps homogeneous we must shift the degree of one (or both) of the modules involved: the module $M(s)$ is identical to M as an R-module, but

$$M(s)_i = M_{s+i}.$$

In other words, degree j elements in M have degree $j - s$ in $M(s)$. Most often one needs shifts that increase the degrees of elements. Therefore they are usually written with a negative sign.

Let K be a field. We say that a graded K-algebra R is *positively graded* if $R = K[x_1, \dots, x_n]$ with homogeneous elements x_i of positive degree. Then $R_i = 0$ for $i < 0$ and $R_0 = K$.

Graded Ext functors

At a certain point in Section 6 it seems inevitable to use derived functors, and we need that $\operatorname{Ext}^i_R(M, N)$ carries a natural grading for finitely generated graded modules M, N over a Noetherian graded ring R. First we endow $\operatorname{Hom}_R(M, N)$ with a natural grading: its homogeneous component of degree i consists of all R-linear maps φ such that $\varphi(M_j) \subset \varphi(N_{i+j})$ for all j. We leave it to the reader to show that $\operatorname{Hom}_R(M, N)$ is indeed the sum of these abelian subgroups.

Next, one constructs a graded free resolution of M,

$$\mathcal{F} : \cdots \to F_n \xrightarrow{\varphi_n} F_{n-1} \to \cdots \to F_1 \xrightarrow{\varphi_1} F_0 \to 0, \qquad M = \operatorname{Coker} \varphi_1,$$

in which the F_i are finitely generated graded free modules and the morphisms $F_{i+1} \to F_i$ are homogeneous. It follows that $\operatorname{Ext}^i_R(M, N) = H^i(\operatorname{Hom}_R(\mathcal{F}, N))$ is certainly a graded R-module, and the grading does not depend on the choice of the homogeneous free (or projective) resolution. Evidently such resolutions exist: if M is graded, generated by elements $x_1, \dots, x_m$ of degrees $g_1, \dots, g_m$, then

$$F_0 = \bigoplus_{i=1}^{m} R(-g_i) \to M, \qquad e_i \mapsto x_i,$$

is a homogeneous R-linear map from a graded free R-module onto M. Its kernel is again a finitely generated graded R-module, onto which we can map a graded free R-module etc.

For numerical applications it is usually best to group the summands $R(-g_i)$ according to their shifts:

$$\bigoplus_{i=1}^{m} R(-g_i) = \bigoplus_{j \in \mathbb{Z}} R(-j)^{\beta_j}, \qquad \beta_j = \#\{i : g_i = j\}.$$

The notion of grading can be generalized by allowing an arbitrary group or even an arbitrary commutative monoid as the set of degrees for R and M.

3. Krull dimension and Noether normalization

Let R be a Noetherian ring. The *Krull dimension* $\dim R$ of R is the supremum of all integers n such that there exists a strictly ascending chain

$$\mathfrak{p}_0 \subset \mathfrak{p}_1 \subset \cdots \subset \mathfrak{p}_n$$

of prime ideals. A priori, the value ∞ is not excluded, but it cannot occur for finitely generated algebras R over a field K. The Krull dimension of a positively graded K-algebra can be described as follows:

Theorem 3.1. *Let $R = \bigoplus_{k=0}^{\infty} R_k$ be a positively graded, finitely generated K-algebra.*

(a) *Then* $\dim R$ *is the smallest integer d for which there exist homogeneous elements $x_1, \dots, x_d \in R$ such that $\mathfrak{m} = \bigoplus_{k=1}^{\infty} R_k$ is the radical of the ideal $(x_1, \dots, x_d)$.*
(b) *Such elements $x_1, \dots, x_d$ are algebraically independent over K, and R is a finitely generated module over $K[x_1, \dots, x_d]$.*
(c) *If K is infinite and $R = K[R_1]$, then $x_1, \dots, x_d$ can be chosen of degree* 1.

Elements $x_1, \dots, x_d$ as in Theorem 3.1 are said to form a *homogeneous system of parameters* for R. More generally, if M is an R-module, then a *homogeneous system of parameters for M* is just a homogeneous system of parameters for $R/\operatorname{Ann} M$. (By $\operatorname{Ann} M$ we denote the annihilator ideal $\{x \in R : xM = 0\}$.) The subalgebra $K[x_1, \dots, x_d]$ is called a *graded Noether normalization* of R. It is isomorphic to a polynomial ring over K, and so Theorem 3.1 shows that every finitely generated graded module over R, in particular R itself, can be understood as such a module over the polynomial ring $K[X_1, \dots, X_d]$. We will intensively use this argument in the next sections.

We cannot include a full proof of Theorem 3.1. The first step is to show that the homogeneous elements in a prime ideal $\mathfrak{p}$ of a graded ring generate again a prime ideal (see [8, 1.5.6]). It follows that all minimal prime ideals of a graded ring are graded. Starting from this basic fact one constructs a chain of graded prime ideals

$$\mathfrak{p}_0 \subset \cdots \subset \mathfrak{p}_d = \mathfrak{m}$$

of length $d = \dim R$. (Since every homogeneous ideal of R, except R itself, is contained in $\mathfrak{m}$, all such chains must end in $\mathfrak{m}$.) Now Krull's principal ideal theorem shows that $\mathfrak{m}$ cannot be the radical of an ideal generated by fewer than d elements.

The existence of $x_1, \dots, x_d$ is shown by induction on d, based on the following lemma:

Lemma 3.2. *Let I be an ideal in a graded ring, generated by elements of positive degree, and let $\mathfrak{p}_1, \dots, \mathfrak{p}_m$ be prime ideals such that $I \not\subset \mathfrak{p}_i$ for $i = 1, \dots, m$.*

(a) *Then I contains a homogeneous element $x \notin \mathfrak{p}_1 \cup \cdots \cup \mathfrak{p}_m$.*
(b) *If R is a graded algebra over an infinite field K and I is generated by elements $x_1, \dots, x_n$ of the same degree g, then there exist $a_1, \dots, a_n \in K$ such that $x = a_1x_1 + \cdots + a_nx_n \notin \mathfrak{p}_1 \cup \cdots \cup \mathfrak{p}_m$.*

For the combinatorial application we are free to choose the field K (and often one can extend K anyway). Therefore it may be enough to prove (b): by hypothesis the intersection $\mathfrak{p}_i \cap V$ is a proper subspace of $V = Kx_1 + \cdots + Kx_n$, and a vector space over an infinite field cannot be the union of finitely many proper subspaces.

For the proof of Theorem 3.1 suppose first that $d = 0$. Then $\mathfrak{m}$ is the only prime ideal in R, and every element in $\mathfrak{m}$ is nilpotent, since the nilradical of R is the intersection of its prime ideals. Therefore $\mathfrak{m}$ is the radical of the ideal generated by the empty set.

Now let $d \geq 1$. To conclude the proof of Theorem 3.1(a) we apply the lemma to $I = \mathfrak{m}$ and the minimal prime ideals $\mathfrak{p}_1, \ldots, \mathfrak{p}_m$ of R, choosing $x_1 = x$. Then $R/(x_1)$ has Krull dimension $d-1$, and we are done by induction, lifting a homogeneous system of parameters of $R/(x_1)$ to R.

This argument shows that we can find a homogeneous system of parameters consisting of degree 1 elements if R is generated over K in degree 1 and K is infinite.

It remains to prove (c). Let I be the ideal in R generated by $x_1, \ldots, x_d$ and $y_1, \ldots, y_m$ be a homogeneous system of generators of the K-algebra R. Since $y_i^e \in I$ for $e \gg 0$, there exists $f \in \mathbb{Z}$ such that $R_i \subset I$ for $i \geq f$. Then a vector space basis of $R_0 \oplus \cdots \oplus R_{f-1}$ generates R as an $K[x_1, \ldots, x_d]$-module. The reader may show this as an exercise or look up [8, p. 38].

For the algebraic independence of $x_1, \ldots, x_d$ it is sufficient that $\dim K[x_1, \ldots, x_d] = \dim R = d$. This follows from the going up theorem for module-finite or, more generally, integral ring extensions.

Krull dimension and transcendence degree

Suppose that R is an integral domain, $y \in R$, and let S be a graded Noether normalization. Since R is a finitely generated S-module, and submodules of finitely generated modules over Noetherian rings are again finitely generated, there exists an exponent q such that y^q is an S-linear combination of the powers $y^0 = 1, y_1, \ldots, y^{q-1}$. It follows that the quotient field $\mathrm{QF}(R)$ is an algebraic extension of $\mathrm{QF}(S)$, and the latter has transcendence degree $d = \dim R$ over K. Therefore d is the transcendence degree of $\mathrm{QF}(R)$ over K.

Let M be an affine monoid. Then $\mathrm{QF}(K[M]) = \mathrm{QF}(K[\mathrm{gp}(M)])$, and the transcendence degree of the latter equals rank M. So $\dim K[M] = \operatorname{rank} M$.

We use this fact to show that $\dim K[\mathcal{M}_n] = (n-1)^2 + 1$. Let $G = \mathrm{gp}(\mathcal{M}_n)$. Let $\mathbf{1}$ be the magic $n \times n$-square with all entries equal to 1, and $A \in \mathbb{Z}^{n\times n}$ a solution of the homogeneous system of equations (2) defining $\mathcal{M}_n$, without sign restriction. Then $A + k\mathbf{1} \in \mathcal{M}_n$ for $k \gg 0$, and it follows that $\mathrm{gp}(\mathcal{M}_n)$ is the group of all solutions to the system (2). The system (2) has rank $2n-2$, so rank $\mathcal{M}_n = n^2 - (2n-2) = (n-1)^2 + 1$.

4. Hilbert functions of graded modules

Suppose that K is a field, considered as a graded ring in which all elements have degree 0, and let $V = \bigoplus_{i\in\mathbb{Z}} V_i$ be a graded K-vector space. Then the *Hilbert function*

$H(M,\_)$ of V is defined as follows:

$$H(M,\_) : \mathbb{Z} \to \mathbb{Z} \cup \{\infty\}, \qquad H(M,i) = \dim_K M_i.$$

A fundamental property of the Hilbert function is its additivity along exact sequences of homogeneous linear maps: If $0 \to V^m \to \cdots \to V^0 \to 0$ is an exact and homogeneous sequence of graded K-vector spaces V^i, then $\sum(-1)^i H(V^i,\_) = 0$. This follows immediately from the additivity of K-dimension, applied to the homogeneous components.

Let R be a K-algebra. If R is graded, then the graded components R_i are automatically K-vector spaces since $K \subset R_0$, and for the same reason the graded components M_i of a graded R-module M are K-vector spaces. If R is a positively graded, finitely generated K-algebra, then the Hilbert function of a finitely generated graded R-module M takes only finite values and moreover $H(M,i) = 0$ for $i \ll 0$. Therefore the *Hilbert* (or *Poincaré*) *series*

$$H_M(t) = \sum_{i \in \mathbb{Z}} H(M,i)t^i$$

is a Laurent series with a finite principal part. It belongs to the field $\mathbb{C}[[t]][t^{-1}]$. Note that $\mathbb{C}[[t]][t^{-1}]$ contains the polynomial ring $\mathbb{C}[t]$ and therefore the field $\mathbb{C}(t)$ of rational functions. The embedding $\mathbb{C}(t) \hookrightarrow \mathbb{C}[[t]][t^{-1}]$ identifies a rational function with its Laurent expansion at 0. In the following we will use the same notation for a rational function and its Laurent expansion at 0.

The fundamental structural result about $H_M(t)$ is the following theorem.

Theorem 4.1 (Hilbert–Serre). *Let R be a graded K-algebra generated by homogeneous elements $x_1, \dots, x_n$ of degrees $g_1, \dots, g_n > 0$. Furthermore let M be a non-zero, finitely generated graded R-module. Then there exists a Laurent polynomial $Q \in \mathbb{Z}[t, t^{-1}]$ such that*

$$H_M(t) = \frac{Q(t)}{\prod_{i=1}^n (1 - t^{g_i})}.$$

Proof. We use induction on n. In the case $n = 0$ we can simply take $Q = H_M$. Let $n > 0$. Then multiplication on M by x_n induces an exact sequence

$$0 \to (0 :_M x_n)(-g_n) \to M(-g_n) \xrightarrow{x_n} M \to M/x_n M \to 0.$$

Both the kernel $U = (0 :_M x_n)$ of the multiplication map and $N = M/x_n M$ are finite modules over the graded algebra $R/(x_n)$ (or over $K[x_1, \dots, x_{n-1}]$). Since the Hilbert function is additive along exact sequences, we obtain

$$(1 - t^{g_n})H_M = H_N - t^{g_n} H_U.$$

It only remains to apply the induction hypothesis. We are allowed to divide by $1 - t^{g_n}$ in the field $\mathbb{C}[[t]][t^{-1}]$, to which H_M, H_U, H_N belong. (Here it is used that $g_i > 0$ for all i.) □

In particular, if $R = K[X_1, \dots, X_n]$ is the polynomial ring over the field K, graded in such a way that $\deg X_i = g_i$, $i = 1, \dots, n$, then

$$H_R(t) = \frac{1}{\prod_{i=1}^{n}(1 - t^{g_i})}. \tag{5}$$

The formula follows by the same induction with $x_i = X_i$. In this case the kernel U is always 0, since X_n is a non-zero-divisor on R.

Algebras generated in degree 1

Note that the Hilbert function of a graded module M depends only on the decomposition of M as a K-vector space. We are therefore free to replace R by any other graded K-algebra as long as the grading of M is not changed. This allows us to refine Theorem 4.1. First we consider the classical case in which $g_1, \dots, g_n = 1$ and the Hilbert function is of polynomial type:

Theorem 4.2. *Under the hypotheses of Theorem 4.1 suppose that* $g_1 = \dots = g_n = 1$ *and let* $d = \dim M$. *Then there exists a Laurent polynomial* $Q \in \mathbb{Z}[t, t^{-1}]$ *such that*

$$H_M(t) = \frac{Q(t)}{(1-t)^d}.$$

Moreover:

(a) *There exists a polynomial* $P_M \in \mathbb{Q}[X]$ *such that*

$$\begin{aligned} H(M,i) &= P_M(i), & i &> \deg H_M, \\ H(M,i) &\neq P_M(i), & i &= \deg H_M. \end{aligned}$$

(b) $e(M) = Q(1) > 0$, *and if* $d \geq 1$, *then*

$$P_M = \frac{e(M)}{(d-1)!} X^{d-1} + \text{terms of lower degree.}$$

Proof. We may assume that K is infinite. Otherwise we choose an infinite extension L of K, replace R by $R \otimes_K L$, and M by $M \otimes_K L$. Next we substitute $R/\operatorname{Ann} M$ for R, and can assume that $d = \dim R$. Then there exists a Noether normalization S of R generated by (necessarily) d homogeneous elements of degree 1. Since R is a finitely generated S-module, so is M, and direct specialization of Theorem 4.1 yields

$$H_M(t) = \frac{Q(t)}{(1-t)^d}.$$

We expand into a Laurent series:

$$\frac{t^j}{(1-t)^d} = \sum_{k=0}^{\infty} \binom{k+d-1}{d-1} t^{k+j} = \sum_{i \in \mathbb{Z}} c_j(i) t^i$$

with

$$c_j(i) = \binom{i+d-1-j}{d-1}, \quad i \geq j, \qquad \text{and} \qquad c_j(i) = 0, \quad i < j.$$

Set

$$p_j(X) = \binom{X+d-1-j}{d-1} = \frac{(X+d-1-j)\cdots(X+1-j)}{(d-1)!}.$$

Then $p_j \in \mathbb{Q}[X]$, and we have

$$c_j(i) = p_j(i), \quad i > j-d, \qquad \text{and} \qquad c_j(i) \neq p_j(i), \quad i = j-d.$$

With $Q(t) = h_a t^a + \cdots + h_b t^b$, $a, b \in \mathbb{Z}$, $a \leq b$, $h_a, h_b \neq 0$, the desired polynomial is

$$P_M = \sum_{j=a}^{b} h_j p_j.$$

That $Q(1) > 0$ will be shown in Proposition 4.5. Then the assertion on the leading coefficient of P_M follows since each of the polynomials p_j has degree $d-1$ (if $d \geq 1$) and leading coefficient $1/(d-1)!$. Therefore P_M has leading coefficient

$$\sum_{j=a}^{b} \frac{h_j}{(d-1)!} = \frac{Q(1)}{(d-1)!}.$$

□

Definition 4.3. The polynomial P_M is called the *Hilbert polynomial* of M, and $e(M)$ is the *multiplicity* of M.

Positively graded algebras

If M cannot be written as a finitely generated module over a K-algebra generated in degree 1, then its Hilbert function need not be of polynomial type. It is however of *quasi-polynomial type*: let $P_0, \ldots, P_{m-1}$ be polynomials. Then the function P,

$$P(km + \ell) = P_\ell(k), \qquad k \in \mathbb{Z}, \ \ell = 0, \ldots, m-1,$$

is called a *quasi-polynomial of period* m.

Theorem 4.4. *Under the hypotheses of Theorem 4.1 suppose that* $d = \dim M$. *Then there exist a Laurent polynomial* $Q \in \mathbb{Z}[t, t^{-1}]$ *and integers* $e_1, \ldots, e_d > 0$ *such that*

$$H_M(t) = \frac{Q(t)}{\prod_{i=1}^{d}(1-t^{e_i})}, \qquad Q(1) > 0.$$

Moreover there exists a quasi-polynomial P_M, *whose period divides* $\operatorname{lcm}(e_1, \ldots, e_d)$, *such that*

$$H(M,i) = P_M(i), \qquad i > \deg H_M,$$
$$H(M,i) \neq P_M(i), \qquad i = \deg H_M.$$

Proof. As above we can assume that K is infinite and replace R by a graded Noether normalization of $R/\operatorname{Ann} M$. Then $e_1, \dots, e_d$ can be chosen as the degrees of the elements in a homogeneous system of parameters. The formula for H_M follows again by direct specialization of Theorem 4.1.

With $m = \operatorname{lcm}(e_1, \dots, e_d)$ we choose $S = R^{(m)} = \bigoplus_{k=0}^{\infty} R_{km}$, the m-th Veronese subalgebra of R. Dividing by m, we normalize the degrees in S. While S need not be generated in degree 1, it is a finite module over $S' = K[S_1] = K[R_m]$, as the reader may check. As an S'-module M splits into the direct sum of its submodules

$$N_\ell = \bigoplus_{k \equiv \ell\ (m)} M_k, \qquad \ell = 0, \dots, m-1.$$

Now we can compose P from the Hilbert polynomials of the N_ℓ. The reader should check that the assertion on the equality of $P(i)$ and $H(M, i)$ follows from the corresponding statement in Theorem 4.2.

Again it remains to show $Q(1) > 0$, but this will be done in the next proposition. □

The theorem contains the statement that the Krull dimension of M is the pole order of H_M at $t = 1$. The proof is not yet complete, but we know at least that the pole order is bounded above by the Krull dimension. We use this fact in the proof of

Proposition 4.5. *Under the hypothesis of Theorem 4.4 suppose that $e_1, \dots, e_d$ are the degrees of the elements in a homogeneous system of parameters $x_1, \dots, x_d$ for M. Let $S = K[x_1, \dots, x_d]$. Then $Q(1) = \operatorname{rank}_S M > 0$.*

Proof. The annihilator of M as a module over S is 0. We can replace R by S, but must keep in mind that M has annihilator 0. The polynomial ring R over K is an integral domain and every R-module has a well-defined rank, given by the vector space dimension of $M \otimes \operatorname{QF}(R)$ over $\operatorname{QF}(R)$. There exist homogeneous elements $y_1, \dots, y_r \in M$ that form a basis of $M \otimes \operatorname{QF}(R)$ over $\operatorname{QF}(R)$. Let F be the R-submodule of M generated by $y_1, \dots, y_r$. Then $y_1, \dots, y_r$ are linearly independent, and there is an exact sequence of graded modules

$$0 \to F \to M \to N \to 0, \qquad N = M/F.$$

Since N has non-zero annihilator, $\dim N < d$. With self-explaining notation we have

$$H_N(t) = \frac{Q_M - Q_F}{\prod_{i=1}^{d}(1 - t^{e_i})}.$$

Since $\dim N < d$, the difference $Q_M - Q_F$ must be divisible by $1 - t$ since the pole order of $H_N(t)$ at $t = 1$ is smaller than d. Therefore $Q_M(1) = Q_F(1)$, and that $Q_F(1) = \operatorname{rank} F = \operatorname{rank} M$ is clear:

$$H_F(t) = \frac{f_a t^a + \cdots + f_b t^b}{\prod_{i=1}^{d}(1 - t^{e_i})}$$

where f_i counts the number of degree i elements in a homogeneous basis of F. □

Remark 4.6.

(a) The module-theoretic argument we have used in the proof of Theorem 4.4 can be replaced by a formal reasoning. It amounts to writing $H_M(t)$ as a rational function with denominator $(1-U)^d$ where $U = T^m$ and to split the numerator according to the residue classes modulo m of the exponents of the T^j. It follows that the assertion on the period of P does not depend on the fact that we can really find a Noether normalization generated in degrees $e_1, \dots, e_d$.

Similarly $Q(1) > 0$ whenever $e_1, \dots, e_d$ admit a representation of $H_M(t)$ with denominator $\prod_{i=1}^{d}(1-t^{e_i})$. In fact two such denominators differ only by polynomial factors $1 + t + \cdots + t^f$, having positive value at 1.

(b) Suppose that there exists a homogeneous element y in R that is a non-zero-divisor of M and whose degree is coprime to m. Then the Hilbert polynomials of the modules N_ℓ have the same degrees and the same leading coefficients. This holds since there exist $u_{ij} \in \mathbb{Z}_+$ for $i, j = 0, \dots, m-1$ such that $y^{u_{ij}} N_i \subset N_j$. This result can be substantially extended: let I be the ideal of R generated by those homogeneous elements that are coprime to m; if $\dim M/IM \le k$, then the coefficients of the degree j terms in the Hilbert polynomials coincide for all $j \ge k$. See [9].

Hilbert functions and homological invariants

The Hilbert series can be calculated from a graded finite free resolution. In the next theorem we use the standard convention of writing the shifts with a negative sign and grouping the terms in a free graded module $R(-s_1) \oplus \cdots \oplus R(-s_m)$ according to their shifts.

Theorem 4.7. *Let M be a finite graded R-module of finite projective dimension, and let*

$$0 \to \bigoplus_j R(-j)^{\beta_{pj}} \to \cdots \to \bigoplus_j R(-j)^{\beta_{0j}} \to M \to 0$$

be a graded free resolution of M. Then

$$H_M(t) = S_M(t) H_R(t)$$

where $S_M(t) = \sum_{i,j} (-1)^i \beta_{ij} t^j$. In particular, if $R = K[X_1, \dots, X_n]$ is the polynomial ring over the field K, graded in such a way that $\deg X_i = g_i$, $i = 1, \dots, n$, *then*

$$H_M(t) = \frac{S_M(t)}{\prod_{i=1}^{n}(1-t^{g_i})}.$$

Proof. This follows immediately from the additivity of Hilbert functions along exact sequences. The formula for the Hilbert function of the polynomial ring has already been proved above. □

If $R = K[X_1, \dots, X_n]$, then every finitely generated module over R has a finite free resolution by *Hilbert's syzygy theorem* (for example, see [8, 2.214]). Under the hypotheses of Theorem 4.1 we can choose $S = K[X_1, \dots, X_n]$ with $\deg X_i = g_i$ and map it onto $R = K[x_1, \dots, x_n]$ via the substitution $X_i \mapsto x_i$. In conjunction with the syzygy theorem we therefore obtain a new (but more difficult) proof of Theorem 4.1. This is Hilbert's original approach [16] to his theorem that the Hilbert function is of polynomial type (in the classical case $\deg X_i = 1$).

The difference between the Hilbert function and the Hilbert polynomial can be expressed in terms of local cohomology. It would take us too far astray to explain this concept here. See [8].

Theorem 4.8 (Serre). *Under the hypotheses of Theorem 4.4 let* $\mathfrak{m}$ *denote the maximal ideal* $(x_1, \dots, x_n)$ *and* $d = \dim M$. *Then*

$$H(M, i) - P_M(i) = \sum_{j=0}^{d} (-1)^j \dim_K H^j_{\mathfrak{m}}(M)_i$$

for all $i \in \mathbb{Z}$.

Here $H^j_{\mathfrak{m}}(M)$ denotes the jth local cohomology of M with support in $\mathfrak{m}$. The proof follows the same pattern as that of Theorem 4.1, but this time one must work in the field $\mathbb{C}[[t^{-1}]][t]$ after one has shown that the generating function of the right hand belongs to it – for the left hand this has been proved above. We refer the reader to [8, 4.4.3] for the details.

5. Cohen–Macaulay rings

Before giving a definition we examine a simple, but very instructive case. Also in the next section it will guide us to a general theorem.

Simplicial normal monoids

An affine monoid M is *simplicial* if the cone $\mathbb{R}_+ M$ is generated by $d = \operatorname{rank} M$ vectors $x_1, \dots, x_d$. Suppose that M is normal, embedded into $\mathbb{Z}^d = \operatorname{gp}(M)$, and let $y \in M$. Then y has a unique representation $y = \sum_{i=1}^{d} a_i x_i$ with $a_i \in \mathbb{Q}_+$. It follows that

$$y = \sum_{i=1}^{d} b_i x_i + \sum_{i=1}^{d} q_i x_i, \qquad b_1, \dots, b_n \in \mathbb{Z}_+, \quad q_1, \dots, q_n \in [0, 1) \cap \mathbb{Q}.$$

In other words, $y = y' + y''$ where y' belongs to the free submonoid $N = \sum \mathbb{Z}_+ x_i$ and y'' is an integral point in the semi-open parallelotope

$$\operatorname{par}(x_1, \dots, x_d) = \left\{ \sum_{i=1}^{d} q_i x_i : q_i \in [0, 1) \right\}.$$

The vectors y', y'' are uniquely determined by y. Since M is normal, $y'' \in M$. Set

$$B = \operatorname{par}(x_1, \dots, x_d) \cap \mathbb{Z}^d.$$

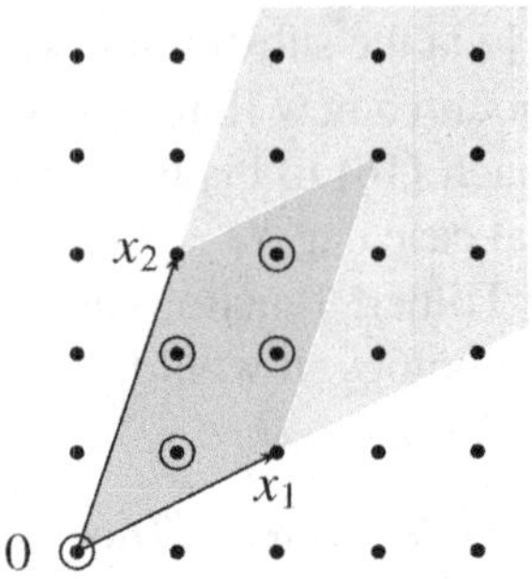

Figure 4. The semi-open parallelotope

If M is an affine monoid with a positive grading, then we can assign a Hilbert function to M directly, namely

$$H(M, k) = \#\{y \in M : \deg y = k\}.$$

Clearly $H(M, k) = H(K[M], k)$ for the induced grading on $K[M]$. Similarly we use the notion of Hilbert series for M.

Proposition 5.1.

(a) *M is the disjoint union of the sets $z + N$, $z \in B$;*
(b) *Let* $\deg : M \to \mathbb{Z}_+$ *be a positive grading on M. Then*

$$H_M(t) = \frac{\sum_{i=0}^{u} h_i t^i}{\prod_{j=1}^{d}(1 - t^{g_j})}, \qquad h_i = \#\{z \in B : \deg z = i\} \geq 0, \quad i = 0, \dots, u.$$

(c) $\deg H_M < 0$*, and* $P_M(i) = H_M(i)$ *for all* $i \geq 0$.

Proof. (a) has been proved above. For (b) it is now enough that $H_N(t) = 1/\prod_{j=1}^{d}(1 - t^{g_j})$. (This follows from equation (5), but can also be proved directly.) For (c) we note that every element in B has degree $< \sum g_j$ so that $u < \sum g_j$. The last assertion follows from Theorem 4.4. □

Proposition 5.1 contains statements of type (ADG-2) and (ADG-4), but it cannot be applied directly since the monoid of magic squares is simplicial only for $n = 1, 2$.

In commutative algebra language 5.1(a) reads as follows: the monoid algebra $S = K[N]$ is a Noether normalization of $K[M]$, and $R = K[M]$ *is a free module over its Noether normalization* S. In fact, S is generated by d elements, $d = \dim R$. Furthermore the finitely many monomials $z \in B$ generate R as a S-module. Their linear independence follows easily from the disjointness of the decomposition in 5.1(a).

Cohen–Macaulay rings

Graded algebras or modules that are free over a Noether normalization deserve (and have) a special name. Its classical and usual definition uses the notion of regular sequence:

Definition 5.2. Let R be a commutative ring and M an R-module. We say that $x_1, \dots, x_m \in R$ form an *M-sequence* or are *M-regular* if

(i) $M/(x_1, \dots, x_m)M \neq 0$;
(ii) x_i is a non-zero-divisor on $M/(x_1, \dots, x_{i-1})M$ for $i = 1, \dots, m$.

Condition (ii) is equivalent to the injectivity of the multiplication by x_i on $M/(x_1, \dots, x_{i-1})M$. Therefore the property of being an M-sequence behaves well under flat ring extensions, like passage to a ring of fractions or to $R \otimes_K L$ if R contains a field K and L is an extension field of K.

The next proposition connects the notion of M-sequence with the freeness of M over a Noether normalization. Moreover, it shows that this property does not depend on the choice of normalization:

Proposition 5.3. *Let R be a positively graded finitely generated K-algebra and M a finitely generated graded R-module. Furthermore, let $x_1, \dots, x_d$ and $y_1, \dots, y_d$ be homogeneous systems of parameters for M. Then the following are equivalent:*

(a) *M is a free module over $S = K[x_1, \dots, x_d]$;*
(b) *$x_1, \dots, x_d$ is an M-sequence;*
(c) *$y_1, \dots, y_d$ is an M-sequence.*

Proof. Both (a) and (b) concern only the structure of M as an S-module. In proving their equivalence we may therefore assume that $R = S = K[X_1, \dots, X_d]$. The implication (a) $\Rightarrow$ (b) is then obvious: M is a direct sum of copies of R, and (b) is trivial for $M = R$.

For the converse we choose a minimal homogeneous system $z_1, \dots, z_m$ of generators of M. "Minimal" means that none of $z_1, \dots, z_m$ can be omitted. We want to show that $z_1, \dots, z_m$ are linearly independent. Choose a presentation

$$F \xrightarrow{\varphi} M \to 0$$

mapping the i-th basis element e_i of the graded free module F to z_i. Our claim amounts to $\operatorname{Ker} \varphi = 0$.

On the contrary, let us assume that $z_1, \dots, z_m$ are linearly dependent. Then there exist $a_1, \dots, a_m \in R$, not all 0, such that $a_1 z_1 + \dots + a_m z_m = 0$. Since $z_1, \dots, z_m$ are homogeneous, we can split this equation into its homogeneous components. Therefore we may assume that the a_i are homogeneous and that we have chosen a non-trivial homogeneous relation of the smallest possible degree.

None of the $a_i \neq 0$ can be in K. Otherwise we could write z_i as a linear combination of the z_j, $j \neq i$. In other words, $\operatorname{Ker} \varphi \subset \mathfrak{m} F$, where $\mathfrak{m} = (X_1, \dots, X_d)$. This implies that the residue classes of $z_1, \dots, z_m$ form a basis of the K-vector space $M/\mathfrak{m}M$.

For $d = 0$, $R = K$, there is nothing to prove. So let $d > 0$. Set $\bar{R} = R/(X_1)$, $\bar{M} = M/X_1 M$. Taking intermediate residue classes with respect to (X_1) does not change the fact that the residue classes of $\bar{z}_1, \dots, \bar{z}_m \in \bar{M}$ form a basis of $M/\mathfrak{m}M$. Moreover $X_2, \dots, X_d$ form an $\bar{M}$-sequence. By the induction hypothesis $\bar{z}_1, \dots, \bar{z}_m$ are linearly independent over $\bar{R}$.

For our relation $a_1 z_1 + \dots + a_m z_m = 0$ this implies $a_i \in (X_1)$, $a_i = b_i X_1$, $i = 1, \dots, m$. Since X_1 is a non-zero-divisor on M, we can factor out X_1 from the equation

$a_1z_1 + \cdots + a_mz_m = 0$, and obtain a contradiction to the minimality of $\deg a_i$. This completes the proof of (b) $\Rightarrow$ (a).

We postpone the somewhat technical proof of the equivalence of (b) and (c) to the end of this section. □

Definition 5.4. Let R be an affine, positively graded K-algebra, and M a finitely generated graded R-module. Then M is *Cohen–Macaulay* if it admits an M-regular homogeneous system of parameters. If R itself is a Cohen–Macaulay module, then it is called a *Cohen–Macaulay ring*.

As stated in Proposition 5.3 (but not yet proved) it follows that every homogeneous system of parameters of a graded Cohen–Macaulay module M is M-regular.

The main combinatorial consequence of the Cohen–Macaulay property is the nonnegativity (or even positivity) of the h-vector.

Theorem 5.5. *Let R be a positively graded affine K-algebra, M a finitely generated graded Cohen–Macaulay R-module, and let $g_1, \dots, g_d$ be the degrees of the elements in a homogeneous system $x_1, \dots, x_d$ of parameters for M. Let*

$$H_M(t) = \frac{\sum_{i=a}^{b} h_i t^i}{\prod_{i=1}^{d}(1 - t^{g_i})}$$

be the Hilbert series of M. Then $h_i \geq 0$ for all i.

If R is a Cohen–Macaulay ring generated in degree 1 and

$$H_R(t) = \frac{\sum_{i=0}^{u} h_i t^i}{(1 - t)^d}, \qquad h_u \neq 0,$$

then $h_i > 0$ for $i = 0, \dots, u$.

Proof. The first assertion follows as in the special case of a simplicial normal monoid algebra considered above: h_i counts the number of degree i basis elements of M over the Noether normalization.

Alternatively we could argue as follows: $\sum_{i=a}^{b} h_i t^i$ is the Hilbert series of $M/(x_1, \dots, x_d)M$. This follows immediately by induction on d (compare the proof of Theorem 4.1).

For the second assertion we first extend the field in order to obtain a degree 1 homogeneous system $x_1, \dots, x_d$ of parameters. The K-algebra $\bar{R} = R/(x_1, \dots, x_d)$ is generated in degree 1 and $\sum_{i=0}^{u} h_i t^i$ is its Hilbert series. If $\bar{R}_i = 0$, then $\bar{R}_j = 0$ for all $i \geq j$. □

Remark 5.6. In the second part of Theorem 5.5 we have used the fact that $\bar{R} = K[\bar{R}_1]$ only in a very weak form. Which h-vectors $(h_0, \dots, h_u)$ can occur for such algebras is determined by *Macaulay's theorem*; see [8, 4.2.15].

In order to apply Theorem 5.5 to $\mathcal{M}_n$ or other normal affine monoids we must prove

Theorem 5.7 (Hochster). *Let M be a normal affine monoid (with a positive grading). Then $K[M]$ is Cohen–Macaulay for every field K.*

The Hochster–Roberts theorem

We want to derive Hochster's theorem from a more general result:

Theorem 5.8 (Hochster–Roberts). *Let K be a field and let S be a an affine graded K-subalgebra of a polynomial ring $R = K[X_1, \dots, X_n]$. If there exists an S-submodule T of R such that $R = S \oplus T$, then S is Cohen–Macaulay.*

Theorem 5.8, in conjunction with Proposition 1.5, immediately implies Hochster's theorem. The theorem of Hochster–Roberts is proved by reduction to characteristic p. However, in general the hypothesis $R = S \oplus T$ in the theorem would not survive the reduction. Therefore one has to prove a more general statement.

Theorem 5.9. *Let K be a field, and let $f_1, \dots, f_s$ be algebraically independent homogeneous elements of positive degree in $R = K[X_1, \dots, X_n]$. Suppose that S is a module-finite graded $K[f_1, \dots, f_s]$-algebra such that there exists a homogeneous homomorphism $\psi : S \to R$ of $K[f_1, \dots, f_s]$-algebras. If $g_{r+1} f_{r+1} = g_1 f_1 + \cdots + g_r f_r$ with $g_1, \dots, g_{r+1} \in S$ for some r, $0 \le r \le s-1$, then $\psi(g_{r+1}) \in (f_1, \dots, f_r)R$.*

For the derivation of the Hochster–Roberts theorem we choose $f_1, \dots, f_s$ as a homogeneous system of parameters for S and $\psi : S \to R$ as the given embedding. We want to show that $f_1, \dots, f_s$ is S-regular. If it should fail, then we have an equation $g_{r+1} f_{r+1} = g_1 f_1 + \cdots + g_r f_r$ with $g_{r+1} \notin (f_1, \dots, f_r)S$ for some r. On the other hand, Theorem 5.9 shows that $g_{r+1} \in (f_1, \dots, f_r)R$, and this is a contradiction: since $R = S \oplus T$, one has $IR \cap S = I$ for every ideal I of S.

Proof of Theorem 5.9. It needs two steps:

(i) the case of positive characteristic;
(ii) the reduction from characteristic 0 to positive characteristic.

Under the general hypothesis of the theorem step (ii) is technically demanding (see [8, p. 294]). Therefore we restrict ourselves to step (i), and do step (ii) only for algebras that, roughly speaking, can be defined over the ring $\mathbb{Z}$ of integers; see Proposition 5.11 below.

So let K be a field of characteristic $p > 0$. We can replace K by its algebraic closure L. The hypothesis certainly survives the base field extension to L, and so does the conclusion since $IL[X_1, \dots, X_n] \cap K[X_1, \dots, X_n] = I$ for every ideal I of $K[X_1, \dots, X_n]$. We may therefore assume that $K = K^p$.

There exist a finitely generated free $K[f_1, \dots, f_s]$-submodule $F \subset S$ and a nonzero element $c \in K[f_1, \dots, f_s]$ such that $cS \subset F$ (just because $K[f_1, \dots, f_s]$ is a domain and S is finitely generated).

Let $q = p^e$ be a power of the characteristic p, take the q-th power of the equation $g_{r+1} f_{r+1} = g_1 f_1 + \cdots + g_r f_r$ and multiply by c to obtain

$$(cg_{r+1}^q) f_{r+1}^q = \sum_{i=1}^{r} (cg_i^q) f_i^q.$$

The elements cg_i^q, $i = 1, \dots, r+1$, are in the free $K[f_1, \dots, f_s]$-module F, and in $K[f_1, \dots, f_s]$ the elements $f_1, \dots, f_s$ behave like indeterminates. Thus an elementary argument yields the existence of $h_{iq} \in F$ with $cg_i^q = h_{iq} f_{r+1}^q$ for $i = 1, \dots, r$. By substituting these expressions into the previous equation and applying $\psi : S \to K[X_1, \dots, X_n]$ one has

$$cf_{r+1}^q \psi(g_{r+1})^q = \sum_{i=1}^r f_i^q f_{r+1}^q \psi(h_{iq}), \quad \text{hence} \quad c\psi(g_{r+1})^q = \sum_{i=1}^r f_i^q \psi(h_{iq}).$$

Let M be the set of monomials $\mu = X_1^{\mu_1} \cdots X_1^{\mu_n}$ with $\mu_i < q$ for $i = 1, \dots, n$. Taking q-th powers in the algebraically closed field K is bijective. Therefore every element $h \in K[X_1, \dots, X_n]$ has a necessarily *unique* representation $h = \sum_{\mu \in M} (h_\mu)^q \mu$; in particular,

$$\psi(h_{iq}) = \sum_{\mu \in M} (h_{iq\mu})^q \mu.$$

Thus

$$\sum_{i=1}^r f_i^q \psi(h_{iq}) = \sum_{\mu \in M} \left(\sum_{i=1}^r h_{iq\mu} f_i \right)^q \mu = \sum_{\mu \in M} (h_{q\mu})^q \mu$$

with $h_{q\mu} \in (f_1, \dots, f_r)R$.

The crucial point is that c does not depend on q. We choose q so large that $c = \sum_{\mu \in M} c'_\mu \mu$ with $c'_\mu \in K$. Let $c'_\mu = (c_\mu)^q$. Then

$$\sum_{\mu \in M} (c_\mu \psi(g_{r+1}))^q \mu = \sum_{\mu \in M} (h_{q\mu})^q \mu.$$

Since $c \neq 0$ there exists μ with $c_\mu \neq 0$, and so

$$\psi(g_{r+1}) = \frac{1}{c_\mu} h_{q\mu} \in (f_1, \dots, f_r)R.$$

□

Remark 5.10.

(a) This proof of the Hochster–Roberts theorem is due to F. Knop. It is a condensation of a tight closure proof by Hochster and Huneke (see [8, 10.1.14]).
(b) The Hochster–Roberts theorem can be strengthened. In characteristic 0 direct summands R of polynomial rings (or the affine varieties defined by them) have rational singularities by a theorem of Boutot [5]; also see Gurjar [15]. In characteristic p Hochster and Huneke showed that R is F-regular; see [8, 10.1.3]. These properties include that R is Cohen–Macaulay.
(c) A general Cohen–Macaulay criterion for affine monoid rings was established by Trung and Hoa [28].

Now Stanley's conjecture (ADG-4) has been completely proved. In fact, for the combinatorial application we are free to choose the coefficient field K, and so we take an infinite field of arbitrary positive characteristic. Moreover, we have shown that every homogeneous system of parameters in a direct summand of a polynomial ring over such a field, for example in $K[\mathcal{M}_n]$, is a regular sequence.

Free $\mathbb{Z}$-algebras

Let M be an affine monoid. Then we can define the monoid algebra $\mathbb{Z}[M]$ in the same way as the monoid algebra $K[M]$ where K is a field: we choose a free $\mathbb{Z}$-module with a basis corresponding to the elements of M, and use the monoid operation for the multiplication table. Then $K[M] = \mathbb{Z}[M] \otimes_{\mathbb{Z}} K$ for every field K.

Proposition 5.11. *Let R be a positively graded, finitely generated and torsionfree $\mathbb{Z}$-algebra. If $R \otimes \mathbb{Z}/(p)$ is Cohen–Macaulay for at least one prime number p, then $R \otimes K$ is Cohen–Macaulay for every field K of characteristic* 0.

Proof. Note that $\bar{R} = R \otimes \mathbb{Z}/(p)$ is a positively graded affine algebra over the field $\mathbb{Z}/(p)$. We choose a homogeneous system of parameters $\bar{x}_1, \dots, \bar{x}_d$ and homogeneous representatives $x_1, \dots, x_d$ in R. Moreover we choose a finite homogeneous system of generators $\bar{y}_1, \dots, \bar{y}_m$ (of positive degree) of $\bar{R}$ over $\mathbb{Z}/(p)[\bar{x}_1, \dots, \bar{x}_d]$ and homogeneous representatives $y_1, \dots, y_m \in R$.

The bridge from characteristic p to characteristic 0 is the ring of fractions $R_{(p)} = R \otimes \mathbb{Z}_{(p)}$ where $\mathbb{Z}_{(p)}$ is the ring of all rational numbers whose denominators are not divisible by p – it is just the localization of $\mathbb{Z}$ with respect to the prime ideal (p). We claim that $R_{(p)}$ is generated over its subalgebra $\mathbb{Z}_{(p)}[x_1, \dots, x_d]$ by $y_1, \dots, y_m$. In fact, consider a single graded component $(R_i) \otimes \mathbb{Z}_{(p)}$. It is a finitely generated $\mathbb{Z}_{(p)}$-module, since R_i is finitely generated over $\mathbb{Z}$. Let N be the submodule of $(R_{(p)})_i$ generated by the elements of type μy_j where μ is a monomial in $x_1, \dots, x_d$. By the choice of the x_i and y_j we have $(R_{(p)})_i = N + p(R_{(p)})_i$, since $(R_{(p)})_i / p(R_{(p)})_i = R_i / pR_i$. By the elementary divisor theorem, applied to the free module $(R_{(p)})_i$ and its submodule N over the Euclidean domain $\mathbb{Z}_{(p)}$ (or Nakayama's lemma), it follows that $(R_{(p)})_i = N$, as desired.

Before we discuss the Cohen–Macaulay property of $R \otimes K$, let us observe that $x_1, \dots, x_d$ is a homogeneous system of parameters for $R \otimes \mathbb{Q}$. The inversion of p leads us from $R_{(p)}$ to $R_{(0)} = R \otimes \mathbb{Q}$, and therefore $R \otimes \mathbb{Q}$ is a finitely generated module over $\mathbb{Q}[x_1, \dots, x_d]$. But $R \otimes \mathbb{Q}$ has the same Krull dimension as $R \otimes \mathbb{Z}/(p)$. For example, this can be concluded from the coincidence of their Hilbert functions:

$$H(R \otimes \mathbb{Z}/(p), i) = \operatorname{rank}_{\mathbb{Z}} R_i = H(R \otimes \mathbb{Q}, i).$$

By hypothesis $\bar{x}_1, \dots, \bar{x}_d$ is an $\bar{R}$-sequence. Moreover p is a non-zero-divisor on R since R is torsionfree. Therefore $p, x_1, \dots, x_d$ is an R-sequence. This property is not lost if we pass to the ring of fractions $R_{(p)} = R \otimes \mathbb{Z}_{(p)}$. Now we can conclude that

(i) $x_1, \dots, x_d$ are algebraically independent over $\mathbb{Z}_{(p)}$ and
(ii) $R_{(p)}$ is a free module over $\mathbb{Z}_{(p)}[x_1, \dots, x_d]$.

Claim (i) holds since $x_1, \dots, x_d$ are algebraically independent over $\mathbb{Q}$, and claim (ii) can be proved by the same induction as the implication (b) $\Rightarrow$ (a) of 5.3. For the case $d = 0$ one uses that $\mathbb{Z}_{(p)}$ is a discrete valuation domain over which a finitely generated module is free if p is a non-zero-divisor on it.

Now it is clear that we can permute $p, x_1, \dots, x_d$ without loosing the property that it is an $R_{(p)}$-sequence. By inversion of p it follows that $x_1, \dots, x_d$ is an $R \otimes \mathbb{Q}$-sequence.

The passage from $\mathbb{Q}$ to an arbitrary field of characteristic 0 is harmless, as observed immediately after Definition 5.2. □

Homogeneous systems of parameters in Cohen–Macaulay rings

It remains to show the equivalence of (b) and (c) in Proposition 5.3. We can assume that $\operatorname{Ann} M = 0$. There is nothing to show in dimension $d = 0$, but $d = 1$ is the critical case. Let x_1 and y_1 be homogeneous system of parameters for M. If x_1 is a non-zero-divisor, then all zero-divisors of M must be contained in minimal prime ideals of R. On the other hand, y_1 cannot be contained in such a prime ideal, and therefore is a non-zero-divisor.

The equivalence 5.3 (a) $\iff$ (b) shows that we can permute a homogeneous system of parameters without destroying the property of being an M-sequence. This will now be used.

For $d > 1$ we argue by induction. Suppose that $x_1, \dots, x_d$ is a homogeneous system of parameters and an M-sequence. We can replace the x_i by suitable powers to obtain elements of the same degree: $K[x_1, \dots, x_d]$ is a free module over $S = K[x_1^{e_1}, \dots, x_d^{e_d}]$, and so M is free over S. We can assume that K is infinite.

The residue class ring $\bar{R} = R/(y_1, \dots, y_{d-1})$ has dimension 1. It is not possible that the residue classes of $x_1, \dots, x_d$ are all contained in a single minimal prime ideal of $\bar{R}$. By the prime avoidance lemma 3.2 we therefore find a linear combination $x' = a_1x_1 + \dots + a_dx_d$ whose residue class is not contained in such a prime ideal. Equivalently, $y_1, \dots, y_{d-1}, x'$ is a homogeneous system of parameters.

Permuting the x_i if necessary we can assume that $a_1 \neq 0$. As follows from 5.3 (a) $\Rightarrow$ (b), also $x', x_2, \dots, x_d$ is an M-sequence. So $x_2, \dots, x_d$ is an $(M/x'M)$-sequence. By induction $y_1, \dots, y_{d-1}$ is an $(M/x'M)$-sequence, too. Thus $x', y_1, \dots, y_{d-1}$ is an M-sequence. Finally, a similar argument allows us to replace x' by y_d (here it is used that $d > 1$).

6. The canonical module and Gorenstein rings

A functional equation for the Hilbert series

Let $R = K[\mathcal{M}_n]$ and $d = \dim R = (n-1)^2 + 1$. The equation $P_n(-k) = (-1)^{d-1}P_n(k-n)$ in (ADG-3) (together with the not yet proved fact that $H(R,r) = P_n(r)$ for all $r \geq 0$) encodes a functional equation between the rational functions $H_R(t)$ and $H_R(t^{-1})$ where the latter is obtained by the substitution of t^{-1} for t. The substitution $t \mapsto t^{-1}$ transforms the Laurent expansion of $H_R(t)$ at 0 into the Laurent expansion of $H_R(t^{-1})$ at ∞, but what we really want to compare are the Laurent expansions at 0:

Lemma 6.1. *Let $P : \mathbb{Z} \to \mathbb{C}$ be a quasi-polynomial. Set $H(t) = \sum_{k=0}^{\infty} P(k)t^k$ and $G(t) = -\sum_{k=1}^{\infty} P(-k)t^k$. Then H and G are rational functions. Moreover*

$$H(t) = G(t^{-1}).$$

Proof. We have to exercise some care and need precise notation for expansions. If R is a rational function, then we let $\mathcal{L}_0(R)$ and $\mathcal{L}_\infty(R)$ denote the Laurent expansions at 0 and ∞ resp. One obtains $\mathcal{L}_\infty(R)$ as follows: in the Laurent expansion of $R(t^{-1})$ at 0 one replaces t by t^{-1}.

Both $\mathcal{L}_0(R)$ and $\mathcal{L}_\infty(R)$ are elements of the abelian group $\mathbb{C}[[t,t^{-1}]]$ of formal Laurent series $\sum_{k\in\mathbb{Z}} a_k t^k$. Note that the multiplication of elements of $\mathbb{C}[[t,t^{-1}]]$ by Laurent polynomials is well-defined. It makes $\mathbb{C}[[t,t^{-1}]]$ a module over $\mathbb{C}[t,t^{-1}]$. Moreover, the maps $\mathcal{L}_0$ and $\mathcal{L}_\infty$ from the field of rational functions to $\mathbb{C}[[t,t^{-1}]]$ are both injective and $\mathbb{C}[t,t^{-1}]$-linear.

Set $J = G(t^{-1})$, $Q = (1-t^e)^{d+1}$ and $F = \mathcal{L}_0(H) - \mathcal{L}_\infty(J)$. Since $\mathcal{L}_\infty(J) = -\sum_{k=1}^{\infty} P(-k)t^{-k}$, we have $F = \sum_{k\in\mathbb{Z}} P(k)t^k$. Let d be the degree of P and e its period. Then

$$QF = (1-t^e)^{d+1}F = 0.$$

In fact, $(1-t^e)F = \sum_{k\in\mathbb{Z}} \tilde{P}(k)t^k$ with the quasi-polynomial $\tilde{P}$ whose components $\tilde{P}_i$, $i = 1,\dots,e$, are given by $\tilde{P}_i(j) = P_i(j) - P_i(j-1)$, $j \in \mathbb{Z}$. Thus $\tilde{P}$ has lower degree than P, and the claim follows by induction on d.

By $\mathbb{C}[t,t^{-1}]$-linearity one has

$$Q\mathcal{L}_0(H) - Q\mathcal{L}_\infty(J) = Q(\mathcal{L}_0(H) - \mathcal{L}_\infty(J)) = QF = 0.$$

The k-th coefficient of $Q\mathcal{L}_0(H)$ vanishes for $k < 0$, and the ℓ-th coefficient of $Q\mathcal{L}_\infty(J)$ vanishes for $\ell > (d+1)e$. So the equation can only hold if $Q\mathcal{L}_0(H)$ and $Q\mathcal{L}_\infty(J)$ are the same Laurent polynomial K.

By the injectivity and $\mathbb{C}[t,t^{-1}]$-linearity of the assignment $R \mapsto \mathcal{L}_0(R)$ we conclude that $H = K/Q$, and similarly $G(t^{-1}) = J = K/Q$. □

Remark 6.2. We have not really used the hypothesis that P is a quasi-polynomial. It is enough that F is annihilated by a non-zero polynomial or, equivalently, that the values $P(k)$ satisfy a linear difference equation with constant coefficients.

We know from Theorem 4.2 that (ADG-2) follows if the Hilbert series of $K[\mathcal{M}_n]$ has degree $-n$. If this should indeed be true, then (ADG-3) is equivalent to the validity of a functional equation for the Hilbert series:

Corollary 6.3. Let R be a positively graded, finitely generated K-algebra of Krull dimension d, and let P be its Hilbert quasi-polynomial. Suppose that the Hilbert series $H_R(t)$ has degree $g < 0$. Then the following are equivalent:

(a) $P(-k) = (-1)^{d-1}P(k+g)$ for all $k \in \mathbb{Z}$;
(b) $(-1)^d H_R(t^{-1}) = t^{-g}H_R(t)$.

This follows from Lemma 6.1 by comparison of coefficients if one takes into account that $H(R,k) = P(k)$ for all $k \geq 0$ and $P(-k) = 0$ for $k = 1,\dots,g-1$ (see Theorem 4.4).

The canonical module and Gorenstein rings

The approach to the proof of (ADG-2) and (ADG-3) that we will take consists of two major steps:

(i) the construction of a graded R-module $\omega = \omega_R$ for a *Cohen–Macaulay* positively graded K-algebra R such that $H_\omega(t) = (-1)^d H_R(t^{-1})$, and
(ii) the exact computation of ω_R for normal affine monoid algebras, and in particular for $K[\mathcal{M}_n]$.

Step (i) is not difficult. Let us first consider the case in which $R = K[X_1, \dots, X_d]$ is the polynomial ring over K with grading given by $\deg X_i = g_i$. Set $e = -\sum_{i=1}^d g_i$ and $\omega = R(e)$. Then

$$H_\omega(t) = t^{-e} H_R(t) = \frac{\prod_{i=1}^d t^{g_i}}{\prod_{i=1}^d (1 - t^{g_i})} = (-1)^d H_R(t^{-1}). \tag{6}$$

If R is a Cohen–Macaulay positively graded K-algebra with graded Noether normalization $S = K[x_1, \dots, x_d] \cong K[X_1, \dots, X_d]$, we set

$$\omega_R = \operatorname{Hom}_S(R, \omega_S).$$

The module ω_R (whose construction so far may depend on the choice of S) has indeed the structure of an R-module defined by $(a\varphi)(b) = \varphi(ab)$ for all $a, b \in R$.

For the computation of its Hilbert series we only need the S-structure. Every direct summand $S(-j)$ of the free S-module R accounts for a summand $\operatorname{Hom}_S(S(-j), S(e)) \cong S(e + j)$ of ω_R. Let

$$H_R(t) = \frac{\sum_{j=0}^u h_j t^j}{\prod_{i=1}^d (1 - t^{g_i})}.$$

Then $R = \bigoplus_j S(-j)^{h_j}$, and so

$$H_{\omega_R}(t) = \frac{\sum_{j=0}^u h_j t^{-e-j}}{\prod_{i=1}^d (1 - t^{g_i})} = \frac{t^{-e}}{\prod_{i=1}^d (1 - t^{g_i})} \sum_{j=0}^u h_j t^{-j} = (-1)^d H_R(t^{-1}). \tag{7}$$

We have reached our goal: ω_R has the right Hilbert series, independently of the choice of the Noether normalization used in its construction. We supplement the combinatorial information on ω_R by

Proposition 6.4. *Let R be a positively graded Cohen–Macaulay K-algebra of Krull dimension d with Hilbert quasi-polynomial P. Then*

(a) $\deg H_R(t) = -\min\{i : (\omega_R)_i \neq 0\}$;
(b) $k \mapsto (-1)^{d-1} P(-k)$ *is the Hilbert quasi-polynomial of* ω_R, *and*

$$H(\omega_R, k) = (-1)^{d-1} P(-k) \qquad \text{for all } k \geq 1.$$

Part (a) follows from equation (7) since the lowest degree appearing in the numerator polynomial of $H_\omega(t)$ is $-\deg H_R(t)$. For part (b) one writes

$$H_R(t) = \sum_{k=0}^{\infty} P(k) t^k + \sum_{k=0}^{\deg H_R(t)} (H(R, k) - P(k)) t^k \tag{8}$$

and applies Lemma 6.1.

As our notation ω_R, in which the Noether normalization does not appear, suggests, this module is uniquely determined by R (up to homogeneous isomorphism).

Definition 6.5. The module ω_R is called the *canonical module* of R. One says that R is a *Gorenstein ring* if ω_R is isomorphic to $R(h)$ for some $h \in \mathbb{Z}$.

Before we indicate why ω_R is uniquely determined, we describe the combinatorial consequences of the Gorenstein property.

Theorem 6.6 (Stanley). *Let R be a positively graded, finitely generated K-algebra of Krull dimension d, let $H_R(t) = (h_0 + \cdots + h_u t^u)/\prod_{i=1}^{d}(1-t^{g_i})$ with $h_u \neq 0$ be the Hilbert series of R, and P the Hilbert quasi-polynomial. Suppose that R is a Gorenstein ring. Then*

(a) $\omega_R \cong R(g)$, $g = \deg H_R(t)$;
(b) $h_i = h_{u-i}$ *for* $i = 0, \dots, u$*: the h-vector is palindromic;*
(c) $H_R(t^{-1}) = (-1)^d t^{-g} H_R(t)$;
(d) $P(-k) = (-1)^{d-1} P(k+g)$ *for all* $k \in \mathbb{Z}$.

Conversely, if R is a Cohen–Macaulay integral domain such that

$$H_R(t^{-1}) = (-1)^d t^{-h} H_R(t)$$

for some $h \in \mathbb{Z}$, then R is Gorenstein.

Proof. Let R be a Gorenstein ring, say $\omega_R \cong R(h)$. Then we have $H_{\omega_R}(t) = t^{-h} H_R(t)$. Thus

$$(-1)^d H_R(t^{-1}) = t^{-h} H_R(t).$$

In particular the numerator polynomials on both sides must be equal:

$$t^{-g} \sum_{j=0}^{u} h_j t^{u-j} = t^{-h} \sum_{j=0}^{u} h_j t^j$$

This is only possible if $h = g$ and the h-vector is palindromic. This proves (a), (b), and (c). The equation in (d) follows from Proposition 6.4 by comparison of coefficients for $k \gg 0$. (However, if $g \geq 0$, (d) does not imply (c).)

For the converse we choose a non-zero element x of degree $-h$ in ω_R. The linear map $\xi : a \mapsto ax$ from R to ω_R is injective, since ω_R is torsionfree and R is an integral domain. Comparing the Hilbert series of the image with that of ω_R, we conclude that ξ is surjective, too. □

The proof that ω_R is unique up to homogeneous isomorphism requires the use of derived functors. First one shows that

$$\operatorname{Ext}_R^j(K, \omega_R) = \begin{cases} K, & j = d, \\ 0, & j \neq d. \end{cases} \tag{9}$$

This is to be understood as a relation between graded modules: $K \cong R/\mathfrak{m}$ (with $\mathfrak{m} = \bigoplus_{k=1}^{\infty} R_k$) lives in degree 0. Equation (9) implies that the localization $(\omega_R)_\mathfrak{m}$

is a canonical module of the local ring $R_{\mathfrak{m}}$. As such it is uniquely determined up to $R_{\mathfrak{m}}$-isomorphism. Finitely generated graded R-modules that become isomorphic after the passage to $R_{\mathfrak{m}}$ are isomorphic up to a shift. Thus graded canonical modules ω_R and ω'_R are isomorphic up to a shift. A non-zero shift is ruled out, since it would show up in $\operatorname{Ext}^d_R(K, \omega_R)$ (or in the Hilbert series). See [8, 3.6.9] for more details.

Equation (9) is proved by induction on d. First let $d = 0$. Then K itself is the Noether normalization, and $\omega_R = R^\vee = \operatorname{Hom}_K(R, K)$. Thus (9) follows from

(i) the exactness of the R-linear functor $V \mapsto V^\vee$ assigning each finitely generated graded R-module M its graded K-dual, and
(ii) the natural isomorphism $M = (M^\vee)^\vee$ if $\dim_K M < \infty$.

The functor $M \mapsto M^\vee$ is defined on the category of graded K-vector spaces: $\bigoplus M_i \mapsto \bigoplus N_i$ with $N_i = \operatorname{Hom}_K(M_{-i}, K)$, and (i) and (ii) follow from the general properties of graded K-duality. For (ii) it is enough that $\dim_K M_i < \infty$ for all i.

The general case is reduced inductively to the case $d = 0$. Let $x_1, \dots, x_d$ be a homogeneous system of parameters as usual, and $S = K[x_1, \dots, x_d]$. We set $\bar{R} = R/x_d R$ and $\bar{S} = S/x_d S$. By construction $\omega_{\bar{S}} = (\omega_S/x_d\omega_S)(\deg x_d)$, and it follows that

$$\omega_{\bar{R}} = (\omega_R/x_d\omega_R)(\deg x_d).$$

Since x_d annihilates K and is a non-zero-divisor on ω_R, the graded Rees lemma (see [8, 3.1.16 and 4.4.20]) yields

$$\operatorname{Ext}^i_{\bar{R}}(K, \omega_R/x_d\omega_R) = \operatorname{Ext}^{i+1}_R(K, \omega_R)(-\deg x_d) \qquad i \in \mathbb{Z},$$

and the two opposite shifts cancel each other in the passage from R to $\bar{R}$.

Remark 6.7.

(a) If we evaluate equation (8) beyond Proposition 6.4(b), then we obtain

$$H(\omega_R, k) = (-1)^d (H(R, -k) - P(-k)), \qquad k \in \mathbb{Z}.$$

The right hand side has appeared before, namely in Theorem 4.8. Since R is Cohen–Macaulay, the local cohomology modules $H^i_{\mathfrak{m}}(R)$ vanish for $i < d$, and so $H(\omega_R, k) = H(H^d_{\mathfrak{m}}(R), -k)$ for all k. The equality of the Hilbert functions shows that ω_R is the graded K-dual of $H^d_{\mathfrak{m}}(R)$, at least as a K-vector space.

This statement can be either viewed as part of the graded local duality theorem [8, 3.6.19] or used for an intrinsic definition of ω_R.

(b) Once it is known that ω_R is the graded K-dual of $H^d_{\mathfrak{m}}(R)$, equation (8) follows immediately from Theorem 4.8.

Simplicial monoids revisited

Now it has become clear that (ADG-3) is equivalent to the Gorenstein property of $K[\mathcal{M}_n]$. In order to prove it, we have to determine the canonical module. We will solve this task for every normal affine monoid ring. As in the previous section, the simplicial normal monoids lead us to the general theorem.

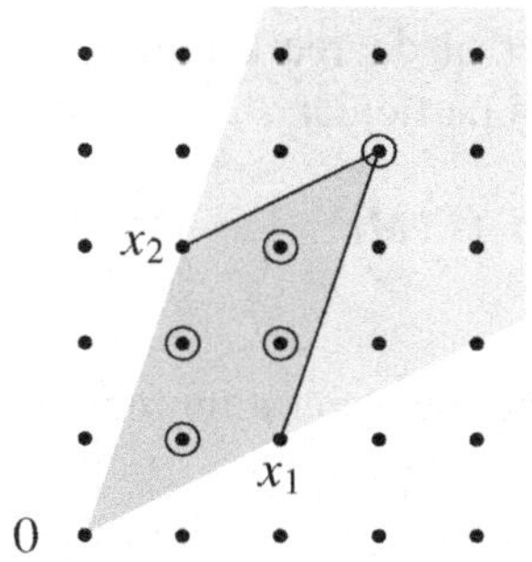

Figure 5. The basis of the interior ideal

Proposition 6.8. *Let M be a simplicial normal affine monoid, and let I be the ideal generated by the monomials in the interior of $\mathbb{R}_+M$. Then I is the graded canonical module of R.*

Proof. We choose $x_i \in M$, $i = 1, \ldots, d$, as in the proof of Proposition 5.1, namely as a linearly independent system of generators of $\mathbb{R}_+M$. They form a homogeneous system of parameters, generating the Noether normalization S of R. As a realization of the module $\omega_S = S(-\sum \deg y_i)$ we choose the S-submodule Sx, $x = x_1 \cdots x_d$, of R.

Let $z_1, \ldots, z_m$ be the monomial basis of the S-module R given by the monomials corresponding to the lattice points in $\operatorname{par}(x_1, \ldots, x_d)$. Then $z_i' = x/z_i$ is a monomial in the interior of $\mathbb{R}_+M$, and, as is easily checked, the z_i' form a basis of I as an S-module. Now we identify $\operatorname{Hom}_S(R, \omega_S)$ with the S-module I as follows: $z_i'(z_j) = x$ if $i = j$, and $= 0$ otherwise.

One must of course check that $\operatorname{Hom}_S(R, \omega_S)$ and I are isomorphic as R-modules. This is left to the reader. □

At this point we should note that Proposition 6.8 opens a purely combinatorial doorway to counting the elements in a normal affine monoid. For the elementary notions of combinatorial topology used in the proof of the next proposition we refer the reader to [6] or [29]. By $\operatorname{relint}(F)$ we denote the relative interior of F, that is the interior of F with respect to the topology on its affine hull.

Proposition 6.9. *Let $M \subset \mathbb{Z}^d$ be a normal simplicial affine monoid with a positive grading* deg. *Set $M_k = \{x \in M : \deg x = k\}$. Then there exists a quasi-polynomial P such that*

$$\#M_k = P(k) \qquad \textit{for all } k \geq 0.$$

Proof. We triangulate the cone $\mathbb{R}_+M$ into a family of simplicial subcones whose rays pass through elements of M. (The extreme rays of $\mathbb{R}_+M$ suffice.) This induces a disjoint decomposition of $\mathbb{R}_+M \setminus \{0\}$ into the relative interiors of the cones $F \in \mathcal{F}$, $F \neq \{0\}$. Thus, for all $k > 0$,

$$\#M_k = \sum_{F \in \mathcal{F}} \#(\operatorname{relint}(F) \cap M_k).$$

By Proposition 5.1 the number of degree k lattice points in F is given by a quasi-polynomial P_F for all $k \geq 0$; in particular $P_F(0) = 1$. Proposition 6.8 shows that

$$\#(\operatorname{relint}(F) \cap M_k) = (-1)^{\dim F - 1} P_F(-k)$$

for all $k \geq 1$.

Let $H_M(t) = \sum_{k=0}^{\infty} \#M_k t^k$ be the generating function. We have shown that

$$H_M(t) = 1 + \sum_{k=1}^{\infty} \sum_{F \in \mathcal{F}} (-1)^{\dim F - 1} P_F(-k) t^k.$$

With $P(k) = \sum_{F \in \mathcal{F}} (-1)^{\dim F - 1} P_F(-k)$ it remains to verify that $P(0) = 1$.

We have not yet used the convexity of $\mathbb{R}_+ M$! Rewriting the sum $\sum (-1)^{\dim F - 1} P_F(0)$ according to the dimension of the faces we obtain

$$\sum_{F \in \mathcal{F}} (-1)^{\dim F - 1} P_F(-k) = \sum_{i=0}^{d-1} (-1)^i f_i,$$

where f_i counts the number of faces in $\mathcal{F}$ that have dimension $i + 1$. Consider a polytopal cross-section Π of $\mathbb{R}_+ M$. An $(i + 1)$-dimensional face $F \in \mathcal{F}$ intersects Π in a simplex of dimension i. Therefore $P(0)$ is the Euler characteristic of Π. But Π is a convex polytope, homotopic to a point, and so $P(0) = 1$. □

For the combinatorial application the algebraic properties of $K[\mathcal{M}_n]$ and its interior ideal are irrelevant. It is enough to show that $H_I(t) = (-1)^d H_R(t^{-1})$. Using more advanced arguments of combinatorics, it is indeed possible to derive this equation from the simplicial case; see [26, 4.6].

The canonical module of a normal affine monoid algebra

Let us return to our main goal, the computation of the canonical module of a normal affine monoid algebra: Proposition 6.8 is a very strong sign that it should always be given by the interior ideal I. In fact, if this turns out to be true, then (ADG-2) and (ADG-3) have been proved simultaneously. A magic square x in the interior of $\mathbb{R}_+ \mathcal{M}_n$ is not contained in any of the coordinate hyperplanes, and we can subtract the square $\mathbf{1}$ with all entries 1 from it: $x - \mathbf{1} \in \mathcal{M}_n$. It follows that the monomial corresponding to $\mathbf{1}$ generates I. It has degree n. So $I \cong K[\mathcal{M}_n](-n)$, the ring $K[\mathcal{M}_n]$ is indeed Gorenstein, and its Hilbert series has degree $-n$.

At this point it seems inevitable to use the monomial structure of $K[M]$ in a deeper way than only through positive gradings. We need a definition of canonical module that respects this structure. So far the grading group of a graded ring has always been the group $\mathbb{Z}$. However, all the notions introduced in Section 2 can be transferred to more general grading groups. For us it is enough to allow the groups (isomorphic to) $\mathbb{Z}^n$ for $n \geq 1$. An affine monoid algebra $R = K[M]$ has a decomposition $\bigoplus_{\mu \in \operatorname{gp}(M)} R_\mu$ by its very definition: $R_\mu = K\mu$ for all $\mu \in M$ and $R_\mu = 0$ for all monomials $\mu \in \operatorname{gp}(M) \setminus M$.

Definition 6.10. Let R be a $\mathbb{Z}^n$-graded Cohen–Macaulay ring such that the homogeneous non-units generate a proper ideal $\mathfrak{p}$ of R. Then $\mathfrak{p}$ is necessarily a prime ideal; we set $d = \dim R_{\mathfrak{p}}$. One says that ω is a *$\mathbb{Z}^n$-graded canonical module* of R if

$$\operatorname{Ext}^j_R(R/\mathfrak{p}, \omega) = \begin{cases} R/\mathfrak{p}, & j = d, \\ 0, & j \neq d. \end{cases} \tag{10}$$

Our discussion of equation (9) and the uniqueness of the $\mathbb{Z}$-graded canonical module shows that the definitions 6.5 and 6.10 are compatible. Moreover, a generalization of the arguments of [8, 3.6.9] (to which we have referred already) shows that the $\mathbb{Z}^n$-graded canonical module is again uniquely determined. In the following we will simply speak of the multigraded canonical module.

In $K[M]$, with its grading by $\operatorname{gp}(M)$, the non-units even generate a maximal ideal so that it is a candidate for the use of Definition 6.10. As we have shown above, $K[M]$ has a multigraded canonical module if M is simplicial. (That equation (10) is satisfied can be shown by varying the positive $\mathbb{Z}$-grading.) However, in the non-simplicial case it is not clear whether the canonical module of R, constructed via a non-monomial homogeneous system of parameters, has a multigraded structure at all. There are various ways to find it; we use a divisorial approach.

The canonical module (with respect to a positive $\mathbb{Z}$-grading) is a Cohen–Macaulay module over R, and therefore a reflexive module: the natural homomorphism

$$\omega \to \operatorname{Hom}_R(\operatorname{Hom}(\omega, R), R)$$

is an isomorphism. Moreover ω has rank 1 as an R-module (since it has the same rank as R over a Noether normalization). Therefore it is isomorphic to a (fractional) divisorial ideal of R. According to equation (4) there exist integers $c_1, \dots, c_s$ such that

$$\omega_R \cong \mathfrak{p}_1^{(c_1)} \cap \cdots \cap \mathfrak{p}_s^{(c_s)}.$$

The prime ideals $\mathfrak{p}_i$ are defined by the support forms of the monoid M; see Section 1. We would like to show that $c_1 = \cdots = c_s = 1$ is a possible choice since the "interior ideal" I is exactly $\mathfrak{p}_1 \cap \cdots \cap \mathfrak{p}_s$.

The numbers $c_1, \dots, c_s$ are not uniquely determined. Therefore let us first make a choice that respects only the module structure and not yet the grading. Denote the chosen ideal by J. At least we have realized the R-module ω_R as a $\mathbb{Z}^d$-graded module J. It follows that $\operatorname{Ext}^d_R(K, J)$ is a $\mathbb{Z}^d$-graded module, too. On the other hand, it is just K: as a multigraded module we can identify it with Kv where v is a monomial in $\operatorname{gp}(M)$. We are free to replace J by $v^{-1}J$. Then $\operatorname{Ext}^d_R(K, v^{-1}J)$ is indeed K in multidegree 0: this is the right choice for the multigraded canonical module. Now $c_1, \dots, c_s$ are uniquely determined, and for the desired generalization of Proposition 6.8 it remains to show that $c_1 = \cdots = c_s = 1$.

Let F_i be the facet of $\mathbb{R}_+M$ that contains the set G of monomials not in $\mathfrak{p}_i$. We invert all of them and pass to the ring $S_i = R[G^{-1}]$. In this ring the non-units among the monomials generate the prime ideal $\mathfrak{P}_i = \mathfrak{p}_i S_i$. Then $S_i/\mathfrak{P}_i$ is the Laurent polynomial ring $K[U]$ where U is the group of monomials living in the support hyperplane of

$\mathbb{R}_+M$ through F_i. The multigraded canonical module ω_{S_i} is defined by the condition that $\operatorname{Ext}^j_{S_i}(S_i/\mathfrak{P}_i, \omega_{S_i}) = S_i/\mathfrak{P}_i$ for $j = 1$ (the Krull dimension of $(S_i)_{\mathfrak{P}_i}$) and $= 0$ otherwise. The ideal $\mathfrak{P}_i$ is a principal ideal generated by every monomial that has value 1 under the support form σ_i, and it is now easy to see that $\mathfrak{P}_i$ is the unique multigraded canonical module of S_i.

The final step is to show that the multigraded canonical module "localizes": $\omega_{S_i} = \omega_R \otimes S_i$. We omit this rather technical argument; it is indicated in Remark 6.12(a) and will be carried out in [6]. Since $\omega_R \otimes S_i = \mathfrak{P}_i^{c_i}$, we obtain $c_i = 1$ as desired:

Theorem 6.11 (Danilov, Stanley). *Let M be a normal affine monoid, and K a field. Then the ideal I generated by the monomials in the interior of $\mathbb{R}_+M$ is the multigraded canonical module of $K[M]$.*

Remark 6.12.

(a) The only case in which one has a multigraded Noether normalization, namely one generated by monomials, is that of a simplicial affine monoid. Consequently there is no simple way for the construction of the multigraded canonical module. A method that always works is to represent R as the residue class ring of a polynomial ring $P = K[X_1, \dots, X_n]$ over K. The grading of R can be pulled back to P by giving each indeterminate the degree of its residue class. The multigraded canonical module of P is $\omega_P = R(-\sum \deg X_i)$. As a P-module, R has a finite multigraded free resolution, and $\operatorname{Ext}^c_P(R, \omega_P)$, $c = \dim P - \dim R$, is a multigraded canonical module of R (see [8, 3.6.12] for the $\mathbb{Z}$-graded case).

   This approach can be used for a proof of the localization of the multigraded canonical module, since it reduces the question to the polynomial ring P.

(b) One can also use local cohomology and graded local duality for the proof of Hochster's theorem and the computation of the canonical module, as in Stanley's original proof [25] or in the version presented in [8]. It uses the topological properties of the face lattice of the cone $\mathbb{R}_+M$.

   Danilov [11] exploits the representation of the canonical module as the module of regular differential d-forms, $d = \dim R$. It can be determined by essentially the same method that we have applied for the divisorial approach.

Remark 6.13. Let us consider a general system of homogeneous linear diophantine equations $\mathcal{L}$, say in m variables. We can assume that there exists a strictly positive solution. Then the monoid M of non-negative solution is normal, all the results above apply, and for a suitable degree function on M (ADG-1)–(ADG-4) remain valid, apart from inevitable adjustments:

(i) in general the number $H(\mathcal{L}, k)$ of degree k solutions is given only by a quasi-polynomial Q for all $k \geq 0$;
(ii) one has $Q(-k) = 0$ for all $k = 1, \dots, s-1$, where s is the smallest degree of a strictly positive solution;
(iii) the general form of (ADG-3) is $Q(-r) = (-1)^{d-1}Q^+(k)$ where $Q^+(k)$ counts the number of the degree k strictly positive solutions and d is the rank of M;
(iv) if $(1, \dots, 1)$ is a solution to $\mathcal{L}$ (necessarily of degree s), then (ADG-3) remains fully valid: $Q(-k) = (-1)^{d-1}Q(k-s)$.

Stanley [25], [27] has extended the commutative algebra approach to the study of the solutions of the inhomogeneous systems associated with $\mathcal{L}$. The K-vector space D whose basis is given by the non-negative solutions to an associated inhomogeneous system is a rank 1 module over $K[M]$. If D is a Cohen–Macaulay module, then a reciprocity law holds similar to that in (iii).

Under certain conditions all these rank 1-modules are divisorial (and a Cohen–Macaulay rank 1 module is automatically divisorial). However, up to isomorphism over $K[M]$ there exist only finitely many Cohen–Macaulay divisorial ideals; see [7].

Ehrhart functions

Let $P \subset \mathbb{R}^n$ be a rational polytope, i. e. the convex hull of finitely many points $x_1, \dots, x_m$. The study of the function

$$E(P,k) = \#(kP \cap \mathbb{Z}^n),$$

counting the lattice points in the multiples $kP = \{kx : x \in P\}$, $k \in \mathbb{Z}_+$, was pioneered by E. Ehrhart [12]. For the commutative algebra approach we let

$$C(P) = \mathbb{R}_+(P,1) = \mathbb{R}_+\{(y,1) \in \mathbb{R}^{n+1} : y \in P\}$$

be the cone over P. The cone is generated by finitely many rational vectors, and therefore $M = C(P) \cap \mathbb{Z}^{n+1}$ is a normal affine monoid. As the grading on M we choose the last coordinate of the elements $x \in M$. Let K be a field. Then clearly

$$E(P,k) = H(K[M],k),$$

and all the results on affine monoid algebras can be exploited for the study of $E(P,k)$. The Hilbert function of the canonical module of $K[M]$ is given by

$$E^+(P,k) = \#(\operatorname{relint}(kP) \cap \mathbb{Z}^n).$$

The corresponding generating functions are

$$E_P(t) = \sum_{k=0}^{\infty} E(P,k)t^k \qquad \text{and} \qquad E_P^+(t) = \sum_{k=0}^{\infty} E^+(P,k)t^k.$$

Theorem 6.14 (Ehrhart). *Let $P \subset \mathbb{R}^n$ be a d-dimensional rational polytope, $d > 0$. Then*

(a) *$E_P(t)$ is a rational function, and there exist quasi-polynomials Q and Q^+ with $E(P,k) = Q(k)$ for all $k \geq 0$ and $E^+(P,k) = Q^+(k)$ for all $k \geq 1$;*
(b) *$E_P^+(t) = (-1)^{d+1} E_P(t^{-1})$, equivalently*

$$Q(k) = (-1)^d Q^+(-k) \qquad \text{for all} \quad k \geq 1.$$

If P is even an integral polytope, i. e. its vertices belong to $\mathbb{Z}^n$, then $E(P, k)$ is given by a polynomial, since $K[M]$ is finite over its subalgebra S generated by the monomials $(x, 1)$, x a vertex of P, and S is an algebra generated in degree 1. Therefore $K[M]$ has a well-defined multiplicity. In order to save us a discussion of volume functions, we restrict the corollary to full-dimensional polytopes:

Corollary 6.15. Let $P \subset \mathbb{R}^d$ be a d-dimensional integral polytope, and let $K[M]$ be the normal monoid algebra constructed above. Then

$$e(K[M]) = d!\ \mathrm{vol}\, P.$$

Proof. Elementary arguments of measure theory show that the volume of P is

$$\mathrm{vol}\, P = \lim_{k\to\infty} \frac{E(P,k)}{k^d}.$$

Being the Hilbert polynomial of the $(d+1)$-dimensional S-module, $E(P, k)$ has degree d. Thus its leading coefficient is given by vol P. On the other hand, it is also given by $e(k[M])/d!$. □

Unimodality of the h-vector

As we have seen, the h-vector $(h_0, \dots, h_u)$ of the Gorenstein ring $K[\mathcal{M}_n]$ is palindromic: $h_{u-i} = h_i$ for $i = 0, \dots, u$. Stanley's conjecture (ADG-5) asks whether it is *unimodal*, i.e. it increases until it reaches its maximum and decreases afterwards. This conjecture has been proved by Athanasiadis [3] and generalized to the Ehrhart functions of compressed Gorenstein polytopes – P is *Gorenstein* if $K[M]$ is Gorenstein, and P is compressed if all its so-called pulling triangulations are unimodular. Meanwhile it has been shown by Römer and the author [10] that unimodality holds for the Ehrhart functions of Gorenstein lattice polytopes, provided they have at least one regular unimodular triangulation. (See [6] for the notion of regular triangulation.)

The method of proof is a reduction to Stanley's g-theorem for simplicial polytopes, for which no access via commutative algebra has been found so far. The simplicial polytope is constructed from the unimodular triangulation and it seems difficult to further weaken the hypothesis on P.

The existence of a unimodular triangulation implies that $K[M]$ is generated by its degree 1 elements. If not even this condition is fulfilled, then the generators of higher degree induce "phenomena of higher period" in the h-vector. Unimodality can no longer be expected, and an example of a Gorenstein polytope P for which unimodality fails has been given by Muşţată and Payne [20].

To the best of our knowledge it is still an open question whether there exist graded Gorenstein integral domains that are generated in degree 1, but whose h-vector is not unimodal.

Acknowledgement

These notes are an expanded version of my lectures at the International Conference on Commutative Algebra and Combinatorics organized by the Bhaskaracharya

Pratishthana, Pune and the Harish-Chandra Research Institute, Allahabad, December 2003. I am grateful to the DFG for a travel grant and to our Indian colleagues that they have made possible a very successful conference in the pleasant environment of the Harish Chandra Research Institute.[1]

References

[1] M. Ahmed, J. De Loera and R. Hemmecke, Polyhedral cones of magic cubes and squares, In: *New directions in computational geometry*, (eds.) B. Aronov et al., (Springer) (2003) 25–41

[2] H. Anand, V. C. Dumir and H. Gupta, A combinatorial distribution problem, *Duke Math. J.* **33** (1966) 757–769

[3] Ch. A. Athanasiadis, Ehrhart polynomials, simplicial polytopes, magic squares and a conjecture of Stanley, *J. Reine Angew. Math.* **583** (2005) 163–174

[4] M. F. Atiyah and I. G. Macdonald, *Introduction to commutative algebra*, Addison–Wesley (1969)

[5] J.-F. Boutot, Singularités rationelles et quotients par les groupes réductifs, *Invent. Math.* **88** (1987) 65–68

[6] W. Bruns and J. Gubeladze, *Polytopes, rings and K-theory*, In preparation. Preliminary version at `http://www.math.uos.de/staff/brunsw`

[7] W. Bruns and J. Gubeladze, Divisorial linear algebra of normal semigroup rings, *Algebr. Represent. Theory* **6** (2003) 139–168

[8] W. Bruns and J. Herzog, *Cohen–Macaulay rings*, (Rev. ed.) University Press, Cambridge (1998)

[9] W. Bruns and B. Ichim, On the coefficients of Hilbert quasipolynomials, *Proc. Amer. Math. Soc.*, to appear. `http://arxiv.org/format/math.AC/0512329`

[10] W. Bruns and T. Römer, h-vectors of Gorenstein polytopes, *J. Comb. Th. Ser. A*, in press

[11] V. I. Danilov, The geometry of toric varieties, *Russian Math. Surveys* **33** (1978) 97–154

[12] E. Ehrhart, *Polynômes arithméthiques et méthode des polyédres en combinatoire*, Birkhäuser (1977)

[13] R. Fossum, *The divisor class group of a Krull domain*, Springer (1973)

[14] W. Fulton, *Introduction to toric varieties*, Princeton University Press (1993)

[15] R. V. Gurjar, On a conjecture of C. T. C. Wall, *J. Math. Kyoto Univ.* **31** (1991) 1121–1124

[16] D. Hilbert, Über die Theorie der algebraischen Formen, *Math. Ann.* **36** (1890) 473–534

[17] M. Hochster, Rings of invariants of tori, Cohen–Macaulay rings generated by monomials, and polytopes, *Ann. of Math.* **96** (1972) 318–337

[18] M. Hochster and J. L. Roberts, Rings of invariants of reductive groups acting on regular rings are Cohen–Macaulay, *Adv. Math.* **13** (1974) 115–175

[19] G. Kempf, F. Knudsen, D. Mumford and B. Saint-Donat, *Toroidal embeddings I*, LNM, **339**, Springer (1973)

[20] M. Mustaţă and S. Payne, Ehrhart polynomials and stringy Betti numbers, *Math. Ann.* **333** (2005) 787–795

[21] R. Y. Sharp, *Steps in commutative algebra*, (2nd ed.) Cambridge University Press, (2001)

[22] R. P. Stanley, Linear homogeneous diophantine equations and magic labelings of graphs, *Duke Math. J.* **40** (1973) 607–632

[23] R. P. Stanley, Magic labelings of graphs, symmetric magic squares, systems of parameters, and Cohen–Macauly rings, *Duke Math. J.* **43** (1976) 511–531

---

[1] It was a special pleasure for the author to lecture about the ADG-conjectures in the presence of his old friend V. C. Dumir. In the academic year 1977/78 the Dumir and Bruns families were neighbors in the married student housing at Orchard Downs, Urbana, Illinois.

[24] R. P. Stanley, Hilbert functions of graded algebras, *Adv. Math.* **28** (1978) 57–83

[25] R. P. Stanley, Linear diophantine equations and local cohomology, *Invent. Math.* **68** (1982) 175–193

[26] R. P. Stanley, *Enumerative combinatorics*, Vol. I, Wadsworth & Brooks/Cole (1986)

[27] R. P. Stanley, *Combinatorics and commutative algebra*, (2nd ed.) Birkhäuser (1996)

[28] Ngô Viêt Trung and Lê Tuân Hoa, Affine semigroups and Cohen–Macaulay rings generated by monomials, *Trans. Amer. Math. Soc.* **298** (1986) 145–167

[29] G. M. Ziegler, *Lectures on polytopes*, Springer (1995)

Proc. Int. Conf. – Commutative Algebra and Combinatorics No. 4, 2007, pp. 39–90.

A Survey of Toric Initial Ideals

Serkan Hoşten

Department of Mathematics, San Francisco State University, San Francisco
e-mail: serkan@math.sfsu.edu

Abstract. We present a broad survey of the work on monomial initial ideals of toric ideals. We start with a characterization of toric Gröbner bases and the corresponding initial ideals in terms of integer programs. Then we present the result that the radical of a given initial ideal is the Stanley-Reisner ideal of a regular triangulation. The case of square-free initial ideals are treated in detail, and we give examples of quadratic initial ideals. Then we introduce a hierarchy of normal toric ideals in terms of the simplicity of their initial ideals. We will prove the chain theorem, the strongest result for the inverse Gröbner problem for toric initial ideals. We will present many examples coming from algebraic geometry, algebraic combinatorics, and algebraic statistics.

1. Introduction

The history of Gröbner bases and initial ideals of toric ideals is not more than 15 years old, but there have been a tremendous amount of work, research, and activity in this subject. A cornerstone of this subject was laid by Bernd Sturmfels in his 1995 book *Gröbner Bases and Convex Polytopes* [77]. Although this article is about toric initial ideals, the mentioned book, which was a summary of the computational algebra and discrete geometry tools for toric and related varieties by 1995, was not only about the toric initial ideals. However, this survey is intended to summarize what happened in the meantime from the perspective of initial ideals. Having said this, disclaimers are needed immediately: We have not included (very interesting) topics which are related to toric initial ideals, such as Gröbner fans [59], [80] and how to compute them efficiently [52], [53], [54], $\mathcal{A}$-graded monomial ideals which are close cousins of toric initial ideals in relation to toric Hilbert scheme [57], [65], [75], the theory of generic toric ideals and their initial ideals [66] etc. *Combinatorial Commutative Algebra* [58] is a new book by Ezra Miller and Bernd Sturmfels which cover some of the more advanced topics on toric ideals such as their free resolutions. The soon to be published book *Polytopes, Rings, and K-Theory* [9] by Winfried Bruns and Joseph Gubeladze also contains material on the algebraic and combinatorial aspects of toric rings. This current survey is by no means a survey on toric varieties; that has been done wonderfully in [15].

2. Affine monoids and toric ideals

An *affine monoid* is a semigroup which is finitely generated and can be embedded in $\mathbb{Z}^d$ for some positive integer d. In this survey, we will present an affine monoid S by giving its generating set $\mathcal{A} = \{a_1, \ldots, a_n\} \subset \mathbb{Z}^d$. This means that

$$S = \mathbb{N}\mathcal{A} = \left\{ \sum_i^n z_i a_i \ : \ z_i \in \mathbb{N} \right\}.$$

There are three more sets we associate to $\mathcal{A}$. The subspace

$$\mathbb{R}\mathcal{A} = \left\{ \sum_i^n r_i a_i \ : \ r_i \in \mathbb{R} \right\}$$

generated by $\mathcal{A}$ will be assumed to be equal to $\mathbb{R}^d$. In other words, the matrix $A = [a_1, \ldots, a_n]$ has rank d. The group

$$\mathbb{Z}\mathcal{A} = \left\{ \sum_i^n z_i a_i \ : \ r_i \in \mathbb{Z} \right\}$$

generated by $\mathcal{A}$ is a sublattice of the lattice $\mathbb{Z}^d$. Since we assume $\mathbb{R}\mathcal{A} = \mathbb{R}^d$, the rank of $\mathbb{Z}\mathcal{A}$ is also d. To simplify the exposition we will also assume that $\mathbb{Z}\mathcal{A} = \mathbb{Z}^d$. Finally, the set

$$C(\mathcal{A}) = \mathbb{R}_+\mathcal{A} = \left\{ \sum_i^n r_i a_i \ : \ r_i \in \mathbb{R}_+ \right\}$$

is the *rational polyhedral cone* generated by $\mathcal{A}$. This cone is rational since the elements of $\mathcal{A}$ are vectors with rational coordinates, and it is a polyhedral cone since

$$C(\mathcal{A}) = \{x \in \mathbb{R}^d \ : \ b_i \cdot x \geq 0, \ \ 1 \leq i \leq m\}$$

where $b_i \in \mathbb{Z}^d$. The fact that every rational cone has a representation as the intersection of finitely many halfspaces $\{x \in \mathbb{R}^d \ : \ b_i \cdot x \geq 0\}$ defined by rational vectors b_i (and vice versa) is a fundamental result of discrete geometry. We will not prove this result, and we refer the reader to [90, Chapter 1].

The fundamental example of an affine monoid is the set of lattice points (= points with integer coordinates) in a rational polyhedral cone C:

$$S(C) := C \cap \mathbb{Z}^d = \{x \in \mathbb{Z}^d \ : \ x \in C\}.$$

It is clear that $S(C)$ is a semigroup. The fact that it is also finitely generated is proved by Gordan. We present a proof below. In fact, we claim and prove a bit more. If the cone C is a pointed cone then a *minimal* generating set of $S(C)$ is unique (a cone is pointed if there exists $h \in \mathbb{R}^d$ such that $h \cdot x > 0$ for all $x \in C \setminus \{0\}$). Such a unique minimal generating set is called the *Hilbert basis* of $S(C)$ (and by abuse of language the Hilbert basis of C).

Theorem 2.1. *If C is a rational polyhedral cone then $S(C)$ is finitely generated. Moreover, if C is a pointed cone then $S(C)$ has a unique minimal generating set.*

Proof. If C is not a pointed cone than $C = C' + V = \{x + v \ : \ x \in C', \ \ v \in V\}$ where C' is a pointed cone and V is a rational subspace [72, Chapter 8]. A generating set of $S(C)$ is then the union of a generating set of $S(C')$ and a basis of the lattice $V \cap \mathbb{Z}^d$. Hence we can assume that C is a pointed cone. Since C is a rational cone it has rational generators $v_1, \ldots, v_m$, i.e., $C = \left\{ \sum_i^m r_i v_i \; : \; r_i \in \mathbb{R} \right\}$. We will assume that each v_i is the first nonzero lattice point on an extreme ray of C (the extreme rays are the halflines that are the one-dimensional faces of C). Let

$$P := \left\{ \sum_i^m \lambda_i v_i \; : \; 0 \leq \lambda_i < 1 \right\}$$

be the open parallepiped generated by $v_1, \ldots, v_m$. We claim that $P \cap \mathbb{Z}^d \cup \{v_1, \ldots, v_m\}$ generates $S(C)$: let u be a lattice point in C. Then we can write $u = \sum_i^m \mu_i v_i$ where $\mu_i \geq 0$ for $i = 1, \ldots, m$. Now if we write this expression in a different way

$$u = \sum_i^m \lfloor \mu_i \rfloor v_i + \sum_i^m (\mu_i - \lfloor \mu_i \rfloor) v_i$$

we see that the first sum is a lattice point generated by $\{v_1, \ldots, v_m\}$, and the second sum is a lattice point in P.

For the second claim, we call a nonzero lattice point u in the above generating set *irreducible* if it cannot be written as the sum of two other such points. Clearly, the set $\mathcal{U}$ of the irreducible points is a minimal generating set, and we claim it is the unique one. Suppose that $\mathcal{U}'$ is a different minimal generating set. Because C is a pointed cone there is a vector h such that $h \cdot x > 0$ for all $x \in C \backslash \{0\}$. Now let $v \in (\mathcal{U}' \backslash \mathcal{U}) \cup (\mathcal{U} \backslash \mathcal{U}')$ be a lattice point with the property that $h \cdot v$ is minimal. Without loss of generality we can assume $v \in \mathcal{U}' \backslash \mathcal{U}$. Then v needs to be a nonnegative integer combination of the generators in $\mathcal{U}$ whose dot product with h are smaller than $h \cdot v$. By our assumption these generators belong to $\mathcal{U}'$, and hence v is irredundant in $\mathcal{U}'$. This is a contradiction to the minimality of $\mathcal{U}'$, and we conclude that $\mathcal{U} = \mathcal{U}'$. □

Definition 2.2. The set $\mathcal{A} = \{a_1, \ldots, a_n\} \subset \mathbb{Z}^d$ is called *normal* if $\mathcal{A}$ is a generating set of the affine monoid $S(C)$ where C is the cone generated by $\mathcal{A}$. In this case, we will call $\mathbb{N}\mathcal{A} = S(C)$ as well as the matrix $A = [a_1, \ldots, a_n]$ also normal.

Example 2.3. We let $\mathcal{A} = \{(1, 0), \ (1, 1), \ (1, 3)\}$ in $\mathbb{Z}^2$. The cone generated by these three vectors is the cone C in Figure 1. The semigroup $\mathbb{N}\mathcal{A}$ is not equal to $S(C)$ and therefore it is not normal. When we add the vector $(1, 2)$ to $\mathcal{A}$ we get a normal monoid $\mathbb{N}\mathcal{A}$.

Example 2.4. We finish our discussion of affine monoids with an important class of examples. Let $P \subset \mathbb{R}^d$ be a *lattice polytope*, i.e., a polytope whose vertices are lattice points, and let $\mathcal{Q}$ be the set of lattice points in P. We form the set $\mathcal{A} = \{(1, u) \; : \; u \in \mathcal{Q}\}$, and we let $C(P) \subset \mathbb{R}^{d+1}$ be the pointed cone generated by $\mathcal{A}$. The affine monoid of the lattice points in $C(P)$ is denoted by $S(P)$. If x_i for $i = 1, \ldots, d+1$ are the coordinates of $\mathbb{R}^{d+1}$ then $C(P) \cap \{x_1 = k\}$ is an isomorphic copy of $kP := \{kv \; : \; v \in P\}$. In particular, $C(P) \cap \{x_1 = 1\}$ is an isomorphic copy of P, and hence $\mathcal{A}$ is contained in

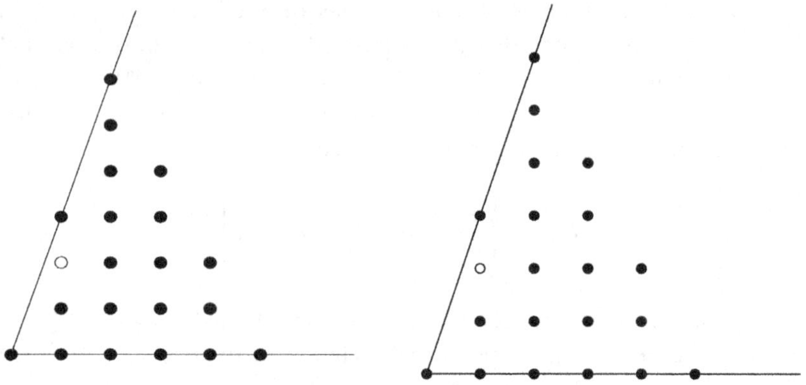

Figure 1. The monoid in Example 2.3 that is not normal

the Hilbert basis of $S(P)$. This containment is strict if the dimension of P is bigger than two. In other words, $S(P)$ is not always normal. For instance, we can take

$$P = \mathrm{conv}\{(0,0,0),\ (1,0,0),\ (0,0,1),\ (2,3,1)\},$$

a simplex that contains only its four vertices as lattice points. Then the affine monoid $S(P)$ contains $(2,1,1,1)$ which cannot be generated by $\{(1,u) : u \in P \cap \mathbb{Z}^3\}$.

Let $R = k[x_1, \dots, x_n]$ be the ring of polynomials with coefficients in the field k. For the purposes of defining and using toric ideals the ground field k does not play a role, and for all practical purposes one can assume that $k = \mathbb{Q}$. A *monomial* of R is a product of powers of variables: $x_1^{u_1} x_2^{u_2} \cdots x_n^{u_n}$. We denote such a monomial by x^u where $u = (u_1, \dots, u_n)$ is the *exponent vector*.

Let $\mathcal{A} = \{a_1, \dots, a_n\}$ be a set of vectors in $\mathbb{Z}^d$, and $A = [a_1, \dots, a_n]$ the corresponding $d \times n$ matrix of rank d. We define the *Laurent polynomial ring* $T = k[t_1, t_1^{-1}, \dots, t_d, t_d^{-1}]$ where monomials are allowed to have exponent vectors with negative coordinates. The ring homomorphism $\psi : R \longrightarrow T$ is defined by the monomial map

$$\psi(x_i) = t^{a_i} = t_1^{a_{i1}} \cdots t_d^{a_{id}}.$$

Definition 2.5. The *toric ideal* I_A is the kernel of the homomorphism ψ. The algebraic variety defined by the ideal I_A is called the *toric variety* V_A. Unless otherwise stated we will assume that V_A is a complex variety defined over $\mathbb{C}$.

Theorem 2.6. *The toric ideal I_A is a prime ideal that is equal to*

$$\langle x^u - x^v \ : \ u, v \in \mathbb{N}^n \text{ and } Au = Av \rangle.$$

The k-algebra R/I_A is isomorphic to the semigroup algebra $k[S] = k[t^m : m \in S] \subset T$. This algebra is of Krull dimension d.

Proof. Because T is an integral domain and I_A is the preimage of the prime zero-ideal of T, the toric ideal is also prime. The polynomials $x^u - x^v$ we claim to generate I_A

are in this ideal since $\psi(x^u - x^v) = t^{Au} - t^{Av} = 0$. Suppose $f = \sum_\alpha c_\alpha x^\alpha$ is a nonzero polynomial in I_A. Because $\psi(f) = 0$, there have to be two terms $c_u x^u$ and $c_v x^v$ of f such that $Au = Av$. Then the polynomial $f - c_u(x^u - x^v)$ is also in I_A and has fewer terms than f. Now induction on the number of terms in f shows that $I_A = \langle x^u - x^v : u, v \in \mathbb{N}^n \text{ and } Au = Av \rangle$.

The image of ψ is the semigroup algebra $k[S]$, and the isomorphism $k[S] = R/I_A$ follows. Since S is generated by $\mathcal{A}$, the algebra $k[S]$ is generated by the monomials $t^{a_1}, \dots, t^{a_n}$. Then the dimension of $k[S]$ is the cardinality of a maximal subset of algebraically independent generators. This is the same as the cardinality of a maximal subset of linearly independent vectors in $\mathcal{A}$ which is equal to $\text{rank}(A) = d$. □

As an immediate consequence of this theorem we get

Corollary 2.7. The toric variety V_A is an irreducible affine variety of dimension d.

Remark 2.8. A polynomial of the form $x^u - x^v$ is called a *binomial* in the literature. Moreover, ideals generated by binomials are called binomial ideals. On the other hand, we like to remark that the term "binomial" and "binomial ideal" could be used in a broader sense to encompass binomial ideals like I_A, prime or not, as well as monomial ideals. The most general definition of a binomial is a polynomial $\alpha x^u - \beta x^v$ where $\alpha, \beta \in k$. Ideals generated by such binomials are called binomial ideals. A thorough treatment was made in [26]. In this survey, we will use binomial and binomial ideal in its original, more restricted sense.

Finally, one little fact we will use many times: if $x^u - x^v$ is in I_A, and a monomial x^a divides both terms then $x^{u-a} - x^{v-a}$ is also in I_A since I_A is a prime binomial ideal, and it does not contain monomials. This means that we can write

$$I_A = \langle x^{u_+} - x^{u_-} : u \in \ker_{\mathbb{Z}}(A) \rangle$$

where $ker_{\mathbb{Z}}(A) = \ker(A) \cap \mathbb{Z}^n$ is the lattice of vectors in the kernel of the matrix A with integer coordinates, and $u = u_+ - u_-$ is the unique way of writing u as the difference of two nonnegative vectors with no common support.

The set of generators of I_A given in Theorem 2.6 is an infinite set of binomials as we have presented it. Nevertheless, the celebrated *Hilbert basis theorem* [23] guarantees that there is a *finite* set of binomial generators.

Corollary 2.9. Every toric ideal has a finite set of generators consisting of binomials:

$$I_A = \langle x^{u_1} - x^{v_1}, \dots, x^{u_t} - x^{v_t} \rangle.$$

Example 2.10. Let $\mathcal{A}$ be the normal set of vectors in Example 2.3, namely

$$\mathcal{A} = \{(1,0), (1,1), (1,2), (1,3)\}.$$

We let $R = k[a, b, c, d]$ and $T = k[s, t]$, and we obtain the monomial map $\psi(a) = s$, $\psi(b) = st$, $\psi(c) = st^2$, and $\psi(d) = st^3$. The toric ideal I_A is then generated by three binomials:

$$I_A = \langle ad - bc, b^2 - ac, c^2 - bd \rangle.$$

These are the equations of the famous twisted cubic curve.

Our discussion so far did not depend on the normality of the affine monoid $S = \mathbb{N}\mathcal{A}$. However, in the "classical" literature on toric varieties [30] only normal affine monoids are allowed. In this case, the toric variety V_A is called an affine toric variety. In fact, an abstract toric variety is constructed by *gluing* a set of affine toric varieties along common open sets. For the details of this development we refer to [27], [30], and [60], as well as to [77, Chapter 13] for the precise connections. We stress that for us an affine toric variety is a toric variety V_A whether $\mathbb{N}\mathcal{A}$ is normal or not. On the other hand, there is a finer notion for affine monoid algebras and hence toric ideals that justifies the term "normal".

Definition 2.11. The affine monoid algebra $k[S]$ and the corresponding toric ideal I_A is called *normal* if $k[S]$ is integrally closed (in its field of fractions).

Proposition 2.12. *The toric ideal I_A and the algebra $k[S] = k[\mathbb{N}\mathcal{A}]$ are normal if and only if the monoid S is normal.*

Proof. We need to show that $S = S(C)$ where C is the cone generated by $\mathcal{A}$ if and only if $k[S]$ is normal. If $u \in S(C)\backslash S$ then one can show that there is a positive integer k such that $ku \in S$. This gives us the identity $(x^u)^k = x^{ku}$. Since x^u is in the field of fractions of $k[S]$ we see that it is integral over $k[S]$. Conversely, suppose $S = S(C)$, and let f be an element in the field of fractions of $k[S]$. We can assume that the cone generated by S together with the monomials in f generate a pointed cone, and if not all of these monomials are in $S(C)$ then we can find a linear functional h that takes a positive value on the exponent of a single monomial x^α of f, and negative values on all of the other monomials of f as well as the elements of S. Now let $f^n + c_{n-1}f^{n-1} + \cdots + c_0 = 0$ be an integral equation with $c_i \in k[S]$. If we evaluate h on the exponents of the monomials on the left hand side we see that the maximum value is attained uniquely at $n\alpha$, and therefore $x^{n\alpha}$ occurs only once. This means the left hand side cannot be zero, and this contradiction shows that $f \in k[S]$. □

Remark 2.13. Finally, we like to comment on when V_A is a *projective toric variety*. It turns out (see for example [30]) that every projective toric variety comes from a polytope P and the affine monoid $S(P)$ associated to this polytope. The set $\mathcal{A} \subset S(P)$ can be taken to be the Hilbert basis of $S(P)$. Each a_i in $\mathcal{A}$ will be of the form (u_i, a_i') where u_i is a positive integer. Then the toric ideal I_A will be in the polynomial ring $k[x_1, \ldots, x_n]$ where $n = |\mathcal{A}|$. When we assign to each variable x_i the *weight* u_i, the binomials in I_A will be homogeneous with respect to these weights.

3. Gröbner bases and initial ideals of toric ideals

The main goal of this survey is to give an overview of monomial initial ideals of toric ideals. In this section we will define Gröbner bases and initial ideals. We assume some knowledge of these concepts as presented in [16] for instance. Hence we will not get into details.

A *term order* on the monomials of $R = k[x_1, \ldots, x_n]$ is a total order of these monomials that is a well-ordering and is compatible with multiplication. In other words, the term order $\prec$ satisfies two conditions: $1 = x^0 \prec x^\alpha$ for any monomial where $\alpha \neq 0$, and if $x^\alpha \prec x^\beta$ then $x^\gamma x^\alpha \prec x^\gamma x^\beta$ for every monomial x^γ. As soon as $n \geq 2$ there exist infinitely many term orders on R. There are a few important examples. The *lexicographic* term order with respect to $x_1 \succ \cdots \succ x_n$ is defined as $x^\alpha \prec x^\beta$ if the leftmost nonzero entry of the vector $\alpha - \beta$ is negative. The *degree reverse lexicographic* term order with respect to $x_1 \succ \cdots \succ x_n$ is in turn defined as $x^\alpha \prec x^\beta$ if the *total degree* of the first monomial is smaller than that of the second monomial, i.e., $\sum_{i=1}^n \alpha_i < \sum_{i=1}^n \beta_i$, and in case of the equality, the rightmost nonzero entry of the vector $\alpha - \beta$ is negative.

For the development of toric initial ideals the following is quiet important: let ω be a vector in $\mathbb{R}^n$. Then we define $x^\alpha \prec_\omega x^\beta$ if $\omega \cdot \alpha < \omega \cdot \beta$. As it stands $\prec_\omega$ does not define a proper term order: the monomial $x^0 = 1$ does not have to be the smallest monomial with respect to $\prec_\omega$, and moreover it is possible that $\omega \cdot \alpha = \omega \cdot \beta$. We can remedy these problems by forcing ω to be strictly positive and with irrational entries. But then how can we make exact symbolic computation ? Plus the reader might have noticed that there will be term orders that cannot be captured by $\prec_\omega$ with a suitable ω, such as the lexicographic term order. In fact, there is a precise answer to all of these worries where instead of a vector ω one needs to use a matrix, and we refer the interested reader to [69]. However, we wish to stick to $\prec_\omega$, and as we will see below there is a nice resolution to all these.

Given a nonzero polynomial $f \in R$, the *initial term* of f with respect to the term order $\prec$ is the biggest term of f according to $\prec$. We will denote this term by $\mathrm{in}_\prec(f)$ (and with $\mathrm{in}(f)$ when $\prec$ is clear). Similarly, $\mathrm{in}_\omega(f)$ will denote the *sum* of the terms $c_\alpha x^\alpha$ of f where $\omega \cdot \alpha$ is maximal. Now let I be an ideal in R. The *initial ideal* $\mathrm{in}_\prec(I)$ of I is the monomial ideal generated by $\mathrm{in}_\prec(f)$ by all $f \in I$:

$$\mathrm{in}_\prec(I) = \langle \mathrm{in}_\prec(f) \,:\, f \in I \rangle.$$

Definition 3.1. The reduced Gröbner basis of I with respect to the term order $\prec$ is $\mathcal{G}_\prec(I) = \{g_1, \ldots, g_t\} \subset I$ such that

i. the coefficient of $\mathrm{in}_\prec(g_i)$ in g_i is one for all i,
ii. $\{\mathrm{in}_\prec(g_1), \ldots, \mathrm{in}_\prec(g_t)\}$ is the unique minimal generating set of $\mathrm{in}_\prec(I)$, and
iii. no nonleading term of $g_1, \ldots, g_t$ is in $\mathrm{in}_\prec(I)$.

The existence and uniqueness of the reduced Gröbner basis $\mathcal{G}_\prec(I)$ are not hard to establish. For this and the description of the celebrated *Buchberger's algorithm* to compute such a basis we refer the reader to [16], [23] etc.

Gröbner bases are extremely useful and form the heart of any computational commutative algebra software these days: they solve the *ideal membership problem*, they are used in a crucial way in computing ideal intersections, ideal quotients, saturations, normalizations, primary decompositions etc. They help to solve systems of polynomial equations. One can generalize them to Gröbner bases of modules, and that opens up the possibility to compute syzygy modules, and hence free resolutions of modules. All these justify the study of Gröbner bases of ideals, and in particular, toric ideals. How about initial ideals? The main motivation there is that for most practical purposes the initial ideal $\mathsf{in}(I)$ can be used as a proxy for I, in fact, a much more manageable object since monomial ideals open up the doors to combinatorial mathematics (something our computers love). When one passes to the initial ideal invariants, such as dimension, degree, and Hilbert function of I are unchanged, and it is much easier to compute these invariants on a monomial ideal. Some other invariants do change but in a predictable or controllable way: the regularity, the arithmetic degree, the projective dimension etc can go up. Knowing a property of an initial ideal helps making inferences about the same property of I: if $\mathsf{in}(I)$ is radical so is I, if $\mathsf{in}(I)$ is Cohen-Macaulay, again, so is I.

There are a few general facts about initial ideals and Gröbner bases we wish to review without proof. We indicate the references for each fact.

(1) The quotient rings R/I and $R/\mathsf{in}_{\prec}(I)$ are isomorphic as k-vector spaces for any term order $\prec$; a vector space basis for both algebras is formed by all monomials *not* in $\mathsf{in}_{\prec}(I)$ [16], and these monomials are called the *standard* monomials of $\mathsf{in}_{\prec}(I)$.
(2) Although there are infinitely many term orders on a polynomial ring R with at least two variables, a fixed ideal I has only finitely many initial ideals, and hence finitely many reduced Gröbner bases [77, Theorem 1.2]; the union of all these reduced Gröbner bases is called the *universal Gröbner basis* of I.
(3) Let $\prec$ be a term order and let $\mathcal{G}_{\prec}(I) = \{g_1, \ldots, g_t\}$ be a Gröbner basis of an ideal I. Then there exists a nonnegative integer vector ω such that $\mathsf{in}_{\prec}(g_i) = \mathsf{in}_{\omega}(g_i)$ for $i = 1, \ldots, t$ [77, Proposition 1.11].

The last fact implies that for theoretical considerations as well as computational purposes we can assume that, in the presence of a *fixed ideal* I, a term order is essentially given by a vector $\omega \in \mathbb{N}^n$. We say "essentially", because, for instance, if we attempt to compute a Gröbner basis using Buchberger's algorithm we might have to break ties between two monomials x^α and x^β when $\omega \cdot \alpha = \omega \cdot \beta$. However, *any* fixed term order could be used as such a tie-breaker, and any such choice will lead to the same Gröbner basis. Let's summarize the first result about reduced Gröbner bases for toric ideals.

Proposition 3.2. *Any toric ideal I_A has a reduced Gröbner basis $\mathcal{G}_{\prec}(I_A) = \{\underline{x^{\alpha_1}} - x^{\beta_1}, \ldots, \underline{x^{\alpha_t}} - x^{\beta_t}\}$ consisting of binomials. Also, there exists $\omega \in \mathbb{N}^n$ such that $\omega \cdot \alpha_i > \omega \cdot \beta_i$ for $i = 1, \ldots, t$. This means that $\mathsf{in}_{\prec}(I_A) = \mathsf{in}_{\omega}(I_A)$.*

Proof. The first statement follows from the fact that Buchberger's algorithm does not destroy binomials: both the S-pair computations and reductions preserve binomials. The second statement is a re-iteration of [77, Proposition 1.11] we mentioned above. □

Definition 3.3. A weight vector ω is called *generic* if for some Gröbner basis $\mathcal{G}_{\prec}(I_A)$ as in Proposition 3.2 we have $\omega \cdot \alpha_i > \omega \cdot \beta_i$ for $i = 1, \ldots, t$.

3.1 *Integer programming and Gröbner bases*

We will characterize the reduced Gröbner bases, initial ideals, and standard monomials of a toric ideal in terms of a very useful notion, namely integer programming. The simple results we present here will be used repeatedly in the rest of the survey.

We let $A = [a_1, \dots, a_n]$ be a $d \times n$ integer matrix of rank d. We assume ω is a generic weight vector. For each vector $b \in \mathbb{N}\mathcal{A}$ we can form the integer program

$$\text{minimize} \quad \omega \cdot x \quad \text{subject to} \quad Ax = b, \quad \text{and} \quad x \in \mathbb{N}^n. \tag{1}$$

We let π be the map $\pi : \mathbb{N}^n \longrightarrow \mathbb{N}\mathcal{A}$ defined by $\pi(x) = Ax$. Given $b \in \mathbb{N}\mathcal{A}$, the set $\pi^{-1}(b) = \{x \in \mathbb{N}^n : Ax = b\}$ is called the b-fiber (of π, or of A). The vectors that belong to the b-fiber are called *feasible solutions* of (1). The b-fibers partition $\mathbb{N}^n$ as b ranges over all vectors in the affine monoid generated by the columns of A. These fibers are either all bounded (and have finitely many points in them) or they are all unbounded: the b-fibers are bounded if and only if the *recession cone* $\mathrm{Rec}(A) = \{x \in \mathbb{R}^n_+ : Ax \leq 0\}$ is trivial [72]. One important case when this condition is met is when the cone generated by the columns of A is a pointed cone. The integer program (1) is an optimization problem that seeks for a vector with the minimum weight in a given b-fiber. Such a feasible solution is called *optimal*. As long as ω is in the interior of $\mathrm{Rec}(A)$, the polar of the recession cone, each integer program (1) has a well defined optimal solution.

Theorem 3.4 ([84, Theorem 1]). *A monomial x^u is in the initial ideal $\mathrm{in}_\omega(I_A)$ if and only if u is not an optimal solution. Equivalently, the monomial x^u is a standard monomial of $\mathrm{in}_\omega(I_A)$ if and only if u is the optimal solution of the b-fiber where $b = Au$.*

Proof. The monomial x^u is in the initial ideal $\mathrm{in}_\omega(I_A)$ if and only if x^u is divisible by the initial term of some reduced Gröbner basis element $x^\alpha - x^\beta$. Then the vector $v = u - \alpha + \beta$ is in the same b-fiber as u, and $\omega \cdot u > \omega \cdot v$, and hence u is not an optimal solution. Equivalently, the exponent vectors of standard monomials are precisely the optimal solutions. □

Corollary 3.5. The reduced Gröbner basis $\mathcal{G}_\omega(I_A)$ consists of binomials $x^u - x^v$ where u is not optimal, but $u - e_i$ is optimal in its own fiber for each i with $u_i > 0$, and v is the optimal solution of the b-fiber where $b = Au = Av$.

Proof. The properties of u mentioned in the statement are equivalent to saying that x^u is a minimal generator of $\mathrm{in}_\omega(I_A)$. By definition, such monomials form the leading terms of reduced Gröbner basis elements. Let $x^u - x^v$ be in $\mathcal{G}_\omega(I_A)$. Suppose x^v is not the optimal solution of the b-fiber where $b = Au = Av$, and let v' be the optimal solution. Then $x^v - x^{v'}$ is a binomial in I_A with leading term x^v. This means that there exists a binomial in $\mathcal{G}_\omega(I_A)$ whose leading term divides x^v. This is a contradiction to the fact that $x^u - x^v$ is in $\mathcal{G}_\omega(I_A)$. □

Example 3.6. Let's continue with the twisted cubic example from Section 1. With respect to the weight vector $\omega = (1, 1, 3, 2)$, the reduced Gröbner basis $\mathcal{G}_\omega(I_A)$ is

$$\{\underline{ac} - b^2,\ \underline{a^2d} - b^3,\ \underline{bc} - ad,\ \underline{c^2} - bd\}$$

where the underlined monomials are initial terms. The second binomial comes from the integer program

$$\begin{aligned} \text{minimize} \quad & x + y + 3z + w \\ \text{subject to} \quad & x + y + z + w = 3 \\ & y + 2z + 3w = 3 x, y, z, w \in \mathbb{N}. \end{aligned}$$

The b-fiber where $b = (3, 3)^t$ consists of three vectors: $(2, 0, 0, 1)$, $(0, 3, 0, 0)$, $(1, 1, 1, 0)$. The ω-weights of these vectors are 4, 3, and 5, respectively, and hence the second vector is the optimal solution. It is easy to check that both $(1, 0, 0, 1)$ and $(2, 0, 0, 0)$ are optimal solutions in their own b-fibers. This shows that indeed $ad^2 - b^3$ is a reduced Gröbner basis element. Once we have this reduced Gröbner basis, and hence the initial ideal, we can describe the standard monomials (and all optimal solutions to all integer programs (1) as b varies) in a compact way. We will look at this compact representation more closely in Section ?. For now, let's list all optimal solutions to the infinitely many integer programs we obtain from the matrix A of the twisted cubic and the weight vector $\omega = (1, 1, 3, 2)$. There are four kinds of optimal solutions. The first two kinds are of the form $(0, y, 0, w)$ and $(1, y, 0, w)$ where $y, w \in \mathbb{N}$. The third kind is $(x, y, 0, 0)$ with $x, y \in \mathbb{N}$, and the last one is $(0, 0, 1, w)$ where $w \in \mathbb{N}$.

Remark 3.7. The connection of integer programming and Gröbner bases has been mined vigorously over the last ten years. The first paper about the subject is by Conti and Traverso [14], and the book by Sturmfels, *Algorithms in Invariant Theory* [76], also contains early references. Thomas gave a geometric interpretation of the Buchberger's algorithm for toric ideals [84] and co-wrote the first application paper [83]. Further theory and algorithms were developed in [86], [87], and [48]. For a recent application of toric Gröbner bases to combinatorial optimization (in particular to the Weak Perfect Graph Theorem) see [64].

3.2 *Computing toric ideals*

How can one get a reduced Gröbner basis of a toric ideal as in Example 3.6? Buchberger's algorithm will produce such a Gröbner basis if we have a generating set to begin with. However, in most cases, the only data concerning the toric ideal we have is the matrix A. Since I_A is defined as the kernel of a polynomial map, the *Elimination Theorem* [16, Thm 3.8] can be used: a) Let $J = \langle x_i - t^{a_i} : i = 1, \dots, n \rangle$ in the polynomial ring $k[x_1, \dots, x_n, t_1, \dots, t_d]$. b) Let $\succ$ be a term order where $\{t_1, \dots, t_d\} \succ \{x_1, \dots, x_n\}$ and compute the reduced Gröbner basis $\mathcal{G}_\succ$ of J with respect to this *elimination term* order. c) The set $\mathcal{G}_\succ \cap k[x_1, \dots, x_n]$ is a reduced Gröbner basis of I_A.

In the above algorithm we assumed that A has only nonnegative entries, and one can modify this algorithm even if there are negative entries (see also [77, Algorithm 4.5]). However, this algorithm is not very practical for matrices already with as small as

10–15 variables since elimination term orders are not very efficient from a computational point of view. There is a different algorithm, first suggested and implemented in [48], that we will present here. It is the superior implementations of this algorithm in various software (see below) which allow us to compute toric ideals with hundreds of variables today. Either variations of this algorithm [2] or new algorithms and implementations [37] have been very successful.

The idea of this *saturation algorithm* which depends on making short successive Gröbner basis computations is based on the following proposition.

Proposition 3.8. *Let* $\mathcal{U} = \{u_1, \dots, u_k\}$ *be a spanning set of the lattice* $\ker_{\mathbb{Z}}(A)$, *and let* $J = \langle x^{u_i^+} - x^{u_i^-} : u_i \in \mathcal{U}\rangle$. *Then*

$$I_A = J : (x_1 \cdots x_n)^\infty.$$

Proof. If a binomial $x^u - x^v$ is contained in the right hand side, then $(x^u - x^v)(x_1 \cdots x_n)^s \in I_A$ for some s. Because I_A is a prime ideal not containing monomials (see Remark 2.8) the binomial $x^u - x^v$ is in I_A. Conversely, let $x^u - x^v$ be a binomial in I_A. Then $u - v = \sum_{i=1}^k s_i u_i$ where $s_i \in \mathbb{Z}$. Replacing u_i with $-u_i$ whenever necessary we can assume that s_i is nonnegative for all i. We observe that if $x^a - x^b$ is a binomial and s is a positive integer, then $x^{sa} - x^{sb} = (x^a - x^b)(x^{(s-1)a} + x^{(s-2)a}x^b + \cdots + x^a x^{(s-2)b}) + x^{(s-1)b})$. Now let $s_1 u_1 = s_1 u_1^+ - s_1 u_1^- = c - d$. The observation we just made implies that $x^c - x^d \in J$. Moreover, let $\sum_{i=2}^k s_i u_i = e - f$. Hence we have $u - v = (c - d) + (e - f)$, and therefore $(x^u - x^v)x^m = x^e(x^c - x^d) + x^d(x^e - x^f)$ where $x^m = \gcd(x^e x^c, x^d x^f)$. In order to show $x^u - x^v$ is in $J : (x_1 \cdots x_n)^\infty$, we need to prove that $x^e - x^f \in I_A$ is in the same ideal. But this we can prove by induction on k. □

Let $J_0 = J$ and $J_i = J_{i-1} : x_i^\infty$ for $i = 1, \dots, n$. Since $J : (x_1 \cdots x_n)^\infty = (\cdots((J : x_1^\infty) : x_2^\infty) \cdots)$, it is enough to show how one can compute $J_i = J_{i-1} : x_i^\infty$ for any i.

Proposition 3.9. *Suppose* I_A *is homogeneous with respect to the weight vector* ω. *Now let* $\prec_\omega^i$ *be a degree reverse lexicographic term order where "degree" means* ω*-degree and where* x_i *is the cheapest variable. Let* $\mathcal{G}_i$ *be the reduced Gröbner basis of* J_{i-1} *with respect to* $\prec_\omega^i$. *Then* J_i *is generated by the reduced Gröbner basis* $\mathcal{G}_i'$ *consisting of binomials in* $\mathcal{G}_i$ *where in each binomial the greatest power of* x_i *shared by both terms is removed.*

Proof. Suppose $x^u - x^v \in J_i$ where the initial term with respect to $\prec_\omega^i$ is the first term. Then there exists $s \in \mathbb{N}$ such that $x_i^s x^u - x_i^s x^v \in J_{i-1}$, and the leading term of this binomial is again the first term. Now this binomial reduced to zero by binomials in $\mathcal{G}_i$. The term order we have chosen guarantees that after removing the relevant powers the initial terms of binomials in $\mathcal{G}_i'$ do not contain x_i. This means that the same sequence of reductions with respect to $\mathcal{G}_i'$ reduces $x^u - x^v$ to zero. This shows that $\mathcal{G}_i'$ is the reduced Gröbner basis of J_i. □

These two propositions together give the algorithm. First a lattice basis of $\ker_{\mathbb{Z}}(A)$ is found, and we form $J = J_0$ generated by the binomials corresponding to the basis vectors. Then n reduced Gröbner bases have to be computed to get a generating set for

I_A. Combined with a special implementation of the Buchberger's algorithm designed to exploit the binomial structure of I_A this algorithm works very well. Further improvements can be achieved since some Gröbner basis computations can be skipped. For instance, after saturating with respect to $x_1, \ldots, x_i$ we might encounter a binomial $x^u - x^v \in J_i$ where x^u consists of variables from $\{x_1, \ldots, x_i\}$. Because of the earlier saturations these variables are invertible modulo I_A (see [23, Exercise 2.3]), and therefore all the variables in x^v are also invertible modulo I_A. The saturations corresponding to these variables can then be skipped. This raises the question of determining a subset X of $\{x_1, \ldots, x_n\}$ of minimum cardinality so that $I_A = J : \left(\prod_{x_i \in X} x_i\right)^\infty$. This question has an algorithmic answer when J is generated by binomials corresponding to a lattice basis of $\ker_{\mathbb{Z}}(A)$.

Definition 3.10. Let $u_1, \ldots, u_m$ be a lattice basis for the lattice $\ker_{\mathbb{Z}}(A)$. The binomial ideal $J = \langle x^{u_i^+} - x^{u_i^-} : i = 1, \ldots, m\rangle$ is called a *lattice basis ideal* [45].

The set of variables we are looking for is derived from the primary decomposition and the minimal primes of the lattice basis ideal J. First, as a consequence of Proposition 3.8 we get the following.

Corollary 3.11. I_A is a primary component in every minimal primary decomposition of J. Furthermore, I_A is a minimal prime and every minimal prime P of J other than I_A has the form $P = \langle x_{i_1}, x_{i_2}, \ldots, x_{i_t}, I'\rangle$ where $t \geq 1$ and I' is the toric ideal corresponding to the lattice generated by the basis vectors $u_{j_1}, \ldots, u_{j_k}$ such that $\mathrm{supp}(u_{j_p}) \cap \{i_1, \ldots, i_t\} = \emptyset$ for $p = 1, \ldots, k$. Thus P is determined by the set of variables contained in it.

The corollary implies that the set X is a minimal set of variables so that at least one variable from each minimal prime of P (other than I_A) is contained in X. We will give a characterization of the variables that appear in a minimal prime. First we need two definitions.

Definition 3.12. Suppose M is an $m \times n$ matrix. Then M is called a *mixed* matrix if every row of M has both a positive and a negative entry; otherwise it is called *unmixed.* The matrix M is called *dominating* if it is mixed, but no $t \times t$ submatrix of M is mixed for $1 \leq t \leq \min(r, n)$ (see [29]).

Let M be the matrix whose rows are the lattice basis vectors $u_1, \ldots, u_m$. Without loss of generality we can assume that a minimal prime P of J which is not I_A contains the first t variables, i.e., $P = \langle x_1, \ldots, x_t, I'\rangle$ where I' is the toric ideal corresponding to the lattice generated by $u_{s+1}, \ldots, u_m$. In other words, the matrix M has the form

$$M = \left[\begin{array}{c|c} N & B \\ \hline 0 & D \end{array}\right], \tag{2}$$

where N is an $s \times t$ matrix with no row of zeros and D is the $(m-s) \times (n-t)$ matrix whose rows come from $u_{s+1}, \ldots, u_m$. We note that N is necessarily mixed, since P is a prime ideal. Conversely, given a decomposition of M as in (2) where N is mixed, the ideal generated by the variables corresponding to the columns of N together with the toric ideal of the lattice generated by rows of D is a prime ideal containing J.

Definition 3.13. A matrix N is called *irreducible* if

((i)) N is a mixed $s \times t$ matrix where $t \leq s$, and
((ii)) one cannot bring N into the following form after permuting its rows and columns,

$$N = \left[\begin{array}{c|c} N' & B' \\ \hline 0 & D' \end{array}\right], \tag{3}$$

where N' is mixed and and D' is of size $r \times p$ where $r < p$.

Remark 3.14. The second condition in the above definition, which is equivalent to saying that N' is "skinnier" than N, also includes the degenerate case when D' is $0 \times p$.

Theorem 3.15 ([45, Theorem 2.5]). *Let P be a prime induced by a decomposition as in (2). Then P is a minimal prime of J if and only if N is irreducible.*

The proof of this theorem is quite technical. We illustrate it on a simple example.

Example 3.16. Let J be the lattice basis ideal whose generators correspond to the rows of the following matrix:

$$M = \begin{bmatrix} 1 & -1 & 2 & 0 & 0 & -2 \\ -3 & 0 & 0 & 1 & -1 & 3 \\ 1 & 1 & 0 & -1 & 1 & -2 \end{bmatrix}.$$

Other than the toric ideal $I_A = J : (x_1x_2x_3x_4x_5x_6)^\infty = \langle x_1x_6 - x_3^2, x_2x_6 - x_1^2, x_1x_4x_6 - x_2^2x_5, x_1x_2x_5 - x_4x_6^2 \rangle$, there are four other minimal primes. We list them together with the decomposition of the matrix M:

$$\begin{bmatrix} 1 & -1 & * & * & * & * \\ -1 & 1 & * & * & * & * \\ 0 & 0 & 1 & -1 & 2 & -2 \end{bmatrix} \qquad P_1 = \langle x_4, x_5, x_1x_3^2 - x_2x_6^2 \rangle,$$

$$\begin{bmatrix} 1 & -2 & * & * & * & * \\ -3 & 3 & * & * & * & * \\ 1 & -2 & * & * & * & * \end{bmatrix} \qquad P_2 = \langle x_1, x_6 \rangle,$$

$$\begin{bmatrix} 2 & 0 & -2 & * & * & * \\ 0 & -1 & 3 & * & * & * \\ 0 & 1 & -2 & * & * & * \end{bmatrix} \qquad P_3 = \langle x_3, x_5, x_6 \rangle,$$

$$\begin{bmatrix} 1 & -1 & 0 & * & * & * \\ -3 & 0 & 1 & * & * & * \\ 1 & 1 & -1 & * & * & * \end{bmatrix} \qquad P_4 = \langle x_1, x_2, x_4 \rangle.$$

In fact, one can check that $\mathrm{rad}(J) = I_A \cap P_1 \cap P_2 \cap P_3 \cap P_4$. This decomposition implies that we can take $X = \{x_1, x_5\}$, and hence two saturations are enough to compute I_A instead of six.

How does one recognize the various irreducible submatrices of M corresponding to the variables of minimal primes of J? At the moment there is no efficient algorithm known for finding the decompositions of M into a form as in (2) where N is irreducible. In fact, it is not very clear how difficult it is to test whether $J = I_A$. This is equivalent to checking whether the matrix M contains no irreducible matrices.

Problem 3.17. Find an efficient algorithm that lists all decompositions of M into a form as in (2) where N is irreducible.

For some new approaches to this problem and more related material we refer the reader to [12]. We close our discussion of computing toric ideals with two remarks. First, it is possible to prove that there exists a set X consisting of at most $\lfloor \frac{n}{2} \rfloor$ variables so that $I_A = J : \left(\prod_{x_i \in X} x_i\right)^{\infty}$, and this set is easy to find [45, Theorem 4.1]. Second, it is enough to do a single saturation provided that one of the vectors a_i of the matrix A is in the interior of the cone generated by the columns of A.

Proposition 3.18. *Suppose a_i is a column of A that is in the interior of the cone generated by A. Then there exists a binomial ideal J such that $I_A = J : x_i^{\infty}$.*

Proof. The hypothesis implies that there is a binomial $x_i^{u_i} - \prod_{j \neq i} x_j^{u_j}$ in I_A. We take J to be the ideal generated by a lattice basis and this binomial. Once J is saturated with respect to x_i all the other variables will be also invertible modulo I_A and this implies the result. □

3.3 *Software*

In this section we present four software programs that are free to download. We will show how one can compute Gröbner bases of toric ideals with the help of these software. We will show how to "plug in" the matrix A and obtain the generators of I_A. We will use the following 3×6 matrix:

$$\begin{bmatrix} 5 & 0 & 0 & 2 & 1 & 0 \\ 0 & 5 & 0 & 1 & 4 & 2 \\ 0 & 0 & 5 & 2 & 0 & 3 \end{bmatrix}$$

4ti2 can be used on a Unix based system, and it is the only one among the four that is specifically created for toric ideals [36]. It also computes Graver bases (see Section 4) and Hilbert bases of rational cones. A new and improved version [38] that works on all platforms has been released recently, and we will use this version. The matrix needs to be written in an input file, say `ourexample`.

```
3 6
5 0 0 2 1 0
0 5 0 1 4 2
0 0 5 2 0 3
```

The positive integers in the first row are the number of rows and number of columns, respectively. We now simply type `groebner -s ourexample`. The option `-s` is for directing 4ti2 to use the saturation algorithm. The option `-p` is an improved *project and lift* algorithm [37]. The output is written to the file `ourexample.gro`:

```
13 6
0 -2 -3 0 0 5
-1 -4 0 0 5 0
0 -1 1 -1 2 -1
-1 -3 -1 1 3 1
0 -3 -2 -1 2 4
-1 0 1 2 1 -3
0 -4 -1 -2 4 3
-1 -2 -2 2 1 2
-1 1 0 3 -1 -2
-2 -3 0 3 4 -2
-2 -2 -1 4 2 -1
-1 -1 -3 3 -1 3
-2 -1 -2 5 0 0
```

There are 13 binomials $x^u - x^v$, and the output file lists the exponent vectors $u - v$. For instance, the first vector corresponds to $x_6^5 - x_2^2 x_3^3$. If the actual binomials are needed, type `output--binomials ourexample.gro`, and the binomials will be stored in the file `ourexample.gro.bin`:

```
x[6]^5-x[2]^2*x[3]^3,
x[5]^5-x[1]*x[2]^4,
x[3]*x[5]^2-x[2]*x[4]*x[6],
x[4]*x[5]^3*x[6]-x[1]*x[2]^3*x[3],
x[5]^2*x[6]^4-x[2]^3*x[3]^2*x[4],
x[3]*x[4]^2*x[5]-x[1]*x[6]^3,
x[5]^4*x[6]^3-x[2]^4*x[3]*x[4]^2,
x[4]^2*x[5]*x[6]^2-x[1]*x[2]^2*x[3]^2,
x[2]*x[4]^3-x[1]*x[5]*x[6]^2,
x[4]^3*x[5]^4-x[1]^2*x[2]^3*x[6]^2,
x[4]^4*x[5]^2-x[1]^2*x[2]^2*x[3]*x[6],
x[4]^3*x[6]^3-x[1]*x[2]*x[3]^3*x[5],
x[4]^5-x[1]^2*x[2]*x[3]^2
```

We could compute a reduced Gröbner basis of the same ideal with respect to a weight (cost) vector ω, say $(22, 7, 2, 1, 1, 1)$. This vector needs to be written into a file called `ourexample.cost`:

```
1 6
22 7 2 1 1 1
```

We again type `groebner -s ourexample`, and then `output -- binomial ourexample.gro` to obtain the following 20 binomials:

```
x[2]^2*x[3]^3-x[6]^5,
x[1]*x[2]^4-x[5]^5,
x[2]*x[4]*x[6]-x[3]*x[5]^2,
x[1]*x[2]^3*x[3]-x[4]*x[5]^3*x[6],
x[2]^3*x[3]^2*x[4]-x[5]^2*x[6]^4,
x[1]*x[6]^3-x[3]*x[4]^2*x[5],
x[2]^4*x[3]*x[4]^2-x[5]^4*x[6]^3,
x[1]*x[2]^2*x[3]^2-x[4]^2*x[5]*x[6]^2,
x[1]*x[5]*x[6]^2-x[2]*x[4]^3,
x[1]^2*x[2]^3*x[6]^2-x[4]^3*x[5]^4,
x[1]^2*x[2]^2*x[3]*x[6]-x[4]^4*x[5]^2,
x[1]*x[2]*x[3]^3*x[5]-x[4]^3*x[6]^3,
x[1]^2*x[2]*x[3]^2-x[4]^5,
x[2]*x[3]^4*x[5]^2-x[4]*x[6]^6,
x[1]*x[3]*x[5]^3*x[6]-x[2]^2*x[4]^4,
x[2]^5*x[4]^3-x[5]^6*x[6]^2,
x[1]*x[3]^4*x[5]^3-x[4]^4*x[6]^4,
x[1]*x[3]^2*x[5]^5-x[2]^3*x[4]^5,
x[1]^2*x[3]^3*x[5]^2-x[4]^6*x[6],
x[3]^5*x[5]^4-x[4]^2*x[6]^7
```

CoCoA is a general purpose algebra system for computational commutative algebra developed by a research group in University of Genoa, Italy [13]. It works on all common operating systems. In a Unix-based system one starts by typing `cocoa`. In order to get useful information about the toric ideal computations the user can type `?toric;`. For instance, by doing this we learn that the main algorithm used is based on [2]. Below is the CoCoA session that computes a Gröbner basis for the same I_A we got above. Note that a ring with six variables have to be defined. The matrix is input by the `Mat` command, and then `Toric` computes the ideal.

```
--------------------------------------------------------
---           ___/       ___/          \             ---
--           /     _ \   /     _ \    , \             --
--           \    |   |  \    |   |  ___ \            --
---          ____, __/  ____,  __/ _/    _\          ---
--------------------------------------------------------
--     Version       : 4.2                            --
--     Online Help   : type  ?   or   ?keyword        --
--     Web site      : http://cocoa.dima.unige.it     --
--------------------------------------------------------


-------------------------------
-- The current ring is R ::= Q[x,y,z];
-------------------------------
```

```
Use R ::= Q[x[1..6]];
A := Mat([[5,0,0,2,1,0],[0,5,0,1,4,2],[0,0,5,2,0,3]]);
IA := Toric(A);
IA;
Ideal(x[3]x[5]^2 - x[2]x[4]x[6], -x[2]x[4]^3
+ x[1]x[5]x[6]^2, -x[3]x[4]^2x[5] + x[1]x[6]^3,
x[1]^2x[2]x[3]^2 - x[4]^5, -x[1]x[2]^4 + x[5]^5,
-x[2]^2x[3]^3 + x[6]^5, x[1]x[2]^2x[3]^2
- x[4]^2x[5]x[6]^2, x[1]x[2]^3x[3] - x[4]x[5]^3x[6],
-x[4]^4x[5]^2 + x[1]^2x[2]^2x[3]x[6],
-x[4]^3x[5]^4 + x[1]^2x[2]^3x[6]^2, x[1]x[2]x[3]^3x[5]
- x[4]^3x[6]^3, -x[2]^4x[3]x[4]^2 + x[5]^4x[6]^3,
-x[2]^3x[3]^2x[4] + x[5]^2x[6]^4)
-------------------------------
QUIT;
```

Macaulay 2 is developed by Dan Grayson and Mike Stillman for computational algebraic geometry computations [31], [24]. It does not have a special implementation for toric ideals. However, the following session is an example about how to use the saturation algorithm. More details about toric Gröbner bases in M2 (with sample code) can be found in [24, pp. 180-181].

```
Macaulay 2, version 0.9.2
--Copyright 1993-2001, D. R. Grayson and M. E. Stillman
--Singular-Factory 1.3c, copyright 1993-2001,
  G.-M. Greuel, et al.
--Singular-Libfac 0.3.3, copyright 1996-2001,
  M. Messollen

i1 : A = {{5,0,0,2,1,0}, {0,5,0,1,4,2}, {0,0,5,2,0,3}};

i2 : B = transpose syz matrix A

o2 = | 2  -1 -1 -5  0  5  |
     | -4 0  -1 10  0  -5 |
     | 5  0  0  -12 -1 8  |

                    3         6
o2 : Matrix ZZ  <--- ZZ

i3 : R = QQ[a,b,c,d,e,f];

i4 : I = ideal(a^2*f^5-b*c*d^5, d^10-a^4*c*f^5,
     a^5*f^8-d^12*e);
```

```
i5 : scan(gens R, g->I=saturate(I,g))

i6 : transpose mingens I

o6 = {-3} | ce2-bdf      |
     {-4} | cd2e-af3     |
     {-4} | bd3-aef2     |
     {-5} | b2c3-f5      |
     {-5} | ab2c2-d2ef2  |
     {-5} | a2bc2-d5     |
     {-5} | ab3c-de3f    |
     {-5} | ab4-e5       |
     {-6} | d4e2-a2b2cf  |
     {-6} | abc3e-d3f3   |
     {-6} | b3c2d-e2f4   |
     {-7} | d3e4-a2b3f2  |
     {-7} | b4cd2-e4f3   |

                 13       1
o6 : Matrix R    <--- R

i7 : quit
```

In the above session, the first two lines compute a basis for the $\ker_{\mathbb{Z}}(A)$. Then we define the polynomial ring with six variables, and the lattice basis ideal corresponding to the rows of the matrix `B`. Line 5 does the successive saturations, and the output is a Gröbner basis of I_A.

Singular is a popular computational algebraic geometry software which could do local ring computations among other things [34], [35]. It has a library that allows one to compute Gröbner bases of toric ideals based on the Proposition 3.9 and other similar ideas. Singular can be used on all platforms, and on a Unix-based system we start by typing `Singular`. At the prompt, we load the library we need by typing `LIB "toric.lib";`

```
> LIB "toric.lib";
// ** loaded /usr/src/Singular/2-0-5/LIB/toric.lib
   (1.11,2001/02/06)
> ring r = 0, (x(1..6)), dp;
> intmat A[3][6] = 5,0,0,2,1,0,
. 0,5,0,1,4,2,
. 0,0,5,2,0,3;
```

In the above lines we have defined a ring of characteristic zero with six variables `x(1), ..., x(6)`, and the 3×6 matrix A. Next we need to input a positive integer vector that is in the row span of our matrix A such as $(1, 1, 1, 1, 1, 1)$. Then we are ready to compute a reduced Gröbner basis of I_A with the command `toric_ideal`. The option `"hs"` refers to "Hoşten-Sturmfels" algorithm based on Proposition 3.9.

```
> intvec ones = 1,1,1,1,1,1;
> ideal T = toric_ideal(A, "hs", ones);
> T = minbase(T);
> T;
T[1]=x(3)*x(5)^2-x(2)*x(4)*x(6)
T[2]=x(3)*x(4)^2*x(5)-x(1)*x(6)^3
T[3]=x(2)*x(4)^3-x(1)*x(5)*x(6)^2
T[4]=x(2)^2*x(3)^3-x(6)^5
T[5]=x(1)*x(2)^2*x(3)^2-x(4)^2*x(5)*x(6)^2
T[6]=x(1)^2*x(2)*x(3)^2-x(4)^5
T[7]=x(1)*x(2)^3*x(3)-x(4)*x(5)^3*x(6)
T[8]=x(1)*x(2)^4-x(5)^5
T[9]=x(4)^4*x(5)^2-x(1)^2*x(2)^2*x(3)*x(6)
T[10]=x(1)*x(2)*x(3)^3*x(5)-x(4)^3*x(6)^3
T[11]=x(2)^3*x(3)^2*x(4)-x(5)^2*x(6)^4
T[12]=x(4)^3*x(5)^4-x(1)^2*x(2)^3*x(6)^2
T[13]=x(2)^4*x(3)*x(4)^2-x(5)^4*x(6)^3
> quit;
```

4. Degrees of Gröbner bases

In this section we construct the *Graver basis* of a toric ideal I_A [84]. The Graver basis contains all reduced Gröbner bases of I_A, and we will use the properties of the Graver basis elements to give bounds on the degree of Gröbner basis elements. In this section $|u|$ will denote $\sum_{i=1}^{n} |u_i|$ for any vector $u \in \mathbb{R}^n$. First we define Graver bases.

We let O_ρ denote an orthant in $\mathbb{R}^n$ where $\rho \in \{+, -\}^n$ and consider $C_\rho = \ker(A) \cap O_\rho$ for $\rho \in \{+, -\}^n$. Every non-zero cone C_ρ is a pointed polyhedral cone in $\mathbb{R}^n$.

Definition 4.1. Let H_ρ be the unique minimal Hilbert basis of the pointed polyhedral cone C_ρ for $\rho \in \{+, -\}^n$. Then $\mathcal{H}_A = \bigcup_\rho H_\rho \backslash 0$ (and the set of the corresponding binomials in I_A) is called the *Graver basis* of I_A.

Lemma 4.2. *Let ω be a generic weight vector. Then if $x^u - x^v \in \mathcal{G}_\omega$ is a reduced Gröbner basis element of I_A then $u - v$ is contained in $\mathcal{H}_A$, the Graver basis of I_A.*

Proof. Suppose that $u - v$ is in C_ρ, but not in H_ρ. This implies that $u - v = \alpha + \beta$ for two vectors in $C_\rho \cap \mathbb{Z}^n$. Moreover either $\omega\alpha > 0$ or $\omega\beta > 0$ since $\omega(u - v) > 0$. Suppose $\omega\alpha$ is strictly positive. Now the initial term of the binomial $x^{\alpha_+} - x^{\alpha_-}$ is x^{α_+} and this divides x^u. A contradiction. □

As a corollary to this lemma we conclude that I_A has finitely many reduced Gröbner bases, a result we have mentioned (but not proved) for general polynomial ideals.

Corollary 4.3. The toric ideal I_A has finitely many reduced Gröbner bases.

Proof. By its definition the Graver basis $\mathcal{H}_A$ is a finite set. Combined with the above lemma we get the result. □

Since we are interested in the maximal degree of Graver basis elements we introduce the following notation.

Definition 4.4. We let $\mathrm{maxg}(I_A)$ to be the maximal degree of any Graver basis element of I_A, i.e.

$$\mathrm{maxg}(I_A) = \max\{|\alpha| + |\beta| : x^\alpha - x^\beta \in \mathcal{H}_A\}.$$

We start with the best known bound on $\mathrm{maxg}(I_A)$. This bound is obtained by many different people from very diverse areas (invariant theory [88], A-C unification [21], linear diophantine equations [5], [67], algebraic geometry [77]) independently. Here we present Theorem 4.7 in [77] since the proof given there leads to ideas which we develop subsequently.

Theorem 4.5. *Let* $D(A) := \max\{|det(a_{i_1}, a_{i_2}, \ldots, a_{i_d})| : 1 \leq i_1 < i_2 < \cdots < i_d \leq n\}$. *If* $n - d \geq 2$ *then* $\mathrm{maxg}(I_A) \leq (n-d-1)(d+1)D(A)$.

For the proof of this theorem we use *circuits* of A and we bound their degrees first.

Definition 4.6. For $v \in \mathbb{R}^n$ let $\mathrm{supp}(v) := \{i : v_i \neq 0\}$. A nonzero vector $u \in \ker_{\mathbb{Z}}(A)$ is called a *circuit* if $\mathrm{supp}(u)$ is minimal with respect to inclusion and the coordinates of u are relatively prime. We will call integral vectors with relatively prime coordinates *primitive* in general.

Lemma 4.7. *If* u *is a circuit of* A *then* $|u| \leq (d+1)D(A)$.

Proof. Let $\mathrm{supp}(u) = \{i_1, \ldots, i_r\}$. Since $\mathrm{supp}(u)$ is minimal, $r \leq (d+1)$. Because A has rank d, we can extend the column vectors $a_{i_1}, \ldots, a_{i_r}$ to a $d \times (d+1)$-submatrix $B = [a_{i_1}, \ldots, a_{i_r}, a_{i_{r+1}}, \ldots a_{i_{d+1}}]$ of rank d. Using Cramer's rule we see that

$$\lambda u = \sum_{j=1}^{d+1} (-1)^j \det\left(a_{i_1}, \ldots, a_{i_{j-1}}, a_{i_{j+1}}, \ldots a_{i_{d+1}}\right) \mathbf{e}_{i_j}$$

where λ is an integer and $\mathbf{e}_i$ is the i-th unit vector. Since the absolute value of each determinant in this expression is bounded by $D(A)$, we conclude $\sum_{i=1}^n |u_i| \leq (d+1)D(A)$. □

Proof of Theorem 4.5. Let $x^\alpha - x^\beta$ be an element of the Graver basis $\mathcal{H}_A$. As in the definition of $\mathcal{H}_A$, let C_ρ be the cone that contains $\alpha - \beta$ as its Hilbert basis element. Since $C_\rho = \{x \in \mathbb{R}^n : Ax = 0,\ \rho_i x_i \geq 0,\ i = 1, \ldots, n\}$ with $\rho_i \in \{+, -\}$, the cone C_ρ is generated by circuits $u^1, \ldots, u^r$. By Carathéodory's theorem [72, Corollary 7.1i], $\alpha - \beta$ is contained in a simplicial subcone generated by at most $(n-d)$ circuits $u^{i_1}, \ldots, u^{i_{n-d}}$. Since $\alpha - \beta$ is also a Hilbert basis element of this subcone,

$$\alpha - \beta = \sum_{j=1}^{n-d} \lambda_j u^{i_j}$$

where $0 \leq \lambda_j < 1$. Therefore $|\alpha - \beta| < \sum_{j=1}^{n-d} \lambda_j |u^{i_j}| \leq (n-d)(d+1)D(A)$, where the last inequality follows from the previous lemma.

The bound can be improved from $(n-d)(d+1)D(A)$ to $(n-d-1)(d+1)D(A)$ in the case when $n-d \geq 2$ as follows. The convex hull of the *nonzero* lattice points in C_ρ is an unbounded polyhedron whose bounded faces are exactly those "visible" from the origin. The lattice points on these bounded faces have to be Hilbert basis elements of C_ρ, and we note that if u is one of these then $|u| \leq (d+1)D(A)$. Now we can subdivide each bounded face into simplices using the lattice points on that face (i.e. we triangulate these faces). The Graver basis element $\alpha - \beta$ is contained in one of the simplicial subcones generated by Hilbert basis elements $h_1, \ldots, h_{n-d}$ that are on one of the bounded facets. The lattice point $\alpha - \beta$ is in the Hilbert basis of this cone and hence $\alpha - \beta = \sum_{j=1}^{n-d} \lambda_j h_j$. We claim that $\sum_{j=1}^{n-d} \lambda_j \leq (n-d-1)$, and this proves the result. Suppose that this sum is bigger than $(n-d-1)$. Then $v := \sum_{j=1}^{n-d} h_j - (\alpha - \beta)$ is in the same cone and $v = \sum_{j=1}^{n-d} \mu_j h_j$ such that $\sum_{j=1}^{n-d} \mu_j < 1$. However this means that v is in the convex hull of the origin and $h_1, \ldots, h_{n-d}$, and this is a contradiction to the fact that these Hilbert basis elements span a facet of the convex hull of the nonzero lattice points in C_ρ. □

Theorem 4.5 together with empirical observations drawn from computations of Gröbner bases of toric ideals leads us to the following conjecture:

Conjecture 4.8. Let A be a $d \times n$ integer matrix with $\text{rank}(A) = d$. The maximum degree $\text{maxg}(I_A)$ of any Graver basis element of the toric ideal I_A is bounded by $(d+1)D(A)$.

Following the proof of Theorem 4.5 the most naive question one can ask is whether a Graver basis element $\alpha - \beta$ as a Hilbert basis element of a cone C_ρ has a representation

$$\alpha - \beta = \sum_{j=1}^{n-d} \lambda_j u_{i_j}$$

with $\sum_{j=1}^{n-d} \lambda_j \leq 1$. We consider this question in a more general setting where we look at general cones:

Definition 4.9. Let $C \in \mathbb{R}^m$ be a full-dimensional pointed cone with $r \geq m$ primitive generators $u_1, \ldots, u_r \in \mathbb{Z}^m$ and let h be a Hilbert basis element of C. The *height* of h is defined to be:

$$\text{height}(h) := \min \left\{ \sum_{j=1}^{m} \lambda_j : h \in \text{cone}(u_{i_1}, \ldots, u_{i_m}) \text{ and } h = \sum_{j=1}^{m} \lambda_j u_{i_j} \right\}.$$

Now the question is whether height$(\alpha - \beta)$ in C_ρ is less than 1 or not. Unfortunately, for general cones the height of a Hilbert basis element can be quite bad: asymptotically $m-1$ is the best possible bound [56]. A more precise bound is given in [39]. For the type of cones we are interested in, the height of a minimal Hilbert basis element could be larger than 1 as the following example shows.

Example 4.10.

$$A = [1\ 3\ 4\ 6], \qquad h = [1, 1, 2, -2].$$

The Graver basis element h is a Hilbert basis element of the cone C_{+++-} generated by $u_1 = [6, 0, 0, -1]$, $u_2 = [0, 2, 0, -1]$, $u_3 = [0, 0, 3, -2]$. Since $h = (1/6)u_1 + (1/2)u_2 + (2/3)u_3$, height$(h) = 1/6 + 1/2 + 2/3 > 1$. However note that degree$(h) = 6 < (d+1)D(A) = 12$ which satisfies the bound in Conjecture 4.8.

Another plausible approach is bounding maxg(I_A) by the degree of the circuits; in other words, we ask whether maxg(I_A) is bounded by

$$\text{maxc}(A) := \max \left\{ \sum_{i=1}^{n} |u_i| \ : \ u \text{ is a circuit of } A \right\}.$$

A positive answer implies an affirmative answer for Conjecture 4.8. Unfortunately, the answer is negative. The following example has also appeared in [78]:

Example 4.11. Let

$$A = \begin{bmatrix} 1 & 3 & 4 & 6 & 0 \\ 0 & 0 & 0 & -5 & 1 \end{bmatrix}$$

The six circuits of A and their degrees are as follows:

circuit	*degree*
$[3,\ -1,\ 0,\ 0,\ 0]$	4
$[4,\ 0,\ -1,\ 0,\ 0]$	5
$[0,\ 4,\ -3,\ 0,\ 0]$	7
$[0,\ 2,\ 0,\ -1,\ -5]$	8
$[6,\ 0,\ 0,\ -1,\ -5]$	12
$[0,\ 0,\ 3,\ -2,\ -10]$	15

However the Graver basis element $[1,\ 1,\ 2,\ -2,\ -10]$ has degree 16.

Remark 4.12. The last circuit in the above list is not the **true** circuit we should consider. When we compute this circuit using Cramer's rule as in Lemma 4.7 we get $[0,\ 0,\ 6,\ -4,\ -20]$ with degree 30. Therefore for the above example maxg $\leq$ maxc if we allow ourselves to consider the circuits as they are computed in Lemma 4.7, and drop the condition of relatively prime coordinates.

In the light of the above example and the remark we will state a conjecture that takes the true circuits into consideration and which implies Conjecture 4.8. But first we formulate this conjecture for general cones (see also [78]).

Let $C \in \mathbb{R}^m$ be a full dimensional pointed cone with r primitive generators $u_1, \ldots, u_r$. Let $b_1, \ldots, b_s$ be the primitive inner normal directions of the facets of C. For each $i = 1, \ldots, r$, let $\mathcal{L}_i := u_i^{\perp} \cap \mathbb{Z}^m$ where $u_i^{\perp}$ denotes the subspace orthogonal to u_i. Also let $\mathcal{L}_i' := \mathbb{Z}\{b_j : b_j \in u_i^{\perp}\}$ be the lattice generated by facet normal vectors of those facets on which u_i lies. Both $\mathcal{L}_i$ and $\mathcal{L}_i'$ are sublattices of $\mathbb{Z}^m$ and moreover $\mathcal{L}_i'$ is a sublattice of $\mathcal{L}_i$ of finite index since rank$(\mathcal{L}_i) = $ rank$(\mathcal{L}_i')$.

Definition 4.13. We define the *index* of u_i as the index of $\mathcal{L}'_i$ in $\mathcal{L}_i$, i.e. $\text{index}(u_i) = [\mathcal{L}_i : \mathcal{L}'_i]$. With this notation we call $\bar{u}_i := \text{index}(u_i)\, u_i$, a *true generator* of C.

Conjecture 4.14. Let C be a cone as above and let h be a Hilbert basis element of C. We conjecture that the

$$\text{true-height}(h) := \min\left\{\sum_{j=1}^{m} \lambda_j : h \in \text{cone}(\bar{u}_{i_1}, \ldots, \bar{u}_{i_m}) \text{ and } h = \sum_{j=1}^{m} \lambda_j \bar{u}_{i_j}\right\}$$

does not exceed 1.

Example 4.15. Let $C \in \mathbb{R}^3$ be the cone generated by the columns u_1, u_2 and u_3 of the matrix

$$U = \begin{bmatrix} 1 & 1 & 2 \\ 0 & 2 & 0 \\ 0 & 0 & 3 \end{bmatrix}.$$

$\text{cone}(u_1, u_2)$, $\text{cone}(u_1, u_3)$, $\text{cone}(u_2, u_3)$ are the facets of C with the corresponding primitive normal vectors $b_1 = [0, 0, 1]$, $b_2 = [0, 1, 0]$ and $b_3 = [6, -3, -4]$. Hence, $\text{index}(u_1) = 1$, $\text{index}(u_2) = 3$ and $\text{index}(u_3) = 2$. The Hilbert basis element $h = [2, 1, 2]$ has $\text{height}(h) = 1/6+1/2+2/3 > 1$ since $h = (1/6)u_1+(1/2)u_2+(2/3)u_3$. On the other hand, $h = (1/6)\bar{u}_1 + (1/6)\bar{u}_2 + (1/3)\bar{u}_3$ so $\text{true-height}(h) = 1/6+1/6+1/3 < 1$.

Remark 4.16. The type of cones we are interested in are of the form $C_\rho = \{x \in \mathbb{R}^n_\rho : Ax = 0\}$. The generators of these cones are circuits of A in $\mathbb{R}^n_\rho$ and the facets are defined by unit vectors and by the rows of A. Hence the index of a circuit of A simply becomes the gcd of the entries of the corresponding true circuit which is computed by Cramer's rule as in Lemma 4.7.

There is not much known about Conjecture 4.14 for general cones. A possible first step for settling this conjecture we prove the statement for simplicial cones (see also [43] where the proof has first appeared).

Theorem 4.17. *Let $C \subseteq \mathbb{R}^m$ be a full dimensional simplicial cone with generators $u_1, \ldots, u_m$. If h is a Hilbert basis element of C, then true-height$(h) \leq 1$.*

Proof. Let $\bar{U} = [\bar{u}_1, \ldots, \bar{u}_m]$ be the $m \times m$ matrix whose columns are the true generators of C. Suppose $b_1, \ldots, b_m$ are the primitive inner normal vectors of the facets of C. We assume that b_i is the unique normal vector with $b_i \notin u_i^\perp$. Let also B be the $m \times m$ matrix with rows $b_1, \ldots, b_m$. With this notation we observe that $C = \{x \in \mathbb{R}^m : Bx \geq 0\}$. From the definition of true generators we conclude that $\bar{U}$ is the adjoint of B; hence

$$B \cdot \bar{U} = \begin{bmatrix} D & & \\ & \ddots & \\ & & D \end{bmatrix},$$

is a diagonal matrix where $D = |\det(B)|$. Similarly let $v = Bh$. In order to prove our claim we just need to show that $\sum_{i=1}^{m} v_i \leq D$. Consider the sequence of vectors $[0, \ldots, 0], [1, \ldots, 0], \ldots, [v_1, \ldots, 0], [v_1, 1, \ldots, 0], \ldots [v_1, v_2, \ldots, 0], \ldots, [v_1, v_2, \ldots, v_m - 1]$. When we view the matrix B as the matrix of a linear transformation, these integral vectors are in $B(C)$. Moreover if true-height$(h) > 1$ then the number of these vectors exceeds D. Now the image of $\mathbb{Z}^m$ under the linear transformation represented by B is a lattice $\mathcal{L} := B(\mathbb{Z}^m)$ with the finite index D; hence if there are more than D vectors in the above sequence then we must have two vectors s_1 and s_2 in this sequence such that $s_1 - s_2 \in \mathcal{L}$. We can assume that $s_1 \geq s_2$, therefore $s_1 - s_2 \in \mathcal{L} \cap B(C)$. In other words we can find an integral vector $t_1 \in C$ such that $Bt_1 = s_1 - s_2$. This says that there exists another integral vector $t_2 \in C$ with $Bt_2 = v - (s_1 - s_2)$, which in turn implies that $h = t_1 + t_2$ contradicting the fact that h is a Hilbert basis element. □

We finish by summarizing a few cases where Conjecture 4.8 is true. Theorem 4.5 implies that the conjecture is true when $n - d = 1$ and $n - d = 2$. However, already the case $n - d = 3$ is open, despite the fact that the cones C_ρ are just three-dimensional cones. The case $d = 1$ is settled in [19] where the Graver basis elements of I_A where $A = [a_1, \ldots, a_n]$ (also known as *primitive partition identities*) have degree at most $a_n + a_{n-1}$. Even the two dimensional case $d = 2$ is still open (see [43, Chapter 2] for some progress).

5. The radical of a toric initial ideal

In this section we study the radical of $\mathrm{in}_\omega(I_A)$ where ω is a generic weight vector. The radical of any monomial ideal is *squarefree*: the minimal generators are monomials not divisible by x_i^2 for $i = 1, \ldots, n$. The variety $V(I)$ of a monomial ideal I is a union of coordinate subspaces:

$$V(I) \quad = \quad \bigcup_{\sigma \in \Delta} V^\sigma \quad \text{where} \quad V^\sigma = \{(a_1, \ldots, a_n) \, : \, a_i = 0 \ \forall i \notin \sigma\}. \qquad (4)$$

Here the members of Δ indexing the union is a set of subsets of $\{1, \ldots, n\}$. Now we can use Hilbert's Nullstellensatz to get a bijection between radical monomial ideals and such unions of coordinate subspaces. Note that the sets $\sigma \in \Delta$ are closed under taking subsets: if $\tau \subset \sigma$ then $\tau \in \Delta$. In other words, there is a bijection between squarefree monomial ideals and such abstract set systems. This brings us to the following important definition.

Definition 5.1. An (abstract) simplicial complex on the vertex set $[n] := \{1, \ldots, n\}$ is a collection Δ of subsets of $[n]$ where if $\sigma \in \Delta$ and $\tau \subset \sigma$ then $\tau \in \Delta$. The radical (squarefree) monomial ideal I_Δ such that $V(I_\Delta) = \bigcup_{\sigma \in \Delta} V^\sigma$ as in (4) is called the *Stanley-Reisner ideal* of Δ.

Proposition 5.2. *Let Δ be simplicial complex on n vertices. Then*

$$I_\Delta = \left\langle \prod_{i \in \sigma} x_i \, : \, \sigma \notin \Delta \right\rangle = \bigcap_{\sigma \in \Delta} \langle x_i \, : \, i \notin \sigma \rangle.$$

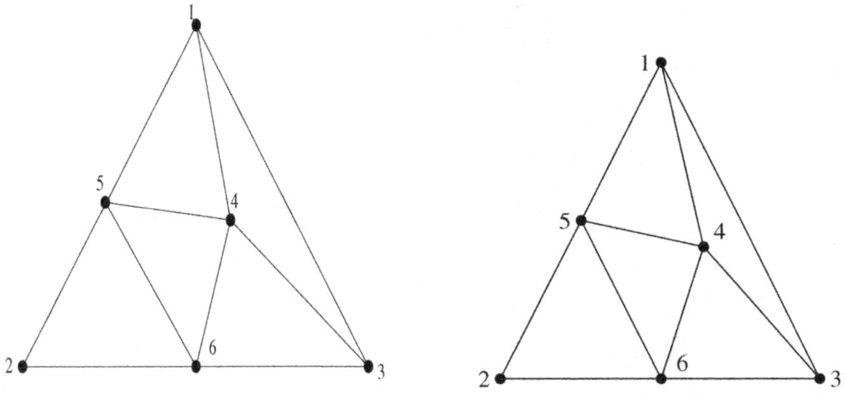

Figure 2. The simplicial complex in Example 5.3

Proof. Since I_Δ is a squarefree monomial ideal, it is characterized by its squarefree standard monomials. Let $x^u \notin I_\Delta$ be such a monomial with $\mathrm{supp}(u) = \sigma$. By (4) we see that $\sigma \in \Delta$. This happens if and only if x^u is not contained in either of the ideals above. □

Example 5.3. Let Δ be the simplicial complex on six vertices whose maximal faces (facets) are the triangles in Figure 2. In other words, Δ consists of $\{1,3,4\}$, $\{1,4,5\}$, $\{2,5,6\}$, $\{3,4,6\}$, $\{4,5,6\}$ and all the subsets of these sets. The corresponding Stanley-Reisner ideal is

$$\begin{aligned} I_\Delta &= \langle x_1x_6,\ x_3x_5,\ x_2x_4,\ x_2x_3,\ x_1x_2\rangle \\ &= \langle x_2, x_5, x_6\rangle \cap \langle x_2, x_3, x_6\rangle \cap \langle x_1, x_3, x_4\rangle \cap \langle x_1, x_2, x_5\rangle \cap \langle x_1, x_2, x_3\rangle. \end{aligned}$$

Next we characterize those simplicial complexes Δ such that $\mathrm{rad}(\mathrm{in}_\omega(I_A)) = I_\Delta$. Given $\sigma \subseteq [n]$, we let $\mathcal{A}_\sigma$ be the subset of $\mathcal{A}$ whose elements are indexed by σ.

Definition 5.4. Let ω be a generic weight vector. For $\sigma \subseteq [n]$, the cone $C(\mathcal{A}_\sigma)$ is a face of the *regular triangulation* Δ_ω of $C(\mathcal{A})$ if and only if there exists a vector $y \in \mathbb{R}^d$ such that $y \cdot a_j = \omega_j$ for all $j \in \sigma$ and $y \cdot a_j < \omega_j$ for all $j \notin \sigma$.

Remark 5.5. The regular triangulation Δ_ω is indeed a triangulation of $C(\mathcal{A})$, i.e., the union of $C(\mathcal{A}_\sigma)$ is equal to $C(\mathcal{A})$, the cones $C(\mathcal{A}_\sigma)$ are all simplicial cones, and the intersection of two such cones are faces of both. In fact, Δ_ω is gotten by taking the cone in $\mathbb{R}^{d+1}$ generated by the lifted vectors $(a_i, \omega_i) \in \mathbb{R}^{d+1}$, and then projecting the *lower* facets of this lifted cone back onto *cone*$(\mathcal{A}) \subset \mathbb{R}^d$ (see [3]). Since ω is generic, the lower facets are simplicial cones, and hence we get a triangulation. Using σ to label $C(\mathcal{A}_\sigma)$, the triangulation Δ_ω becomes naturally a simplicial complex.

Example 5.6. Suppose $A = \begin{pmatrix} 1 & 2 & 2 & 3 & 2 \\ 3 & 3 & 1 & 1 & 0 \end{pmatrix}$ and $\omega = (9, 9, 4, 0, 9)$.

Figure 3 illustrates how the regular triangulation Δ_ω of $C(\mathcal{A})$ is obtained via Remark 5.5. The columns of A are the five vectors in $\mathbb{R}^2$ shown in the top picture. These five vectors are then replaced by the lifted vectors (a_i, ω_i) in $\mathbb{R}^3$. The cone

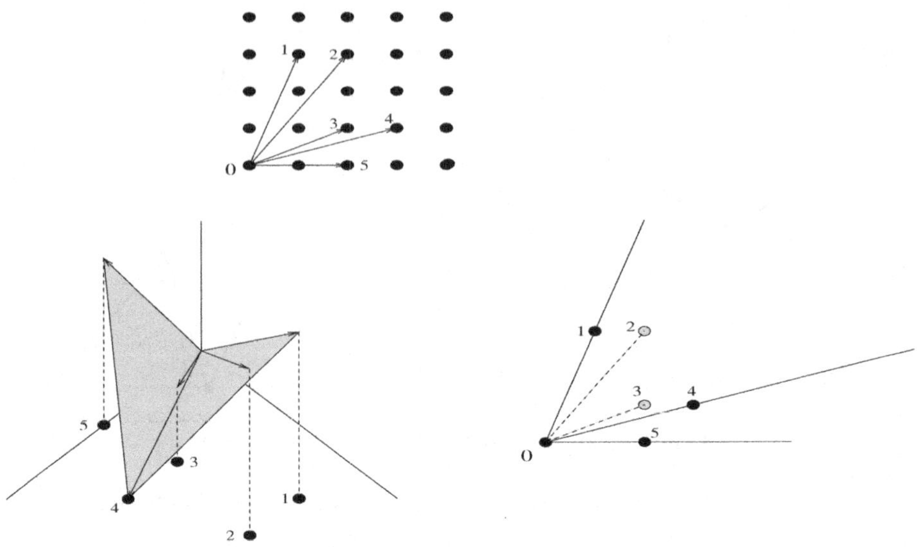

Figure 3. The regular triangulation in Example 5.6.

generated by the lifted vectors is shown in the bottom left of Figure 3. The shaded facets are the *lower* facets of this cone. Note that the liftings of columns 2 and 3 are in the interior of the cone. The bottom right picture shows the regular triangulation Δ_ω.

Theorem 5.7 ([77, Theorem 8.3]). *Let* ω *be generic. Then* $\mathrm{rad}(\mathrm{in}_\omega(I_A)) = I_{\Delta_\omega}$.

Proof. Let $\prod_{i\in\sigma} x_i$ be a squarefree monomial not in $\mathrm{rad}(\mathrm{in}_\omega(I_A))$. This is equivalent to the existence of a standard monomial x^u of $\mathrm{in}_\omega(I_A)$ such that $\mathrm{supp}(u) = \sigma$ and every power of x^u is standard. By Theorem 3.4 the vector u is the optimal solution to the integer program (1) over the b-fiber where $b = Au$. We consider the *linear program*

$$\text{minimize} \quad \omega \cdot x \quad \text{subject to} \quad Ax = b, \quad \text{and} \quad x \in \mathbb{R}^n_+. \tag{5}$$

We claim that u is also an optimal solution to this program. The optimal solution v to (5) does not have to be an integer vector in general. However, an appropriate integer multiple tv will be an integer vector and it is necessarily the optimal solution for the integer program over the tb-fiber. By our assumption, the vector tu is also an optimal solution for the same integer program. Now the genericity of ω implies that $tv = tu$ and hence $u = v$. Now by linear programming duality and complementary slackness [72, §7.9] the vector u with $\mathrm{supp}(u) = \sigma$ is an optimal solution to (5) if and only if there exits $y \in \mathbb{R}^d$ such that $a_i \cdot y = \omega_i$ for $i \in \sigma$ and $a_i \cdot y < \omega_i$ otherwise. By Definition 5.4, the subset σ is a face of Δ_ω, and this happens if and only if $\prod_{i\in\sigma} x_i \notin I_{\Delta_\omega}$. □

Example 5.8. The reduced Gröbner basis $\mathcal{G}_\omega(I_A)$ for toric ideal in Example 5.6 with respect to $\omega = (9, 9, 4, 0, 9)$ consists of the following binomials with the underlined initial terms:

$$\{\underline{x_2x_3} - x_1x_4,\ \underline{x_1^2x_5} - x_2^2,\ \underline{x_1x_3x_5} - x_2x_4,\ \underline{x_3^2x_5} - x_4^2,\ \underline{x_2x_5^2} - x_3^3,\ \underline{x_1x_5^3} \\ - x_3^2x_4,\ \underline{x_1x_4x_5^2} - x_3^4,\ \underline{x_2^3x_5} - x_1^2x_3^3,\ \underline{x_2^2x_4x_5} - x_1x_3^4,\ \underline{x_3^5} - x_2x_4^2x_5,\ \underline{x_1^4x_3^3} \\ - x_2^5,\ \underline{x_1^3x_3^4} - x_2^4x_4,\ \underline{x_2^6} - x_1^5x_3^2x_4\}.$$

The radical of $\mathrm{in}_\omega(I_A)$ is equal to $\langle x_2, x_3, x_1x_4, x_1x_5\rangle$ which is the same as the Stanley-Reisner ideal of the regular triangulation Δ_ω in Example 5.6.

Remark 5.9. Definition 5.4 and the proof of Theorem 5.7 imply that the cones $C(\mathcal{A}_\sigma)$ where σ ranges over the facets of Δ_ω are the optimal bases of the linear programs (5) as the right hand side vector b ranges over the cone $C(\mathcal{A})$. In other words, computing Δ_ω is equivalent to solving all such linear programs: given $b \in C(\mathcal{A})$ one solves $A_\sigma x^\sigma = b$ for each facet σ of Δ_ω. The optimal solution x^* is given by that σ where $b \in C(\mathcal{A}_\sigma)$, i.e. when $x^\sigma \geq 0$. In this case, $x_i^* = x_i^\sigma$ if $i \in \sigma$ and $x_i^* = 0$ otherwise.

The structure of *all* regular triangulations of $\mathcal{A}$ as well as *all* triangulations of $\mathcal{A}$ have lead to the construction of the *secondary polytope* (see [33], [3], [80]), and the *universal polytope* (see [3], [17]). The software TOPCOM [68] provides an efficient tool for computing and analyzing (regular) triangulations. The deepest structural result is the fact that the graph of all triangulations of $\mathcal{A}$ (the vertices of this graph are the triangulations and two such triangulations are connected by an edge if there is a *bistellar flip* between them) is disconnected [71].

6. Squarefree initial ideals

In this section we discuss the toric ideals which happen to possess at least one squarefree (radical) initial ideal. We will look at these kinds of toric ideals when I_A has low (co-)dimension, and list some important known families of examples. The key word of this section is *unimodular* triangulations of $\mathcal{A}$ which we first define.

Definition 6.1. A triangulation Δ of $\mathcal{A}$ is called a unimodular triangulation if $\mathcal{A}_\sigma$ forms a basis of the lattice $\mathbb{Z}\mathcal{A}$ for all facets $\sigma \in \Delta$. When $\mathbb{Z}\mathcal{A} = \mathbb{Z}^d$, this means $|\det(A_\sigma)| = 1$ for all such σ.

Theorem 6.2 ([77, Corollary 8.9]). *The initial ideal* $\mathrm{in}_\omega(I_A)$ *is squarefree if and only if the regular triangulation* Δ_ω *is unimodular.*

Proof. If Δ_ω is unimodular then the integer programs (1) are solved by the corresponding linear programs (5). See Remark 5.9. This means x^u is a standard monomial of $\mathrm{in}_\omega(I_A)$ if and only if $\mathrm{supp}(u) = \sigma$ where σ is a face of Δ. If $\mathrm{in}_\omega(I_A)$ is not radical, then it has a minimal generator x^v that is not squarefree. The squarefree monomial $\prod_{i \in \mathrm{supp}(v)} x_i$ is a standard monomial, and hence x^v should also be a standard monomial. This contradiction shows the if-direction. For the converse, suppose $\mathrm{in}_\omega(I_A)$ is squarefree. As above, x^u is a standard monomial if and only if $\mathrm{supp}(u) = \sigma$ where σ is a face of Δ_ω. Now assume that Δ_ω is not unimodular. This implies that there is a facet $\sigma \in \Delta$ such that $|\det(A_\sigma)| > 1$. Below in Theorem 5.7 we will see that the multiplicity of the prime ideal $\langle x_i \; : \; i \notin \sigma \rangle$ in $\text{in}_\omega(I_A)$ is equal to $|\det(A_\sigma)|$. But since $\text{in}_\omega(I_A) = I_{\Delta_\omega} = \bigcap_{\tau \in \Delta} \langle x_i \; : \; i \notin \tau \rangle$ by Proposition 5.2, this multiplicity is equal to one. Again a contradiction. □

Corollary 6.3. If I_A has a squarefree initial ideal then $\mathcal{A}$ is normal.

Proof. By Theorem 6.2 $\mathcal{A}$ has a unimodular triangulation Δ_ω. If h is in the affine monoid $S(C(\mathcal{A}))$, then it will be in one of $C(\mathcal{A}_\sigma)$ where $\sigma \in \Delta_\omega$ is a facet. Now Cramer's rule implies that h is a nonnegative integer combination of the vectors in $\mathcal{A}_\sigma$. Hence $h \in \mathbb{N}\mathcal{A}$ and this shows that $\mathcal{A}$ is normal. □

6.1 *Squarefreeness in low (co-)dimension*

Now we like to identify $\mathcal{A}$ that have unimodular triangulations when the dimension d is ≤ 3 and when the codimension $n - d$ is ≤ 2.

Dimension 1: In this case $\mathcal{A} = \{a_1, \ldots, a_n\}$ where we can assume that $a_i's$ are nonnegative integers (recall that we are assuming that $\mathbb{Z}\mathcal{A} = \mathbb{Z}^d$ and $C(\mathcal{A})$ is a pointed cone). Hence $\mathcal{A}$ will have a (regular) unimodular triangulation if and only if $a_i = 1$ for some i. In other words, $\mathcal{A}$ has a unimodular triangulation if and only if $\mathcal{A}$ is normal.

Dimension 2: Like in the case of Dimension 1, we claim that I_A has a squarefree initial ideal if and only if $\mathcal{A}$ is normal. By Corollary 6.3 we just need to show the if-direction. If $\mathcal{A}$ is normal, by definition of normality $\mathcal{A}$ has to contain a Hilbert basis of the cone $C(\mathcal{A})$. The proof of Theorem 4.5 implies that the Hilbert basis of $C(\mathcal{A})$ consists of the vectors $a_1, \ldots, a_k$ which are the lattice points on the bounded faces of the polyhedron $\text{conv}(u \in C(\mathcal{A}) \cap \mathbb{Z}^2 \; : \; u \neq 0)$. We can assume that $a_1, \ldots, a_k$ are ordered clockwise, and it is not hard to see that $\{a_1, a_2\}, \{a_2, a_3\}, \ldots, \{a_{k-1}, a_k\}$ form a regular unimodular triangulation of $\mathcal{A}$.

Dimension 3: By a result of Sebö [73] what is true in Dimension 1 and Dimension 2 is also true here: every normal $\mathcal{A}$ has a regular unimodular triangulation and hence I_A has a squarefree initial ideal if and only if $\mathcal{A}$ is normal. One special case of Sebö's result is easy to see, namely the case where $C(\mathcal{A}) = C(P)$ where P is a lattice polygon in $\mathbb{R}^2$ (see Example 2.4). We can triangulate P into *empty* triangles (i.e. triangles whose only lattice points are its vertices) as follows: Pick a vertex v of P, and let Q be the convex hull of the lattice points in $P \backslash \{v\}$. By induction we can obtain a desired triangulation of Q, and by adding the empty triangles formed by v and the boundary of Q "visible" from v, we extend this triangulation to a triangulation of P. We point out that this is an instance of a *lexicographic triangulation* that are known to be regular triangulations [77, Proposition 8.6]. Such a triangulation induces a regular triangulation of $\mathcal{A} \subset \mathbb{R}^3$ that is also unimodular.

Dimension 4 and beyond: Already in dimension 4 there are normal $\mathcal{A}$ which do not have unimodular triangulations. See [6] and [28] for examples. Examples have been constructed where $\mathcal{A}$ is normal and consists of 0/1 vector but does not have regular unimodular triangulations [41]. Finally, an impressive breakthrough showed that there are normal $\mathcal{A}$ in dimension 6 and more which do not possess *unimodular Hilbert covers* [7].

Codimension 1: A toric ideal I_A of codimension one is a principal ideal $\langle x^u - x^v \rangle$ where $u - v$ generates the lattice $\ker_{\mathbb{Z}}(A)$. So I_A has only two initial ideal generated by either x^u or x^v, and therefore I_A has a squarefree initial ideal if and only if the binomial $x^u - x^v$ has a squarefree term. A sufficient condition is that $\mathcal{A}$ is normal. The proof of the following result was communicated to us by Winfried Bruns and the details will appear in [22].

Proposition 6.4. *If $\mathcal{A}$ is normal then every minimal generator of I_A has a squarefree term.*

A corollary to Proposition 6.4 is that if $\mathcal{A}$ is normal of codimension one then it has a squarefree initial ideal.

Codimension 2: The strongest result we can cite is again when $\mathcal{A}$ is normal. This new result will appear in [22], and its proof heavily depends on Proposition 6.4 and Hochster's result Theorem 8.1 that I_A is Cohen-Macaulay if $\mathcal{A}$ is normal. The proof treats the complete intersection and the non-complete intersection case separately. Here we will present a sketch of the first case.

If I_A is a complete intersection of codimension two then $I_A = \langle x^p - x^q,\ x^r - x^s \rangle$. We let B to be the $2 \times n$ matrix whose rows are $g_1 = p - q$ and $g_2 = r - s$. The columns of B form a *Gale diagram* of the matrix A. Such a Gale diagram is called *imbalanced* if either $B_{1j} = 0$ or $B_{2j} \geq 0$.

Lemma 6.5 ([29, Theorem 3.9]). *I_A is a complete intersection if and only if there is an imbalanced Gale diagram B.*

In the light of this lemma Figure 4 illustrates the imbalanced Gale diagram of a complete intersection. The arrows represent a *class* of columns of B and not just an individual vector. For instance the class "D" consists of vectors with $B_{1j} < 0$ and $B_{2j} > 0$.

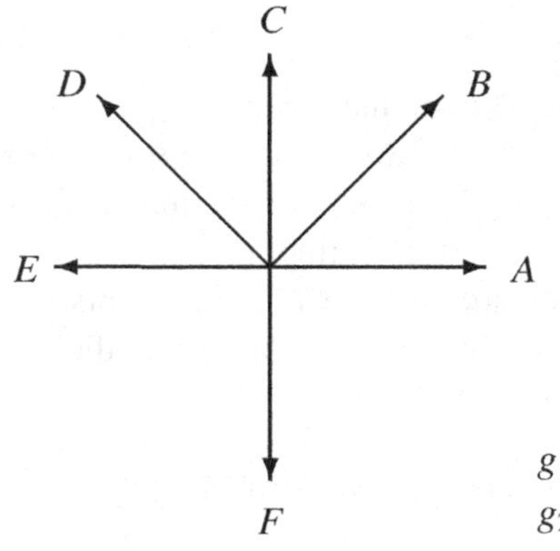

	A	B	C	D	E	F
g_1	+	+	0	−	−	0
g_2	0	+	+	+	0	−

Figure 4. Sign patterns in the (imbalanced) Gale diagram of a complete intersection.

Theorem 6.6. *Let $\mathcal{A}$ be normal such that I_A is a complete intersection of codimension two. Then I_A has a squarefree initial ideal.*

Before we prove this theorem we will mention a result we need.

Proposition 6.7. *The set $\mathcal{A}$ (and hence the associated matrix A) is normal if and only if for each vector $u \in \ker(A)$ there exists an integer vector $v \in \ker(A)$ such that $v \leq \lceil u \rceil$.*

Proof of Theorem 6.6. By Proposition 6.4 both generators $g_1 = x^p - x^q$ and $g_2 = x^r - x^s$ have a squarefree term. Note that the class of vectors F must exist in the Gale diagram since otherwise $C(\mathcal{A})$ would not be pointed. If the term x^s of g_2 corresponding to F is squarefree then we can use a term order so that x^s is the initial term of g_2 and the squarefree term, say x^p, of g_1 is the initial term of this binomial. Since x^p and x^s are relatively prime the two binomials form a reduced Gröbner basis of $I_{\mathcal{A}}$. So we assume that x^s is not squarefree. This means there is $i \in F$ with $(g_2)_i = -c$ where $c \geq 2$, and consequently the positive components of g_2 are all one. Without loss of generality we assume also the squarefree term of g_1 to be the A and B components. We choose representatives from the D and E classes, labeling them $-a$ and $-b$ where $a, b \geq 1$.

$$\begin{array}{rcccccccccccc} & & A & & B & & C & & D & & E & & F \\ g_1 & = & 1 & \ldots & 1 & \ldots & 0 & \ldots & -a & \ldots & -b & \ldots & 0 \\ g_2 & = & 0 & \ldots & 1 & \ldots & 1 & \ldots & 1 & \ldots & 0 & \ldots & -c \end{array}$$

Now we consider $u = -\frac{1}{2}(p-q) + \frac{1}{2}(r-s)$ and hence

$$u = \left(-\frac{1}{2}, \ldots, 0, \ldots, \frac{1}{2}, \ldots, \frac{a}{2} + \frac{1}{2}, \ldots, \frac{b}{2}, \ldots, -\frac{c}{2}\right) \in \ker(A).$$

Note that $-c/2 \leq -1$. By Proposition 6.7 there exists an integral $v \in \ker(A)$ with $v \leq \lceil u \rceil$.

$$\begin{array}{rcccccccccccc} & & A & \ldots & B & \ldots & C & \ldots & D & \ldots & E & \ldots & F \\ g_1 & = & 1 & \ldots & 1 & \ldots & 0 & \ldots & -a & \ldots & -b & \ldots & 0 \\ g_2 & = & 0 & \ldots & 1 & \ldots & 1 & \ldots & 1 & \ldots & 0 & \ldots & -c \\ \lceil u \rceil & = & 0 & \ldots & 0 & \ldots & 1 & \ldots & \lceil \frac{a+1}{2} \rceil & \ldots & \lceil \frac{b}{2} \rceil & \ldots & \lceil -\frac{c}{2} \rceil \end{array}$$

The vector v is an integral combination $v = \alpha(p-q) + \beta(r-s)$. If $v_1 = 0$, then v must be a multiple of $(r-s)$. Furthermore v must be a *positive* multiple of $(r-s)$ to ensure that $v_F = -c \leq \lceil -c/2 \rceil = \lceil u \rceil_F$. This implies the contradiction $v_B = \beta > 0$.

Next suppose that $v_1 \leq -1$. This implies that $\alpha \leq -1$ (considering the A components) and $\beta \geq 0$ (considering the F components). The D components require $a + 1 \leq \lceil \frac{a+1}{2} \rceil$. But this implies that $a = 0$, a contradiction. This brings us to the case where x^s is squarefree. □

6.2 *Total unimodularity and unimodularity*

A large class of toric ideals that are guaranteed to have squarefree initial ideals are given by *(totally) unimodular* $\mathcal{A}$. These kinds of toric ideals provide the best possible such examples since *every* initial ideal is squarefree. We first give the definitions and then present two examples: one classical and one brand new.

Definition 6.8. A $0/\pm 1$-matrix A is called totally unimodular if eery square submatrix of A has determinant $0/\pm 1$. Similarly, a $d \times n$ integer matrix A of rank d is called unimodular if all $d \times d$ submatrices hace determinant $0/\pm 1$.

Proposition 6.9. *If $\mathcal{A}$ is totally unimodular then all initial ideals of I_A are squarefree. Moreover, $\mathcal{A}$ is unimodular if and only if all initial ideals are squarefree.*

Proof. Except the if-direction of the second statement the proof is clear. For this, we observe that every $\sigma \subset [n]$ of cardinality d such that $|\det(A_\sigma)| > 0$ appears as the facet of some regular triangulation Δ_ω. Hence, if all regular triangulations of $\mathcal{A}$ are unimodular then $\mathcal{A}$ is itself unimodular. □

Corollary 6.10. If $\mathcal{A}$ is unimodular then the Graver basis of $\mathcal{A}$ is equal to the set of circuits of $\mathcal{A}$.

Proof. By definition, the set of circuits of $\mathcal{A}$ are a subset of the Graver basis. Now the proof of Lemma 4.7 implies that if h is Graver basis element then $h_i = 0/\pm 1$ for all i. Since h is supposed to be a Hilbert basis element of some C_ρ, and since circuits are the primitive generators of such cones we conclude that h has to be a circuit. □

Segre Products: Combinatorial optimization literature is rich with examples of (totally) unimodular matrices. We refer the reader to the Chapters 19–21 of [72] for a thorough treatment. We use the following little gem.

Theorem 6.11 ([72, §19.3]). *Let G be an undirected graph with vertex set V and edge set E, and let A be the $|V| \times |E|$ incidence matrix of G (i.e. A is the $0/1$-matrix with rows and columns indexed by the vertices and edges of G, respectively, where $A_{v,e} = 1$ if and only if v is an endpoint of e). Then A is totally unimodular if and only if G is bipartite.*

Corollary 6.12. All initial ideals of the defining ideal of the Segre product $\mathbb{P}^{n-1} \times \mathbb{P}^{m-1}$ are squarefree.

Proof. The Segre embedding is defined by the monomial map $\psi(x_{ij}) = s_i t_j$ for $i = 1, \ldots, n$ and $j = 1, \ldots, m$. Hence it is a toric variety defined by I_A where A is the incidence matrix of the complete bipartite graph $K_{n,m}$. Now Theorem 6.11 implies the result. □

Cyclotomic Polytopes: This section is taken from a recent paper [1] and we refer to this paper for the proofs. Let ζ_m be a primitive mth root of unity. The ring of integers in the cyclotomic field of order m, $\mathbb{Z}[\zeta_m]$, is a lattice of full rank in $\mathbb{Z}[\zeta_m] \otimes_{\mathbb{Z}} \mathbb{R} \cong \mathbb{R}^{\phi(m)}$ and hence isomorphic to $\mathbb{Z}^{\phi(m)}$. The set of all mth roots of unity generates $\mathbb{Z}[\zeta_m]$ as a monoid.

We define the mth *cyclotomic polytope* $\mathcal{C}_m$ as the convex hull of the m roots of unity in $\mathbb{R}^{\phi(m)}$. When $m = p$ is a prime number we consider the $\mathbb{Z}$-basis $1, \zeta_p, \zeta_p^2, \ldots, \zeta_p^{p-2}$ of the lattice $\mathbb{Z}[\zeta_p]$. Since $\zeta_p^{p-1} = -\sum_{i=0}^{p-2} \zeta_p^i$, these p elements form a monoid basis for $\mathbb{Z}[\zeta_p]$. We identify them with $e_0, e_1, \ldots, e_{p-2}, -\sum_{i=0}^{p-2} e_i$ in $\mathbb{Z}^{p-1}$.

Proposition 6.13. *The cyclotomic polytope $\mathcal{C}_p \subset \mathbb{R}^{p-1}$ is the simplex*

$$\mathcal{C}_p = \operatorname{conv}\left(e_0,\ e_1, \ldots, e_{p-2},\ -\sum_{i=0}^{p-2} e_i\right).$$

The only interior point of $\mathcal{C}_p$ is the origin.

In order to describe $\mathcal{C}_m$ for general m we need two operations. The first one is the *direct sum*. Let $P \subset \mathbb{R}^{d_1}$ and $Q \subset \mathbb{R}^{d_2}$ two polytopes which contain the origin in their interior. Then we define

$$P \circ Q := \operatorname{conv}(P \times \mathbf{0}_{d_2}, \mathbf{0}_{d_1} \times Q) \subset \mathbb{R}^{d_1+d_2}.$$

Here $\mathbf{0}_d$ denotes the zero vector in $\mathbb{R}^d$. The polytope $P \circ Q$ contains $\mathbf{0}_{d_1+d_2}$ in its interior and its dimension is the sum of the dimensions of P and Q. We denote the k-fold direct sum $P \circ \cdots \circ P$ by $P^{\circ k}$.

The second polytope operation we need is the *tensor product*. Let $P \subset \mathbb{R}^{d_1}$ and $Q \subset \mathbb{R}^{d_2}$ be two polytopes with vertices $v_1, \ldots, v_s$ and $w_1, \ldots, w_t$, respectively. Then $P \otimes Q \subset \mathbb{R}^{d_1 d_2}$ is the polytope

$$P \otimes Q := \operatorname{conv}(v_i \otimes w_j \ :\ 1 \leq i \leq s,\ \ 1 \leq j \leq t).$$

Theorem 6.14. *Let $m = p_1^{\alpha_1} \cdots p_t^{\alpha_t}$ be a positive integer with its unique prime decomposition, and let $\sqrt{m} = p_1 \cdots p_t$ be its squarefree part. Then the mth cyclotomic polytope is*

$$\mathcal{C}_m = \left(\bigotimes_{i=1}^{t} \mathcal{C}_{p_i}\right)^{\circ(m/\sqrt{m})}.$$

It is a $0/\pm 1$-polytope with only one lattice point other than its vertices. This lattice point is the origin and it is in the interior of $\mathcal{C}_m$.

All of this culminates in the following result which is used to describe the form of the *growth series* of *cyclotomic lattices*.

Theorem 6.15. *Suppose the positive integer m is one of the following:*

(1) *$m = p^\alpha$ where p is prime,*
(2) *$m = p^\alpha q^\beta$ where p and q are distinct primes, or*
(3) *$m = 2^\alpha p^\beta q^\gamma$ where p and q are distinct odd primes.*

Let $\mathcal{A}_m$ be the set of the vertices of the cyclotomic polytope $\mathcal{C}_m$ in these cases. Then $\mathcal{A}_m$ is totally unimodular.

Sketch of Proof. It is clear that $\mathcal{A}_p$ is totally unimodular and the direct sum operation preserves total unimodularity. It requires a bit of work to show that $\mathcal{A}_p \otimes \mathcal{A}_q$ is totaly unimodular for two distinct primes. Once this is established the last case also follows easily because if a matrix A is totally unimodular so is $[A\ -A]$. □

7. Quadratic initial ideals

After the squarefree initial ideals the "simplest" initial ideals are the ones that are generated in lowest (nontrivial) degree, namely in degree two. If such an initial ideal is also squarefree, i.e., it is the Stanley-Reisner ideal of a simplicial complex Δ, then the simplicial complex Δ is a *flag* complex. We will give examples of such quadratic initial ideals.

7.1 *Normal toric surfaces, lattice polygons, and multiples of polytopes*

A normal toric surface corresponds to the dimension two case we studied in Section 6.1. There we showed that there exists a squarefree initial ideal corresponding to the unimodular triangulation consisting of the cones $\{a_1, a_2\}, \ldots, \{a_{n-1}, a_n\}$ where $\mathcal{A} = \{a_1, \ldots, a_n\} \subset \mathbb{Z}^2$ is ordered clockwise. The corresponding initial ideal $\mathsf{in}_{\succ}(I_{\mathcal{A}})$ is given by the lexicographic term order $x_1 \succ \cdots \succ x_n$. A Gröbner basis consists of the binomials $x_i x_j - x_k^{j_k} x_{k+1}^{j_{k+1}}$ where $1 \leq i < k < j \leq n$.

When the polytope $P \subset \mathbb{R}^2$ is a *lattice polygon* with integral vertices, the defining toric ideal $I_{\mathcal{A}}$ of the semigroup $S(P)$ has nice initial ideals. Whether one gets quadratic initial ideals depends on the number of lattice points on the boundary of P.

Theorem 7.1 ([8, Corollary 3.2.5]). *Let P be a lattice polygon containing at least four lattice points. Then the corresponding toric ideal has a quadratic squarefree initial ideal if and only if P has at least four lattice points on its boundary.*

One of the strongest and most general results about the existence of quadratic initial ideals is about the toric ideals of multiples of polytopes. For $c > 0$ and the polytope P we define $cP = \{cx : x \in P\}$. A result of Bruns, Gubeladze and Trung [8, Theorem 1.3.3] together with a more general theorem of Eisenbud, Reeves, and Totaro [25] shows that if we blow up the polytope P by a big enough factor c then the toric ideal $I_{\mathcal{A}}$ corresponding to cP has a quadratic initial ideal.

Theorem 7.2. *Let P be a d-dimensional lattice polytope in $\mathbb{R}^d$. Then for $c \geq d$ the toric ideal $I_{\mathcal{A}}$ corresponding to the affine monoid $S(cP)$ has a quadratic initial ideal.*

We note that it is not known whether the initial ideal in Theorem 7.2 is squarefree.

7.2 *Toric ideals of distributive lattices*

Let L be a finite distributive lattice. The Birkhoff theorem for distributive lattices assert that L is isomorphic to $J(P)$, the lattice of all order ideals of some poset P [74]. If $|L| = n$ we construct a configuration $\mathcal{A}_L$ as follows: for each $x \in L$ corresponding to the order ideal J_x of P we let a_x to be the characteristic vector of J_x, i.e. $(a_x)_i = 1$ if $i \in J_x$ and $(a_x)_i = 0$ otherwise. We let $I_{\mathcal{A}_L}$ be the toric ideal of $\mathcal{A} = \{(1, a_x) : x \in L\}$. The following theorem is proved in [40].

Theorem 7.3. *Let $\prec$ be a reverse lexicographic ordering of the variables $x \in L$ such that $x \prec y$ whenever $x <_L y$. Then the binomials*

$$\{xy - (x \wedge y)(x \vee y) \ : x \text{ and } y \text{ are incomparable in } L\}$$

is a Gröbner basis of I_{A_L} with respect to $\prec$. In particular, this toric ideal has a squarefree quadratic initial ideal.

We illustrate this theorem with an example. We let P be the poset of two disjoint chains: $1 < 2 < 3$ and $1' < 2'$. The lattice $L = J(P)$ consists of the order ideals

x_{00}	$x_{01'}$	$x_{02'}$	x_{10}	$x_{11'}$	$x_{12'}$
$\emptyset$	$\{1'\}$	$\{1', 2'\}$	$\{1\}$	$\{1, 1'\}$	$\{1, 1', 2'\}$
x_{20}	$x_{21'}$	$x_{22'}$	x_{30}	$x_{31'}$	$x_{32'}$
$\{1, 2\}$	$\{1, 2, 1'\}$	$\{1, 2, 1', 2'\}$	$\{1, 2, 3\}$	$\{1, 2, 3, 1'\}$	$\{1, 2, 3, 1', 2'\}$

with their corresponding labels. So the toric ideal I_{A_L} is given by the matrix

$$A_L = \begin{pmatrix} 1&1&1&1&1&1&1&1&1&1&1&1 \\ 0&0&0&1&1&1&1&1&1&1&1&1 \\ 0&0&0&0&0&0&1&1&1&1&1&1 \\ 0&0&0&0&0&0&0&0&0&1&1&1 \\ 0&1&1&0&1&1&0&1&1&0&1&1 \\ 0&0&1&0&0&1&0&0&1&0&0&1 \end{pmatrix}$$

where the columns are labeled by $x_{00}, \ldots, x_{32'}$. The columns of this matrix is affinely isomorphic to the columns of

$$B_L = \begin{pmatrix} 1&1&1&0&0&0&0&0&0&0&0&0 \\ 0&0&0&1&1&1&0&0&0&0&0&0 \\ 0&0&0&0&0&0&1&1&1&0&0&0 \\ 0&0&0&0&0&0&0&0&0&1&1&1 \\ 1&0&0&1&0&0&1&0&0&1&0&0 \\ 0&1&0&0&1&0&0&1&0&0&1&0 \\ 0&0&1&0&0&1&0&0&1&0&0&1 \end{pmatrix}$$

whose columns are labeled as A_L and whose rows are labeled as $0, 1, 2, 3$ and $0', 1', 2'$. The matrix A_L is obtained from B_L by adding each row $i = 1, 2, 3$ of B_L to all the earlier rows, and each row $i' = 1', 2'$ also to the earlier rows, and then deleting the row $0'$. The second configuration defines the ideal of the Segre product $\mathbb{P}^3 \times \mathbb{P}^2$. This ideal is generated by the 2×2 minors of a 4×3 matrix

$$\begin{pmatrix} x_{00} & x_{01'} & x_{02'} \\ x_{10} & x_{11'} & x_{12'} \\ x_{20} & x_{21'} & x_{22'} \\ x_{30} & x_{31'} & x_{32'} \end{pmatrix}.$$

With respect to the reverse lexicographic term order $x_{00} \prec \cdots \prec x_{32'}$ induced by the ordering of the columns of both A_L and B_L these 2×2 minors form a Gröbner basis where the anti-diagonal terms of each minor is the initial term. This is precisely the initial ideal constructed in Theorem 7.3.

7.3 *Quadratic initial ideals for root systems*

We let Γ to be one of the classical root systems $\mathbf{A}_{n-1}$, $\mathbf{B}_n$, $\mathbf{C}_n$, or $\mathbf{D}_n$, and we let Γ^+ be the configuration obtained from the origin and the positive roots of Γ:

$$\begin{aligned}
\mathbf{A}_{n-1}^+ &= \{0\} \cup \{e_i - e_j \,:\, 1 \le i < j \le n\},\\
\mathbf{B}_n^+ &= \mathbf{A}_{n-1}^+ \cup \{e_1, \ldots, e_n\} \cup \{e_i + e_j \,:\, 1 \le i < j \le n\},\\
\mathbf{C}_n^+ &= \mathbf{A}_{n-1}^+ \cup \{2e_1, \ldots, 2e_n\} \cup \{e_i + e_j \,:\, 1 \le i < j \le n\},\\
\mathbf{D}_n^+ &= \mathbf{A}_{n-1}^+ \cup \{e_i + e_j \,:\, 1 \le i < j \le n\}.
\end{aligned}$$

For each Γ^+ we get the toric ideal I_{Γ^+} corresponding to the configutarion $\{1\} \times \Gamma^+$. These toric ideals have been extensively studied in [61] where it is shown that the toric ideal in all of these four families have quadratic squarefree initial ideals. To illustrate this we give the example for the case $\mathbf{B}_n^+$. For this and the other positive root systems the proofs can be found in [61].

We use a variable a for $\{0\}$, the variable x_{ij} for $e_i - e_j$, the variable y_i for e_i, and the variable z_{ij} for $e_i + e_j$. We use a reverse lexicographic order where

$$\begin{aligned}
\{y_1 \prec \cdots \prec y_n\} \prec \{&x_{ij} \prec x_{kl} : i < k, \text{ or } i = k \text{ and } j > l\}\\
&\prec \{z_{ij} \prec z_{kl} : i < k, \text{ or } i = k \text{ and } j > l\} \prec a.
\end{aligned}$$

We use the convention that $z_{ji} = z_{ij}$ whenever $i < j$.

Theorem 7.4. *When we use the above reverse lexicographic term order the following binomials form a Gröbner basis of $I_{\mathbf{B}_n^+}$.*

$$\begin{aligned}
z_{i,j}z_{k,\ell} - z_{i,\ell}z_{j,k}, &\quad i < j < k < \ell,\\
z_{i,k}z_{j,\ell} - z_{i,\ell}z_{j,k}, &\quad i < j < k < \ell,\\
x_{i,k}x_{j,\ell} - x_{i,\ell}x_{j,k}, &\quad i < j < k < \ell,\\
x_{i,j}x_{j,k} - ax_{i,k}, &\quad i < j < k,\\
z_{i,j}x_{k,\ell} - x_{i,\ell}z_{j,k}, &\quad i < j < k < \ell,\\
z_{i,k}x_{j,\ell} - x_{i,\ell}z_{j,k}, &\quad i < j < k < \ell,\\
z_{i,\ell}x_{j,k} - x_{i,k}z_{j,\ell}, &\quad i < j < k < \ell,\\
x_{i,j}z_{j,k} - y_iy_k, &\quad i < j, \ \ j \neq k,\\
z_{i,j}y_k - y_iz_{j,k}, &\quad i < j < k,\\
z_{i,k}y_j - y_iz_{j,k}, &\quad i < j < k,\\
x_{i,k}y_j - y_ix_{j,k}, &\quad i < j < k,\\
x_{i,j}y_j - y_ia, &\quad i < j,\\
az_{i,j} - y_iy_j, &\quad i < j.
\end{aligned}$$

7.4 *Toric ideals of matroids*

Let M be a matroid of rank d on the ground set $[n] = \{1, \dots, n\}$. For each basis (= an independent set of d elements) B of M we let $a_B \in \{0, 1\}^n$ be the characteristic vector of B, i.e. $(a_B)_i = 1$ if $i \in B$ and $(a_B)_i = 0$ otherwise. We let $\mathcal{A}_M$ be the corresponding vector configuration and I_M be the toric ideal defined by $\mathcal{A}_M$. Given two bases B_1 and B_2, for every element $b_1 \in B_1$ there exists $b_2 \in B_2$ such that $B_1' = B_1 - b_1 \cup b_2$ and $B_2' = B_2 - b_2 \cup b_1$ are bases. This is known as the *symmetric exchange property*. If B_1' and B_2' are obtained from B_1 and B_2 in this way we say that the first pair is obtained from the second one by a *double swap*. The following conjecture is the algebraic version of Neil White's conjecture in [89] which was first translated into Gröbner basis language in [78].

Conjecture 7.5. The toric ideal I_M is generated by the quadratic binomials $x_{B_1}x_{B_2} - x_{B_1'}x_{B_2'}$ where B_1' and B_2' are obtained from B_1 and B_2 by a double swap. More strongly, these binomials form a Gröbner basis.

This conjecture is still wide open, and until only very recently there was just one general result known concerning Conjecture 7.5. This result of Sturmfels [77, Chapter 14] concerns *uniform matroids*. The uniform matroid of rank d on $[n]$ is the matroid where all $\binom{n}{d}$ subsets are bases.

Theorem 7.6. *If M is a uniform matroid the toric ideal I_M has a squarefree quadratic initial ideal.*

The proof of this Theorem is ethereal: although an explicit description of a Gröbner basis is given, the term order that makes this set into a Gröbner basis is not explicit. All one knows is that such a term order must exist!

The newest result has been obtained in Summer 2005 by Jonah Blasiak [4] on *graphic matroids*: Let G be a graph. Then the graphic matroid $M(G)$ is the matroid whose ground set is the edges of G and whose bases are the spanning trees of G.

Theorem 7.7. *The toric ideal of a graphic matroid $M(G)$ is generated by quadratic binomials.*

The proof of Theorem 7.7 is a wonderful (but involved) combinatorial argument. We refer the reader to [4] for the proof of this theorem and for pointers to other related results.

8. A hierarchy of normal toric ideals

Throughout this section we will assume that $\mathcal{A}$ is normal. The following theorem is due Hochster [42] and forms one of the hallmarks of toric algebra.

Theorem 8.1. *If I_A is a normal toric ideal then it is Cohen-Macaulay.*

In this section we are motivated by the following question: given a normal toric ideal I_A does there exists a generic weight vector ω such that $\mathrm{in}_\omega(I_A)$ is also Cohen-Macaulay? The previous section provides us a rich class of cases where the answer is "yes".

Proposition 8.2. *If I_A has a squarefree initial ideal $\mathsf{in}_\omega(I_A)$ then this initial ideal is Cohen-Macaulay.*

Proof. A quick-and-dirty proof depends on the following fact (see [10, Chapter 5] for definition and proofs): If a simplicial complex Δ is *shellable* then its Stanley-Reisner ideal I_Δ is Cohen-Macaulay. Now if $\mathsf{in}_\omega(I_A)$ is squarefree Theorem 5.7 implies that it is equal to the Stanley-Reisner ideal I_{Δ_ω}. The triangulation Δ_ω viewed as a simplicial complex is equal to the lower hull of the boundary of a simplicial polytope by its construction. One can use a *line shelling* of this simplicial polytope to show that indeed Δ_ω is shellable. This shows that the squarefree initial ideal $\mathsf{in}_\omega(I_A)$ is Cohen-Macaulay. □

It is a standard commutative algebra fact (see for instance [23]) that a Cohen-Macaulay ideal I is *unmixed*: it has only minimal primes that are equidimensional with dimension equal to $\dim(I)$. Hence a refinement of the main question we are asking here is the following: if I_A is normal, is there an initial ideal $\mathsf{in}_\omega(I_A)$ without embedded primes (the equidimensionality of the minimal primes is mentioned after Theorem 5.7 below). This brings us to a pair of interrelated concepts.

8.1 Δ_ω-*normality and supernormality*

Definition 8.3. The set of vectors $\mathcal{A}$ (and the corresponding $d \times n$ integer matrix A) is Δ_ω-normal if $C(\mathcal{A})$ has a regular triangulation Δ_ω such that for *every* maximal face $\sigma \in \Delta_\omega$, the vectors of $\mathcal{A}$ in $C(\mathcal{A}_\sigma)$ form a Hilbert basis for $C(\mathcal{A}_\sigma)$.

If $\mathcal{A}$ is Δ_ω-normal for some triangulation then it is normal: Note that every lattice point in $C(\mathcal{A})$ lies in $C(\mathcal{A}_\sigma)$ for some maximal face $\sigma \in \Delta_\omega$. Since $\mathcal{A}$ is Δ_ω-normal, this lattice point also lies in the monoid generated by the columns of A in $C(\mathcal{A}_\sigma)$ and hence in $\mathbb{N}\mathcal{A}$.

It is clear that $\mathcal{A}$ is Δ_ω-normal with respect to all the unimodular triangulations of $\mathcal{A}$. Hence triangulations Δ_ω with respect to which $\mathcal{A}$ is Δ_ω-normal generalize unimodular triangulations of $\mathcal{A}$.

Example 8.4. We have already mentioned examples of normal $\mathcal{A}$ without unimodular triangulations [6], [28]. So it is also easy to construct examples that do not have unimodular triangulations but which are Δ_ω-normal. For instance, consider the following example taken from [28]:

$$A = \begin{pmatrix} 1 & 0 & 0 & 1 & 1 & 1 & 1 & 1 \\ 0 & 1 & 0 & 1 & 1 & 2 & 2 & 2 \\ 0 & 0 & 1 & 1 & 2 & 2 & 3 & 3 \\ 0 & 0 & 0 & 1 & 2 & 3 & 4 & 5 \end{pmatrix}.$$

Here $C(\mathcal{A})$ has 77 regular triangulations none of which are unimodular. Since $C(\mathcal{A})$ is the simplicial cone generated by a_1, a_2, a_3, a_8, $\mathcal{A}$ is Δ_ω-normal with respect to its coarsest regular triangulation.

Example 8.5. There are normal $\mathcal{A}$ that are not Δ_ω-normal with respect to any triangulation of $C(\mathcal{A})$. To see such an example, consider the following modification of the matrix in Example 8.4 that appears in [77, Example 13.17]:

$$A = \begin{pmatrix} 0 & 1 & 0 & 0 & 1 & 1 & 1 & 1 & 1 \\ 0 & 0 & 1 & 0 & 1 & 1 & 2 & 2 & 2 \\ 0 & 0 & 0 & 1 & 1 & 2 & 2 & 3 & 3 \\ 0 & 0 & 0 & 0 & 1 & 2 & 3 & 4 & 5 \\ 1 & 1 & 1 & 1 & 1 & 1 & 1 & 1 & 1 \end{pmatrix}.$$

This matrix is again normal and each of its nine columns generate an extreme ray of $C(\mathcal{A})$. Hence the only way for this matrix to be Δ_ω-normal would be if Δ_ω is a unimodular triangulation of $C(\mathcal{A})$. However, $\mathcal{A}$ has no unimodular triangulations.

Definition 8.6. The set $\mathcal{A}$ is *supernormal* if for every subset $\mathcal{A}'$ of $\mathcal{A}$, the vectors of $\mathcal{A}$ that lie in $C(\mathcal{A}')$ form a Hilbert basis for $C(\mathcal{A}')$.

Proposition 8.7. *The following are statement are equivalent:*

(i) *$\mathcal{A}$ is supernormal,*
(ii) *$\mathcal{A}$ is Δ_ω-normal for every regular triangulation Δ_ω of $\mathcal{A}$,*
(iii) *Every triangulation of $C(\mathcal{A})$ in which every element of $\mathcal{A}$ generates a one-dimensional face, is unimodular.*

Proof. The equivalence of (i) and (iii) was established in [44, Proposition 3.1]. Definition 8.6 shows that (i) $\Rightarrow$ (ii). Hence we just need to show that (ii) $\Rightarrow$ (i). Suppose that $\mathcal{A}$ is Δ-normal for every regular triangulation of $\mathcal{A}$. In order to show that $\mathcal{A}$ is supernormal we only need to check submatrices $\mathcal{A}'$ where the dimension of $C(\mathcal{A}')$ is d. It is easy to get a polyhedral subdivision of $C(\mathcal{A})$ in which $C(\mathcal{A}')$ is a maximal face. There are also standard procedures that will refine this subdivision to a regular triangulation Δ_ω of $cone(A)$. Let T be the set of maximal faces σ of Δ_ω such that $C(\mathcal{A}_\sigma)$ lies in $C(\mathcal{A}')$. Since $\mathcal{A}$ is Δ_ω-normal, the elements of $\mathcal{A}$ that lie in $C(\mathcal{A}_\sigma)$ form a Hilbert basis for $C(\mathcal{A}_\sigma)$ for each $\sigma \in T$. However, since their union is the set of vectors of $\mathcal{A}$ that are in $C(\mathcal{A}')$, this union forms a Hilbert basis for $C(\mathcal{A}')$. □

Remark 8.8. For small values of the dimension d it is not difficult to classify supernormality. In dimension one, $\mathcal{A}$ is normal if and only if it is supernormal. If $d = 2$, by our discussion in Section 6.1 $\mathcal{A}$ normal implies $\mathcal{A}$ supernormal since it is Δ-normal for every triangulation Δ — the Hilbert basis of a maximal subcone of Δ is precisely the set of vectors of $\mathcal{A}$ in that subcone. If $d = 3$ then as mentioned before, $\mathcal{A}$ has a unimodular triangulation with respect to which $\mathcal{A}$ is Δ_ω-normal. However, not every such $\mathcal{A}$ needs to be supernormal: see Example 5 in [51]. Or discussion of the case $d = 3$ in Section 6.1 when $\mathcal{A}$ comes from the lattice points of a lattice polygon in $\mathbb{R}^2$ show that every triangulation of $\mathcal{A}$ that uses all the vectors in $\mathcal{A}$ is unimodular. Hence, by Proposition 8.7, $\mathcal{A}$ is supernormal, and therefore Δ_ω-normal for any triangulation of $\mathcal{A}$.

We finish this subsection by pointing out the hierarchy of normal $\mathcal{A}$ that we have established so far in Figure 5. In this figure Δ-unimodular means that I_A has a unimodular regular triangulation. We will explain the class of *Gomory* normal toric ideals and

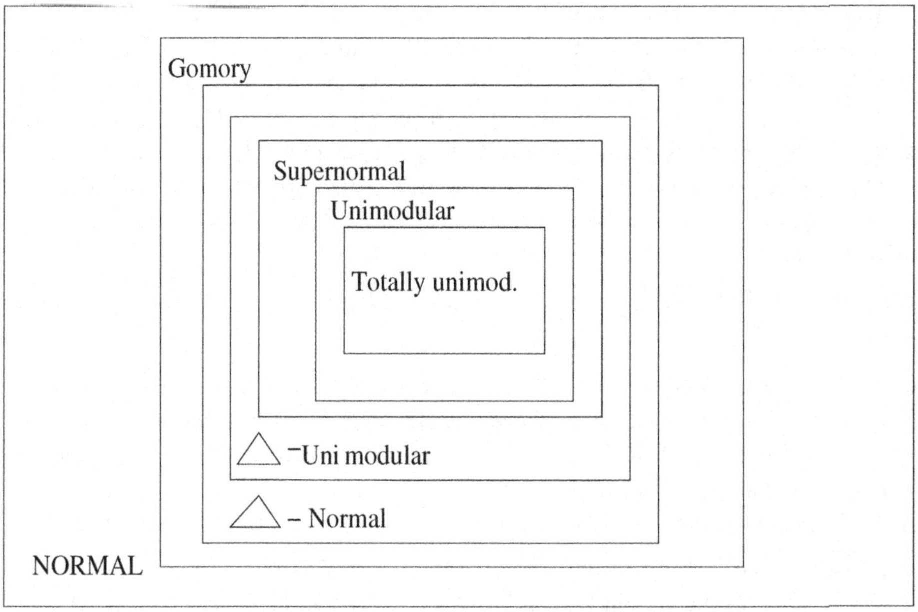

Figure 5. The hierarchy of normal $\mathcal{A}$

its connection to Δ_ω-normality in the next subsection. Note that we have seen that all the inclusions are strict.

8.2 *Standard pair decompositions*

In this section we will study a certain decomposition of the standard monomials of a general monomial ideal. This was first introduced in [81]. After some general facts we will concentrate on the case of toric initial ideals. This study was initiated in [77, §12.D] and then was extended and used in [49], [50], and [51]. We start with definitions.

Definition 8.9. For a monomial x^m and $\sigma \subseteq [n]$, we say that (x^m, σ) is an admissible pair of a monomial ideal M if (i) $\text{supp}(m) \cap \sigma = \emptyset$ and (ii) every monomial in $x^m \cdot k[x_j : j \in \sigma]$ is a standard monomial of M.

We partially order admissible pairs of M by $(x^m, \sigma) \leq (x^{m'}, \sigma')$ if and only if x^m divides $x^{m'}$ and $supp(x^{m'}/x^m) \cup \sigma' \subseteq \sigma$.

Definition 8.10. An admissible pair (x^m, σ) of M is called a standard pair of M if it is a minimal element in the poset of all admissible pairs with respect to the above partial order.

The standard pairs of M induce a unique covering of the set of standard monomials of M which we refer to as the standard pair decomposition of M. There are two different algorithms for computing the standard pair decomposition of a monomial ideal M; see [49] and [70, Section 3.2]. For implementations of both algorithms in `Macaulay 2` see [46].

It was shown in [81] that the *arithmetic degree* of M, denoted *arithdeg(M)*, equals the number of standard pairs of M. Algebraically, arithdeg $(M) = \sum$ mult (P) deg(P) where the sum is over all homogeneous prime ideals P, and mult (P) denotes the length of the largest ideal of finite length in the ring $k[x]_P/Mk[x]_P$, and deg(P) is the geometric degree of the variety of P. The integer mult (P) is positive if and only if P is an associated prime of M. In our situation, since M is a monomial ideal, every associated prime of M has the form $P_\sigma := \langle x_j : j \notin \sigma \rangle$ for some $\sigma \subseteq [n]$ and $\deg(p_\sigma) = 1$.

Theorem 8.11 ([81]). *Given a monomial ideal M and $P_\sigma = \langle x_j : j \notin \sigma \rangle$ for some $\sigma \subseteq [n]$:*

(i) *P_σ is an associated prime of M if and only if M has a standard pair of the form $(\cdot, \sigma)$. If P_σ is an associated prime of M then mult (P_σ) equals the number of standard pairs of M of the form $(\cdot, \sigma)$. Hence, arithdeg $(M) = \sum_{\sigma \subseteq [n]}$ mult (P_σ) is the number of standard pairs of M.*
(ii) *P_σ is a minimal prime of M if and only if $(1, \sigma)$ is a standard pair of M.*

Now we return to toric initial ideals. Recall that π is the map $\pi : \mathbb{N}^n \longrightarrow \mathbb{N}\mathcal{A}$ defined by $\pi(x) = Ax$. Given $b \in \mathbb{N}\mathcal{A}$, the set $\pi^{-1}(b) = \{x \in \mathbb{N}^n : Ax = b\}$ is called the b-fiber (of π, or of A). The vectors that belong to the b-fiber are called *feasible solutions* of (1). Given such a fiber and $v \in \pi^{-1}(b)$ we can identify this fiber with the lattice points in the polytope

$$P_v := \{u \in \mathbb{R}^m : Bu \leq v\}, \tag{6}$$

via the map $P_v \cap \mathbb{Z}^m \to \mathbb{N}^n$ such that $u \to v - Bu$. Here B is a $n \times m$ matrix (where $m = n - d$) whose columns form a basis of the lattice $\ker_{\mathbb{Z}}(A)$. In particular, $0 \in P_v$ maps to $v \in \pi^{-1}(b)$. If v and v' are both in the same finer then P_v and $P_{v'}$ are lattice translates of each other and hence we may represent this fiber by any of the polytopes P_v where $v \in \pi^{-1}(b)$.

Proposition 8.12. *Let v be an element of a fiber $C = \pi^{-1}(b)$. Then the monomial $x^{(v-Bu^*)}$ is a standard monomial of $\mathrm{in}_\omega(I_A)$ if and only if u^* is the optimal solution to the integer program*

$$\textit{minimize}\{-(\omega B) \cdot u : u \in P_v \cap \mathbb{Z}^m\}. \tag{7}$$

Proof. By Theorem 3.4, the monomial x^{v^*} is a standard monomial of $\mathrm{in}_\omega(I_A)$ if and only if the lattice point $v^* \in C$ minimizes $\omega \cdot x$ over C. This happens
$\Leftrightarrow$ there exists $u^* \in \mathbb{Z}^m$ such that $v^* = v - Bu^* \geq 0$ and $\omega(v - Bu^*) < \omega(v - Bu)$ for all $u \neq u^* \in \mathbb{Z}^m$ with $v - Bu \geq 0$
$\Leftrightarrow$ there exist $u^* \in P_v \cap \mathbb{Z}^m$ such that $-(\omega B) \cdot u^* < -(\omega B) \cdot u$ for all $u \in P_v \cap \mathbb{Z}^m$ and $u \neq u^*$
$\Leftrightarrow$ u^* is the optimal solution of the integer program (7). □

Using the optimal solution u^* to the program (7), we define the following subpolytope of P_v:

$$P_v(u^*) := \{u \in \mathbb{R}^m : Bu \leq v, -(\omega B) \cdot u \leq -(\omega B) \cdot u^*\}. \tag{8}$$

Theorem 8.13. *Let $v \in \mathbb{N}^n$ be a lattice point in a fiber $C = \pi^{-1}(b)$. Then x^{v-Bu^*} is a standard monomial of $\mathrm{in}_\omega(I_A)$ if and only if u^* is the unique lattice point in $P_v(u^*)$. In particular, x^v is a standard monomial if and only if 0 is the unique lattice point in $P_v(0) = \{u \in \mathbb{R}^m : Bu \leq v, -(\omega B) \cdot u \leq 0\}$.*

Proof. Since ω is assumed to be generic, each integer program (7) has a unique optimal solution. By Proposition 8.12, $u^* \in \mathbb{Z}^m$ is the optimal solution to (7) if and only if u^* is in P_v and there is no $u \neq u^*$ in $P_v \cap \mathbb{Z}^m$ such that $-(\omega B) \cdot u \leq -(\omega B) \cdot u^*$. This is equivalent to u^* being the unique lattice point in $P_v(u^*)$. The second statement follows immediately. □

For $\tau \subseteq [n]$, let $P_v^{\bar{\tau}} := \{u \in \mathbb{R}^m : B^{\bar{\tau}} u \leq v^{\bar{\tau}}\}$ be the polyhedron obtained from P_v by removing the inequalities indexed by τ. Here we use $v^{\bar{\tau}}$ to denote the truncation of v obtained by restricting to the coordinates indexed by $\bar{\tau}$ ($v^{\bar{\tau}}$ lies in $\mathbb{N}^{|\bar{\tau}|}$). Let $P_v^{\bar{\tau}}(0) := \{u \in \mathbb{R}^m : (-\omega B) \cdot u \leq 0, B^{\bar{\tau}} u \leq v^{\bar{\tau}}\}$ which is gotten from $P_v^{\bar{\tau}}$ by adding the additional constraint $(-\omega B) \cdot u \leq 0$. For a polyhedron $P = \{x \in \mathbb{R}^p : T_i x \leq t_i, i = 1, \ldots, q\}$ we say that an inequality $T_i x \leq t_i$ is *essential* if the relaxation of the polyhedron obtained by removing $T_i x \leq t_i$ contains a new lattice point.

Theorem 8.14. *An admissible pair (x^v, τ) of $\mathrm{in}_\omega(I_A)$ is a standard pair if and only if 0 is the unique lattice point in $P_v^{\bar{\tau}}(0)$ and all of the inequalities in the system $B^{\bar{\tau}} u \leq v^{\bar{\tau}}$ are essential.*

Proof. An admissible pair (x^v, τ) of $M := \mathrm{in}_\omega(I_A)$ is a standard pair if and only if all monomials in $x^v \cdot k[x_j : j \in \tau]$ are standard monomials, and for every $i \in \bar{\tau}$ there exists $m_i > 0$ such that $x_i^{m_i} \cdot x^v \in M$. By Theorem 8.13, this is the case if and only if for *every* v' with support in τ, the origin is the unique lattice point in $P_{(v+v')}(0) = \{u \in \mathbb{R}^m : B^{\bar{\tau}} u \leq v^{\bar{\tau}}, B^{\tau} u \leq v'^{\tau}, (-\omega B) \cdot u \leq 0\}$ and for each $i \in \bar{\tau}$ there exists $m_i > 0$ such that $P_{(v_i+v')}(0)$ where $v_i = v + (0, \ldots, 0, m_i, 0, \ldots, 0)$ contains an extra lattice point. This is equivalent to saying that $P_v^{\bar{\tau}}(0)$ contains the origin as the unique lattice point, and each inequality in this polytope is essential. □

The following theorem relates the faces of the simplicial complex Δ_ω to the associated primes of $\mathrm{in}_\omega(I_A)$. Its proof can be found in [50].

Theorem 8.15. *The prime ideal P_σ is a minimal prime of $\mathrm{in}_\omega(I_A)$ if and only if σ is a maximal face of Δ_ω. The multiplicity of such a minimal prime is $|\det(B^{\bar{\sigma}})| = |\det(A_\sigma)|$. Moreover, if P_τ is an associated prime of $\mathrm{in}_\omega(I_A)$, then τ is a face of Δ_ω.*

This theorem also shows the well known fact that $\mathrm{in}_\omega(I_A)$ is an equidimensional ideal (all minimal primes have the same dimension d). To illustrate Theorem 8.14 and Theorem 8.15 we give the following example.

Example 8.16. Consider $A = \begin{pmatrix} 1 & 1 & 1 & 1 & 1 \\ 0 & 4 & 5 & 2 & 0 \\ 0 & 1 & 4 & 5 & 2 \end{pmatrix}$ and $\omega = (17, 21, 0, 0, 0)$. We let a, b, c, d, e be the variables indexing the five columns of A. Then $\ker_{\mathbb{Z}}(A)$ is generated by the rows of $B^T = \begin{pmatrix} -8 & 7 & -6 & 1 & 6 \\ -13 & 10 & -8 & 0 & 11 \end{pmatrix}$, and $-(\omega B) = (-11, 11)$. Since the columns of A lie on the hyperplane $x_1 = 1$, we can draw (the regular triangulation) Δ_ω as a triangulation of a pentagon as in Figure 6. The minimal primes of $\mathrm{in}_\omega(I_A)$ are in bijection with the

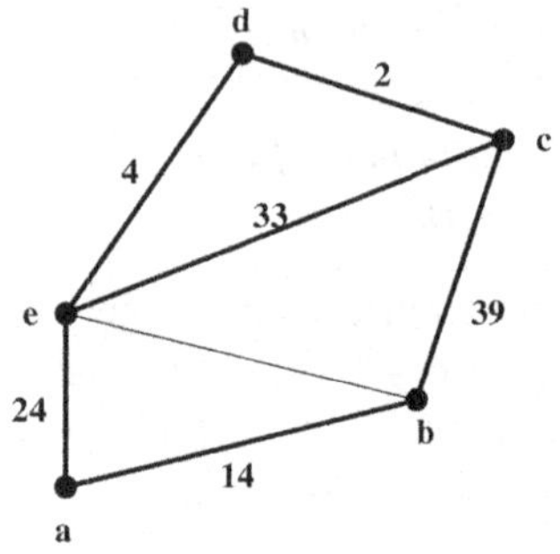

Figure 6. The bold faces of Δ_ω index the embedded primes of $\mathsf{in}_\omega(I_A)$.

three maximal faces of Δ_ω. The simplices $\tau \in \Delta_\omega$ such that P_τ is an embedded prime are the faces of Δ_ω indicated by bold lines. The multiplicity of each embedded prime is given next to the corresponding face. In particular, all embedded primes of $\mathsf{in}_\omega(I_A)$ have dimension two and $P_{\{2,5\}}$ is not an embedded prime.

Consider $\tau = \{3, 5\}$ in Δ_ω. There are a total of 33 standard monomials of the form $(\cdot, \{3, 5\})$, and they appear in two groups: the the first group the variable d has exponent one, and in the second group it has exponent two.

The shaded polytope in Figure 7 is $P_v^{\bar{\tau}}(0)$ for the monomial a^2bd. Each inequality is labeled by a monomial in the variable indexing the inequality. The exponent of this monomial is the component of v in the slot indexed by this variable. In Figure 7, the inequality indexed by d is fixed at right hand side value one. Notice that 0 is the unique lattice point in the shaded polytope and that all the three inequalities (indexed

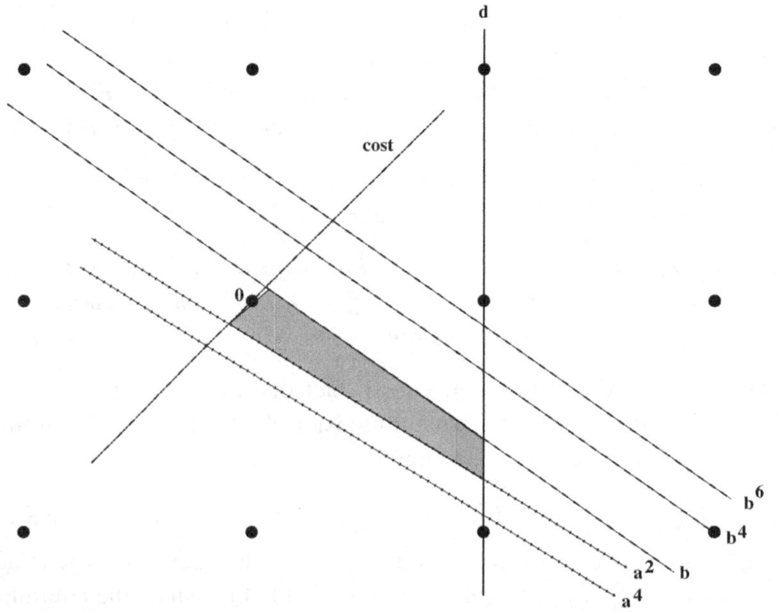

Figure 7. The polytopes $P_v^{\bar{\tau}}(0)$ for the monomials in Group 1.

by a, b, d) are essential for the shaded polytope (if you remove inequality d, the lattice point $(3, -2)$ enters the relaxation).

8.3 *Gomory initial ideals*

An important consequence of Theorem 8.15 is that $\mathrm{in}_\omega(I_A)$ has no embedded primes if and only if the polytopes $P_v^{\bar{\tau}}(0)$ are simplices (in other words, the only standard pairs are of the form $(\cdot, \tau)$ where τ is a maximal face of Δ_ω. This comment brings us back to our original question in this section. We noted earlier that if $\mathrm{in}_\omega(I_A)$ is Cohen-Macaulay then it does not have embedded primes. Since the minimal primes of all initial ideals are equidimensional it is worth considering when $\mathrm{in}_\omega(I_A)$ does not have embedded primes. We will call an initial ideal without embedded primes a *Gomory initial ideal.*

Remark 8.17. The choice of the terminology comes from the theory of integer programming. Ralph Gomory has been an influential figure with many contributions, in particular with his idea of *group relaxations* in integer programming. For these connections see the articles [49] and [51].

The next result illustrates the connection between Δ_ω-normality and Gomory initial ideals.

Theorem 8.18. *If $\mathcal{A}$ is Δ-normal for some regular triangulation Δ then there exists a generic weight vector $\omega \in \mathbb{Z}^n$ such that $\Delta = \Delta_\omega$ and $\mathrm{in}_\omega(I_A)$ is a Gomory initial ideal. Moreover, this initial ideal is Cohen-Macaulay.*

Proof. We only sketch the ideas of the proof. The details can be found in [51] and [63]. Since Δ is a regular triangulation, $\Delta = \Delta_c$ for some $c \in \mathbb{R}^n$. The maximal faces of Δ are the projections of the lower facets of the cone C generated by the lifted vectors (a_i, c_i). In the first step of the proof, we will define a new vector c' such that $\Delta_c = \Delta_{c'}$. We will choose c' so that each (a_i, c_i') (the i-th lifted column of A) will lie on a lower facet of C. In the second step, we will use the fact the $\mathcal{A}$ is Δ-normal, and we will show that if $b \in \mathbb{N}\mathcal{A}$ lies in $C(A_\sigma)$ for a maximal face $\sigma \in \Delta_{c'}$ then there is an optimal solution to the integer program (1) – here we use c' as the weight vector and b as the right hand side – whose support is contained in the set of indices corresponding to the vectors of $\mathcal{A}$ that are in $C(\mathcal{A}_\sigma)$. In fact, by perturbing c' to a generic ω, we will make sure that there is a unique such optimal solution. The last step of the proof will show that this unique optimal solution is covered by a standard pair of the form $(\cdot, \sigma)$. We will conclude that only maximal faces of Δ are associated to $\mathrm{in}_\omega(I_A)$, and hence this initial ideal is a Gomory initial ideal. □

Corollary 8.19. Let $\mathcal{A}$ be normal such that $C(\mathcal{A})$ is simplicial, and let Δ be the coarsest triangulation whose single maximal face has support $C(\mathcal{A})$. Then there exists a weight vector $\omega \in \mathbb{Z}^n$ such that $\Delta = \Delta_\omega$ and $\mathrm{in}_\omega(I_A)$ is a Gomory initial ideal.

Remark 8.20. The normal simplicial $\mathcal{A}$ in the above corollary is a natural example of Δ_ω-normal configurations. However, it would be nice to find other examples. For an inductive construction of infinite families of Δ_ω-normal configurations see [63].

9. The chain theorem

We have gone a long way in understanding toric initial ideals. Yet, they are still elusive. We do not have a precise characterization of them: which monomial ideals could be toric initial ideals? Theorem 5.7 gives a strong necessary condition by characterizing the radicals of initial ideals. The next result we will present was proved in [50] and probably the second strongest necessary condition to date one can cite. This is a structural result for $Ass(\mathrm{in}_\omega(I_A))$, the poset of associated primes of $\mathrm{in}_\omega(I_A)$ where for $P, Q \in Ass(\mathrm{in}_\omega(I_A)$, we write $P \leq Q$ if $P \supseteq Q$. Theorem 8.15 implies that $Ass(\mathrm{in}_\omega(I_A))$ is a subposet of the face lattice of Δ_ω, both posets having the same maximal elements.

Theorem 9.1. *Let τ be a non-maximal face of Δ_ω such that P_τ is an embedded prime of $\mathrm{in}_\omega(I_A)$. Then there exists a $\tau' \in \Delta_omega$ such that $\tau' \supset \tau$, $|\tau'| = |\tau| + 1$ and $P_{\tau'}$ is an associated prime of $\mathrm{in}_\omega(I_A)$.*

Proof. Since P_τ is an embedded prime, there is a standard pair of the form (x^v, τ) where $v \neq 0$ and $supp(v) \subseteq \bar{\tau}$. By Theorem 8.14, the origin is the unique lattice point in $P_v^{\bar{\tau}}(0) = \{u \in \mathbb{R}^m \,:\, B^{\bar{\tau}}u \leq v^{\bar{\tau}}, -(\omega B)\cdot u \leq 0\}$ and all the inequalities of $B^{\bar{\tau}}u \leq v^{\bar{\tau}}$ are essential. For each $i \in \bar{\tau}$, let R^i be the relaxation of $P_v^{\bar{\tau}}(0)$ obtained by removing the ith inequality $b_i \cdot u \leq v_i$, i.e., $R^i := \left\{u \in \mathbb{R}^m \,:\, B^{\overline{\tau\cup\{i\}}}u \leq v^{\overline{\tau\cup\{i\}}},\ -(\omega B)\cdot u \leq 0\right\}$. Let $E^i := R^i \backslash P_v^{\bar{\tau}}(0) = \left\{u \in \mathbb{R}^m \,:\, B^{\overline{\tau\cup\{i\}}}u \leq v^{\overline{\tau\cup\{i\}}}, b_i\cdot u > v_i,\ -(\omega B)\cdot u \leq 0\right\}$. Clearly, $E^i \cap P_v^{\bar{\tau}}(0) = \emptyset$, and since the removal of each inequality introduces at least one lattice point into R^i we have $E^i \cap \mathbb{Z}^m \neq \emptyset$. Let u_i^* be the optimal solution to the integer program $\min\{-(\omega B)\cdot u : u \in E^i \cap \mathbb{Z}^m\}$. This integer program is always feasible since $E^i \cap \mathbb{Z}^m \neq \emptyset$, but it may not have a finite optimal value. However, there exists at least one $i \in \bar{\tau}$ for which the above integer program is bounded. To see this, pick a maximal simplex $\sigma \in \Delta_\omega$ such that $\tau \subset \sigma$. The polytope $\{u \in \mathbb{R}^m : B^{\bar{\sigma}}u \leq v^{\bar{\sigma}}, -(\omega B)\cdot u \leq 0\}$ is a simplex and hence bounded. This polytope contains all E^i for $i \in \sigma\backslash\tau$ and hence all these E^i are bounded and have finite optima with respect to $-(\omega B)$. We may assume that the inequalities in $B^{\bar{\tau}}u \leq v^{\bar{\tau}}$ are labeled so that the finite optimal values are ordered as $-(\omega B)\cdot u_1^* \geq -(\omega B)\cdot u_2^* \geq \cdots \geq -(\omega B)\cdot u_p^*$ where $\bar{\tau} \supseteq \{1, 2, \ldots, p\}$. □

Claim. Let $N^1 := \left\{u \in \mathbb{R}^m \,:\, B^{\overline{\tau\cup\{1\}}}u \leq v^{\overline{\tau\cup\{1\}}}, -(\omega B)\cdot u \leq -(\omega B)\cdot u_1^*\right\}$. Then u_1^* is the unique lattice point in N^1 and all inequalities are essential.

Proof. Since u_1^* lies in R^1, $0 = -(\omega B)\cdot 0 \geq -(\omega B)\cdot u_1^*$. However, $0 > -(\omega B)\cdot u_1^*$ since otherwise, both u_1^* and 0 would be optimal solutions to $\min\{-(\omega B)\cdot u : u \in R^1\}$ contradicting that ω is generic. Therefore, $N^1 = R^1 \cap \{u \in \mathbb{R}^m : -(\omega B)\cdot u \leq -(\omega B)\cdot u_1^*\} = (E^1 \cup P_v^{\bar{\tau}}(0)) \cap \{u \in \mathbb{R}^m : -(\omega B)\cdot u \leq -(\omega B)\cdot u_1^*\} = (E^1 \cap \{u \in \mathbb{R}^m : -(\omega B)\cdot u \leq -(\omega B)\cdot u_1^*\}) \bigcup (P_v^{\bar{\tau}}(0) \cap \{u \in \mathbb{R}^m : -(\omega B)\cdot u \leq -(\omega B)\cdot u_1^*\})$. Since ω is generic, u_1^* is the unique lattice point in the first polytope and the second polytope is free of lattice points. Hence u_1^* is the unique lattice point in N^1. The relaxation of N^1 got by removing $b_j \cdot u \leq v_j$ is the polyhedron $N^1 \cup (E^j \cap \{u \in \mathbb{R}^m : -(\omega B)\cdot u \leq -(\omega B)\cdot u_1^*\})$ for $j \in \bar{\tau}$ and $j \neq 1$. Either this is unbounded, in which case there is a lattice point u in this relaxation such that $-(\omega B)\cdot u_1^* \geq -(\omega B)\cdot u$ or (if $j \leq p$), we have $-(\omega B)\cdot u_1^* \geq -(\omega B)\cdot u_j^*$ and u_j^* lies in this relaxation which shows that $b_j \cdot u \leq v_j$ is essential for N^1. ◇

Translating N^1 by $-u_1^*$ we get $P_{v'}^{\overline{\tau\cup\{1\}}}(0) := \{u \in \mathbb{R}^m : -(\omega B)\cdot u \leq 0,\ B^{\overline{\tau\cup\{1\}}}\cdot u \leq v'\}$ where $v' = v^{\overline{\tau\cup\{1\}}} - B^{\overline{\tau\cup\{1\}}}\cdot u_1^* \geq 0$ since u_1^* is feasible for all inequalities except the first one. Now $P_{v'}^{\overline{\tau\cup\{1\}}}(0)$ contains only the origin as an integral vector and hence $(x^{v'}, \tau \cup \{1\})$ is a standard pair of $\mathrm{in}_\omega(I_A)t$. □

Remark 9.2. Theorem 9.1 fails for general monomial ideals. The associated primes of the monomial ideal $\langle x^2yz, xy^2z, xyz^2\rangle \subset k[x, y, z]$ are $\langle x\rangle$, $\langle y\rangle$, $\langle z\rangle$, and $\langle x, y, z\rangle$ and each has multiplicity one. However, there is no associated prime generated by a pair of variables.

Remark 9.3. Theorem 9.1 may fail when the weight vector is not generic. David Eisenbud and Sorin Popescu have generated the following example. Consider the toric ideal I_A where

$$A = \begin{pmatrix} 3 & 2 & 2 & 1 & 1 & 0 & 0 & 0 & 0 \\ 0 & 1 & 0 & 2 & 0 & 3 & 2 & 1 & 0 \\ 0 & 0 & 1 & 0 & 2 & 0 & 1 & 2 & 3 \end{pmatrix} \text{ and the weight vector } \omega = (1, 1, 1, 1, 1, 1, 1, 1, 0).$$

Then $\mathrm{in}_\omega(I_A) = \langle h^2, gh, ch, eg, e^2, ce, aeh, g^2 - fh, cg - bh, ef - dh, cf - bg, de - bh, be - ah, d^2 - bf, cd - ag, bd - af, c^2 - ae, b^2 - ad\rangle$ which shows that ω is not generic. Here, the variables $a, b, c, d, e, f, g, h, i$ have been associated with the columns of A. The associated primes of $\mathrm{in}_\omega(I_A)$ are $P_1 = \langle c, e, g, h, d^2 - bf, bd - af, b^2 - ad\rangle$ of dimension three and $P_2 = \langle a, b, c, d, e, f, g, h\rangle$ of dimension one. The embedded prime P_2 contains the minimal prime P_1. However, there is no embedded prime of dimension two for this initial ideal.

10. Toric ideals in statistics

In this last section we touch upon the rapidly growing field of algebraic statistics. Toric ideals emerge in many different application areas such as phylogenetic analyis [79]. Here we will look at those which appear in contingency tables of multivariate analysis. The chapter by Ohsugi and Hibi in this volume [62] handles the "no n-way interaction" model in detail. We will give a more general presentation and show a quadratic initial ideal for *decomposable models*. Here we will not give the motivation for statistics. The reader can find more in-depth information and motivation in [20] and [47], as well as in the chapter we mentioned.

An n-dimensional contingency table Γ is a $d_1 \times \cdots \times d_n$ array of nonnegative integers where $d_i \geq 2$. We view such tables as elements of the vector space $\bigotimes_{i=1}^n \mathbb{R}^{d_i} = \mathbb{R}^D$ with $D = \prod d_i$. The coordinate functions and the standard unit vectors of this vector space will be denoted by x_K and e_K respectively where $K = (k_1, \ldots, k_n)$ is an indexing vector with $0 \leq k_i \leq d_i - 1$. A *hierarchical model* [55] consists of a simplicial complex Δ on n vertices and a linear map

$$\pi_\Delta : \bigotimes_{i=1}^{n} \mathbb{R}^{d_i} \longrightarrow \bigoplus_{F} \bigotimes_{j\in F} \mathbb{R}^{d_j} \quad e_K \mapsto \bigoplus_{F} e_{K_F}$$

where the direct sums are taken over all facets F of Δ and K_F denotes the restriction of $K = (k_1, \ldots, k_n)$ to F. The image $\pi_\Delta(\Gamma)$ is the Δ-marginal (or just marginal) of the table Γ with respect to Δ. For instance, the very popular $m \times n$ contingency tables where the marginals are the row and columns sums correspond to the simplicial complex of two isolated vertices $\Delta = \{\{1\}, \{2\}\}$ with $d_1 = m$ and $d_2 = n$.

Example 10.1. For a more complicated example let Δ be the one-dimensional complex of a four-cycle on four vertices, and let $d_i = 2$ for $i = 1, 2, 3, 4$. If we order the indices of the unit vectors $e_{i,j,k,\ell}$ in $\bigotimes_{i=1}^4 \mathbb{R}^2$ where $0 \leq i, j, k, \ell \leq 1$ in increasing lexicographic order then π_Δ is given by the following $0/1$-matrix. The rows correspond (in blocs of four) to the simplices $\{1,2\}, \{2,3\}, \{3,4\}, \{1,4\}$ respectively.

$$A_\Delta = \begin{pmatrix}
1&1&1&1&0&0&0&0&0&0&0&0&0&0&0&0\\
0&0&0&0&1&1&1&1&0&0&0&0&0&0&0&0\\
0&0&0&0&0&0&0&0&1&1&1&1&0&0&0&0\\
0&0&0&0&0&0&0&0&0&0&0&0&1&1&1&1\\
1&1&0&0&0&0&0&0&1&1&0&0&0&0&0&0\\
0&0&1&1&0&0&0&0&0&0&1&1&0&0&0&0\\
0&0&0&0&1&1&0&0&0&0&0&0&1&1&0&0\\
0&0&0&0&0&0&1&1&0&0&0&0&0&0&1&1\\
1&0&0&0&1&0&0&0&1&0&0&0&1&0&0&0\\
0&1&0&0&0&1&0&0&0&1&0&0&0&1&0&0\\
0&0&1&0&0&0&1&0&0&0&1&0&0&0&1&0\\
0&0&0&1&0&0&0&1&0&0&0&1&0&0&0&1\\
1&0&1&0&1&0&1&0&0&0&0&0&0&0&0&0\\
0&0&0&0&0&0&0&0&1&0&1&0&1&0&1&0\\
0&1&0&1&0&1&0&1&0&0&0&0&0&0&0&0\\
0&0&0&0&0&0&0&0&0&1&0&1&0&1&0&1
\end{pmatrix}$$

The toric ideal $I_\Delta = I_{A_\Delta}$ is generated by the following 16 binomials:

$$\begin{aligned}
I_\Delta = \langle\, & x_{1011}x_{1110} - x_{1010}x_{1111},\; x_{0111}x_{1101} - x_{0101}x_{1111},\\
& x_{1001}x_{1100} - x_{1000}x_{1101},\; x_{0110}x_{1100} - x_{0100}x_{1110},\\
& x_{0011}x_{1001} - x_{0001}x_{1011},\; x_{0010}x_{1000} - x_{0000}x_{1010},\\
& x_{0011}x_{0110} - x_{0010}x_{0111},\; x_{0001}x_{0100} - x_{0000}x_{0101},\\
& x_{0000}x_{0011}x_{1101}x_{1110} - x_{0001}x_{0010}x_{1100}x_{1111},\\
& x_{0000}x_{0111}x_{1001}x_{1110} - x_{0001}x_{0110}x_{1000}x_{1111},\\
& x_{0000}x_{0110}x_{1101}x_{1101} - x_{0010}x_{0100}x_{1001}x_{1111},\\
& x_{0001}x_{0110}x_{1010}x_{1101} - x_{0010}x_{0101}x_{1001}x_{1110},\\
& x_{0000}x_{0111}x_{1011}x_{1100} - x_{0011}x_{0100}x_{1000}x_{1111},\\
& x_{0010}x_{0101}x_{1011}x_{1100} - x_{0011}x_{0100}x_{1010}x_{1101},\\
& x_{0001}x_{0111}x_{1010}x_{1100} - x_{0011}x_{0101}x_{1000}x_{1110},\\
& x_{0100}x_{0111}x_{1001}x_{1010} - x_{0101}x_{0110}x_{1000}x_{1011}\,\rangle
\end{aligned}$$

For the rest of this section we will denote the monomial $x_{K_1}x_{K_2}\cdots x_{K_t}$ where $K_j = (k_{j1}, \dots, k_{jn})$ by a $t \times n$ tableau

$$\begin{bmatrix} k_{11} & \cdots & k_{1n} \\ \vdots & \ddots & \vdots \\ k_{tn} & \cdots & k_{tn} \end{bmatrix}.$$

If a certain variable appears to power p in the monomial, the corresponding row in the tableau will be repeated p times. Also when we write $T = T'$ for two tableaus this means that these arrays are equal up to a permutation of the rows. Note that if $F = \{j_1, \dots, j_t\} \subset [n]$ then the F-marginal of a contingency table Γ is a $d_{j_1} \times \cdots \times d_{j_t}$ table. The tableau $T|_F$ corresponding to this table is obtained from T by deleting columns of T not in F.

Now let Δ be a simplicial complex with exactly two facets F_1 and F_2. Let $S = F_1 \cap F_2$ and assume that $F_1 \backslash S = \{1, \dots, a\}$, $S = \{a+1, \dots, b\}$, and $F_2 \backslash S = \{b+1, \dots, n\}$. We define $\mathcal{F}(F_1, F_2)$ to be the set of binomials

$$\begin{bmatrix} p_1 & q & r_1 \\ p_2 & q & r_2 \end{bmatrix} - \begin{bmatrix} p_1 & q & r_2 \\ p_2 & q & r_1 \end{bmatrix}$$

where $p_1 \neq p_2$ ranges over $[0 \dots d_1 - 1] \times \cdots \times [0 \dots d_a - 1]$, q ranges over $[0 \dots d_{a+1} - 1] \times \cdots \times [0 \dots d_b - 1]$, and $r_1 \neq r_2$ ranges over $[0 \dots d_{b+1} - 1] \times \cdots \times [0 \dots d_n - 1]$. The next result is an easy exercise.

Proposition 10.2. *Let Δ be a simplicial complex with two facets F_1 and F_2. Then $\mathcal{F}(F_1, F_2)$ is a quadratic Gröbner basis of I_Δ with respect to a lexicographic term order.*

Definition 10.3. A simplicial complex Δ is called reducible if there is a nontrivial decomposition (Δ_1, S, Δ_2) such

(1) $\Delta = \Delta_1 \cup \Delta_2$,
(2) S is a face of both Δ_1 and Δ_2, and
(3) when S is removed from $|\Delta_1|$ and $|\Delta_2|$ the remaining sets do not share a vertex.

Now if Δ is reducible, and let H be a subset of binomials in I_{Δ_1} represented as differences of tableaus $T - T'$. Then by $Ext(H \to I_\Delta)$ we will denote all possible binomials of I_Δ in tableau form obtained by appending all possible index vectors supported on $|\Delta_2|\ S$ to the rows of T and T'. Next is the main result in this section.

Theorem 10.4. *Let G_1 and G_2 be Gröbner bases of I_{Δ_1} and I_{Δ_2} with respect to the term orders $\prec_1$ and $\prec_2$. Then there exists a lexicographic term order $\prec$ as in Proposition 10.2 which restricts to $\prec_1$ and $\prec_2$ so that*

$$Ext(G_1 \to I_\Delta) \cup Ext(G_2 \to I_\Delta) \cup \mathcal{F}(|\Delta_1|, |\Delta_2|)$$

is a Gröbner basis of I_Δ.

The proof of this theorem can be found in [47]. We invite the reader to observe that since the degree of a binomial in G_1 or G_2 stays the same when it is "extended" to a binomial in I_Δ, and since the binomials in $\mathcal{F}(|\Delta_1|, |\Delta_2|)$ are quadratic, the degrees of the binomials in the Gröbner basis of I_Δ constructed in Theorem 10.4 is at most the degrees of those in G_1 and G_2.

Finally, we have come to the construction of quadratic initial ideals of a nice class of hierarchical models known as *decomposable models*.

Definition 10.5. A simplicial complex Δ is called decomposable if it has a decomposition (Δ_1, S, Δ_2) as in Definition 10.3 where Δ_1 and Δ_2 are reducible or simplices.

This definition says that a decomposable simplicial complex is one which could recursively be decomposed into simplices of various sizes.

Theorem 10.6. *If Δ is a decomposable simplicial complex then I_Δ has a quadratic squarefree initial ideal*

Proof. We do induction of the number of facets of Δ. If there are two facets then Proposition 10.2 gives the result. For $r > 2$ facets, we look at the decomposition (Δ_1, S, Δ_2). Since the two simplicial complexes have fewer than r facets, induction provides us with quadratic squarefree initial ideals of I_{Δ_1} and I_{Δ_2}. Now Theorem 10.4 proves the result. □

The surprising result is that the converse of the last theorem is also true: if I_Δ has a quadratic squarefree (reverse lexicographic) initial ideal then Δ gives rise to a decomposable model, see [32]. We also refer the reader to [18] for the toric initial ideals of hierarchical models when Δ is one-dimensional.

References

[1] M. Beck and S. Hoşten, Growth series of cyclotomic lattices and cyclotomic polytopes, *Math. Res. Lett.*, `math.CO/0508136`, Preprint

[2] A. M. Bigatti, R. La Scala and L. Robbiano, Computing toric ideals, *J. Symbolic Comput.* **27** (1999) 351–365

[3] L. J. Billera, P. Filliman and B. Sturmfels, Constructions and complexity of secondary polytopes, *Adv. Math.* **83** (1990) 155–179

[4] J. Blasiak, The toric ideal of a graphical matroid is generated by quadrics. `math.CO/0511223`, Preprint

[5] I. Borosh and L.B. Treybig, Bound on positive integral solutions of linear diophantine equations, *Proc. Amer. Math. Soc.* **55** (1976) 299–304

[6] C. Bouvier and G. Gonzalez-Springberg, Système generateurs minimal, diviseurs essentiels et G-désingularizations de variétés toriques, *Tôhoku Math. J.*, **46** (1994) 125–149

[7] W. Bruns and J. Gubeladze, Normality and covering properties of affine semigroups, *J. Reine Angew. Math.* **510** (1999) 161–178

[8] W. Bruns, J. Gubeladze and N. V. Trung, Normal polytopes, triangulations, and Koszul algebras, *J. Reine Angew. Math.* **485** (1997) 123–160

[9] W. Bruns and J. Gubeladze, *Polytopes, Rings and K-Theory*, Preliminary version at `http://www.math.uos.de/staff/brunsw`

[10] W. Bruns and J. Herzog, *Cohen-Macaulay rings,* (Rev. ed.) University Press, Cambridge, (1998)

[11] W. Bruns and R. Koch, Normaliz, *A program to compute normalizations of semigroups*. Available by anonymous ftp from
`ftp.mathematik.Uni-Osnabrueck.DE/pub/osm/kommalg/software/`
[12] E. Cattani and A. Dickenstein, *Counting solutions to binomial complete intersections*. `math.AC/0510520`, Preprint
[13] CoCoATeam, CoCoA: *A system for doing Computations in Commutative Algebra*, available at `http://cocoa.dima.unige.it`
[14] P. Conti and C. Traverso, Gröbner bases and integer programming. *Proc. AAECC*-9 (New Orleans), Springer-Verlag, LNCS **539** (1991) 130–139
[15] D.A. Cox, Update on toric geometry, in, *Geometry of toric varieties*, Sémin. Congr. **6** Soc. Math. France (2002) 1–41
[16] D.A. Cox, J. Little and D. O'Shea, *Using Algebraic Geometry*, Graduate Texts in Mathematics, **185** Springer-Verlag, New York, (1998)
[17] J. A. de Loera, S. Hoşten, F. Santos and B. Sturmfels, The polytope of all triangulations of a point configuration, *Doc. Math.* **1** (1996) 103–119
[18] M. Develin and S. Sullivant, Markov bases of binary graph models, *Ann. Comb.* **7** (2003) 441–466
[19] P. Diaconis, R. L. Graham and B. Sturmfels, Primitive partition identities. In, *Paul Erdös is Eighty* [V. T. Sós, D. Miklós, and T. Szönyi, eds.], pages 173–192, Janós Bolyai Mathematical Society, Hungary, (1996)
[20] P. Diaconis and B. Sturmfels, Algebraic algorithms for sampling from conditional distributions, *Ann. Stat.* **26** (1998) 363–397
[21] E. Domenjoud, Solving systems of linear Diaphontine equations: An algebraic approach, *J. Symbolic Comput.* **8** (1989) 173–182
[22] P. Dueck and S. Hoşten, Hilbert triangulations in low codimension. In preparation
[23] D. Eisenbud, *Commutative algebra. With a view toward algebraic geometry*, Graduate Texts in Mathematics **150** Springer-Verlag, New York, (1995)
[24] D. Eisenbud, D. Grayson, M. Stillman and B. Sturmfels, *Math. Comp.* with `Macaulay2`, Algorithms and Computation in Mathematics, Vol. 8, Springer-Verlag, Heidelberg, (2001)
[25] D. Eisenbud, A. Reeves and B. Totaro, Initial ideals, Veronese subrings, and rates of algebras, *Adv. Math.* **109** (1994) 168–187
[26] D. Eisenbud and B. Sturmfels, Binomial ideals, *Duke Math. J.* **84** (1996) 1–45
[27] G. Ewald, *Combinatorial convexity and algebraic geometry*, Graduate Texts in Mathematics **168**, Springer-Verlag, New York, (1996)
[28] R. Firla and G. Ziegler, Hilbert bases, unimodular triangulations, and binary covers of rational polyhedral cones, *Discrete Comput. Geom.* **21** (1999) 205–216
[29] K. G. Fischer and J. Shapiro, Mixed matrices and binomial ideals, *J. Pure Appl. Algebra* **113** (1996) 39–54
[30] W. Fulton, *Introduction to Toric Varieties*. Princeton University Press, Princeton, (1993)
[31] D. Grayson and M. Stillman, Macaulay 2. *A Software System for Algebraic Geometry*, `http://www.math.uiuc.edu/Macaulay2/`
[32] D. Geiger, C. Meek and B. Sturmfels, On the toric algebra of graphical models, to appear in, *Ann. Stat.*
[33] I. M. Gel'fand, M. Kapranov and A. Zelevinsky, *Multidimensional Determinants, Discriminants and Resultants*. Birkhäuser, Boston, (1994)
[34] G.-M. Greuel, G. Pfister and H. Schönemann, Singular 2.0. *A Computer Algebra System for Polynomial Computations*, University of Kaiserslautern, (2001),
`http://www.singular.uni-kl.de`
[35] G.-M. Greuel and G. Pfister, *A Singular Introduction to Commutative Algebra*, Springer-Verlag, (2002)
[36] R. Hemmecke and R. Hemmecke, 4ti2 Version 1.1 — *Computation of Hilbert bases, Graver bases, toric Gröbner bases*, and more. Available at `www.4ti2.de` (2003)

[37] R. Hemmecke and P. Malkin, *Computing generating sets of lattice ideals.* `math.CO/0508359`, Preprint

[38] R. Hemmecke, R. Hemmecke and P. Malkin, 4ti2 Version 1.2 — *Computation of Hilbert bases, Graver bases, toric Gröbner bases*, and more, available at `www.4ti2.de` (2005)

[39] M. Henk and R. Weismantel, *On Hilbert bases of polyhedral cones*. Technical Report SC 96-12, Konrad-Zuse Zentrum, Berlin, (1996)

[40] T. Hibi, Distributive lattices, affine semigroup rings and algebras with straightening laws, in Commutative Algebra and Combinatorics (Kyoto, 1985), 93–109, *Adv. Stud. Pure Math.* **11** North-Holland, Amsterdam, (1987)

[41] T. Hibi and H. Ohsugi, A normal (0,1)-polytope none of whose regular triangulations is unimodular, *Discrete Comput. Geom.* **21** (1999) 201–204

[42] M. Hochster, Rings of invariants of tori, Cohen-Macaulay rings generated by monomials, and polytopes, *Ann. Math.* **96** (1972) 318–337

[43] S. Hoşten, *Degrees of Gröbner bases of integer programs*. Ph. D Thesis, Cornell University, (1997)

[44] S. Hoşten, D. Maclagan and B. Sturmfels, Supernormal vector configurations, *J. Algebraic Combin.* **19** (2004) 297–313

[45] S. Hoşten and J. Shapiro, Primary decomposition of lattice basis ideals, *J. Symbolic Comput.* **29** (2000) 625–639

[46] S. Hoşten and G. Smith, Monomial ideals, in "Computations in Algebraic Geometry with Macaulay 2" [D. Eisenbud, D. Grayson, M. Stillman and B. Sturmfels, eds.], Springer, *Algorithms Comput. Math.* **8** (2001), 73–100

[47] S. Hoşten and S. Sullivant, Gröbner bases and polyhedral geometry of reducible and cyclic models, *J. Combin. Theory Ser. A.* **100** (2002) 277–301

[48] S. Hoşten and B. Sturmfels, GRIN: An implementation of Gröbner bases of for integer programming, in Integer programming and combinatorial optimization (Copenhagen, 1995) 267–276, *Lecture Notes in Comput. Sci.* **920** Springer, Berlin, (1995)

[49] S. Hoşten and R. Thomas, Standard pairs and group relaxations in integer programming, *J. Pure Appl. Algebra* **139** (1999) 133–157

[50] S. Hoşten and R. Thomas, The associated primes of initial ideals of lattice ideals, *Math. Res. Lett.* **6** (1999) 83–97

[51] S. Hoşten and R. Thomas, Gomory integer programs, *Math. Program. Ser. B* **96** (2003) 271–292

[52] B. Huber and R. Thomas, Computing Gröbner fans of toric ideals, *Experiment. Math.* **9** (2000) 321–331

[53] A. N. Jensen, `CaTS` - *A computer program for computing the Gröbner fan of a toric ideal*, `http://www.daimi.au.dk/~u950710/cats.html`

[54] A. N. Jensen, `Gfan`, *A software system for Gröbner fans*, `http://home.imf.au.dk/ajensen/software/gfan/gfan.html`

[55] S. Lauritzen, *Graphical Models*. Oxford University Press, New York, (1996)

[56] J. Liu, L. E. Trotter and G. Ziegler, On the height of the minimal Hilbert basis, *Results Math.* **23** (1993) 374–376

[57] D. Maclagan and R. Thomas, Combinatorics of the toric Hilbert scheme, *Discrete Comput. Geom.* **27** (2002) 249–264

[58] E. Miller and B. Sturmfels, *Combinatorial Commutative Algebra*. Graduate Texts in Mathematics **227**, Springer-Verlag, New York, (2005)

[59] T. Mora and L. Robbiano, The Gröbner fan of an ideal, *J. Symbolic Comput.* **6** (1988) 183–208

[60] T. Oda, *Convex bodies and algebraic geometry. An introduction to the theory of toric varieties*. Results in Mathematics and Related Areas (3) **15**, Springer-Verlag, Berlin, (1988)

[61] H. Ohsugi and T. Hibi, Quadratic initial ideals of root systems, *Proc. Amer. Math. Soc.* **130** (2002) 1913–1922

[62] H. Ohsugi and T. Hibi, *Toric ideals arising from contingency tables*. This volume.
[63] E. O'Shea and R. Thomas, Toric initial ideals of Δ-normal configurations: Cohen-Macaulayness and degree bounds, *J. Algebraic Combin.* **21** (2005) 247–268
[64] E. O'Shea, *Sufficient conditions for total dual integrality*, Preprint
[65] I. Peeva and M. Stillman, Toric Hilbert schemes, *Duke Math. J.* **111** (2002) 419–449
[66] I. Peeva and B. Sturmfels, Generic lattice ideals, *J. Amer. Math. Soc.* **11** (1998) 363–373
[67] L. Pottier, Minimal solutions to linear Diaphontine systems: Bound and algorithms. In, *Proc. RTA'91*, pages 162–173, LNCS **488**, Springer-Verlag, (1991)
[68] J. Rambau, TOPCOM: triangulations of point configurations and oriented matroids, in, *Mathematical software* (Beijing, 2002), 330–340, World Science Publishing, River Edge, NJ, (2002)
[69] L. Robbiano, Term orderings on the polynomial ring. EUROCAL '85, Vol. 2 (Linz, 1985) 513–517, *Lecture Notes in Comput. Sci.* **204**, Springer, Berlin, (1985)
[70] M. Saito, B. Sturmfels and N. Takayama, *Gröbner deformations of hypergeometric differential equations*. Springer-Verlag, Berlin, (2000)
[71] F. Santos, A point configuration whose space of triangulations is disconnected, *J. Amer. Math. Soc.* **13** (2000) 611–637
[72] A. Schrijver, *Theory of Linear and Integer Programming*, Wiley Interscience, Chichester, (1986)
[73] A. Sebö, Hilbert bases, Caratheodory's theorem and combinatorial optimization, in, *Integer Programming and Combinatorial Optimization* [R. Kannan and W. Pulleyblank, eds.], 431–456, University of Waterloo Press, Waterloo, (1990)
[74] R. Stanley, *Enumerative Combinatorics* Vol. I. Cambridge Studies in Advanced Mathematics **49**, Cambridge University Press, Cambridge, (1997)
[75] M. Stillman, B. Sturmfels and R. Thomas, Algorithms for the toric Hilbert scheme in "Computations in Algebraic Geometry with Macaulay 2" [D. Eisenbud, D. Grayson, M. Stillman and B. Sturmfels, eds.], Springer, *Algorithms Comput. Math.* **8** (2001) 179–213
[76] B. Sturmfels, *Algorithms in Invariant Theory*. Texts and Monographs in Symbolic Computation, Springer-Verlag, Vienna, (1993)
[77] B. Sturmfels, *Gröbner Bases and Convex Polytopes*, University Lecture Series **8**, American Mathematical Society, (1995)
[78] B. Sturmfels, Equations defining toric varieties. Algebraic geometry—Santa Cruz 1995, 437–449, Proc. Sympos. Pure Math., 62, Part 2, *Amer. Math. Soc.*, Providence, RI, (1997)
[79] B. Sturmfels and S. Sullivant, Toric ideals of phylogenetic invariants, *J. Comput. Biology* **12** (2005) 204–228
[80] B. Sturmfels and R. Thomas, Variation of cost functions in integer programming, *Math. Program.* **77** (1997) 357–387
[81] B. Sturmfels, N. V. Trung and W. Vogel, Bounds on degrees of projective schemes, *Math. Ann.* **302** (1995) 417–432
[82] B. Sturmfels, R. Weismantel and G. Ziegler, Gröbner bases of lattices, corner polyhedra, and integer programming, *Beiträge Algebra Geom.* **36** (1995) 281–298
[83] S. R. Tayur, R. Thomas and N. R. Natraj, An algebraic geometry algorithm for scheduling in presence of setups and correlated demands, *Math. Program.* **69** (1995) 369–401
[84] R. Thomas, A geometric Buchberger algorithm for integer programming, *Math. Oper. Res.* **20** (1995) 864–884
[85] R. Thomas, The structure of group relaxations, to appear in, *Handbook of Discrete Optimization* [eds: K. Aardal, G. Nemhauser, R. Weismantel], (2002), `http://www.math.washington.edu/~thomas/articles.html`
[86] R. Thomas and R. Weismantel, Test sets and inequalities for integer programs, *Proceedings of the 5th International IPCO conference*, Vancouver, Lecture Notes in Computer Science **1084** (1996) 16–30

[87] R. Thomas and R. Weismantel, Truncated Gröbner bases for integer programming, *Appl. Algebra Engrg. Comm. Comput.* **8** (1997) 241–257

[88] D. L. Wehlau, Constructive invariant theory of tori, *Ann. Inst. Fourier* **43** (1993) 1055–1066

[89] N. White, A unique exchange property for bases, *Linear Algebra Appl.* **31** (1980) 81–91

[90] G. Ziegler, *Lectures on Polyopes*. Graduate Texts in Mathematics **152**, Springer-Verlag, New York, (1995)

Proc. Int. Conf. – Commutative Algebra and Combinatorics No. 4, 2007, pp. 91–115.

Toric Ideals Arising from Contingency Tables

Hidefumi Ohsugi[1] and Takayuki Hibi[2]

[1]Department of Mathematics, Rikkyo University, Toshima, Tokyo 171-8501, Japan
e-mail: ohsugi@rkmath.rikkyo.ac.jp
[2]Department of Pure and Applied Mathematics, Graduate School of Information Science and Technology, Osaka University, Toyonaka, Osaka 560-0043, Japan
e-mail: hibi@math.sci.osaka-u.ac.jp

Abstract. Fundamental questions on semigroup rings and toric ideals arising from contingency tables will be studied. In addition to discussing recent developments on such the topic, the algebraic background of contingency tables and basic notions related with Gröbner bases will be also explained clearly.

Background

An n-way contingency table is an n dimensional array whose entries are nonnegative integers. In the present paper, we study "no n-way interaction models." For example, the following table is given in [7, Table 2].

Example 0.1. The following 2-way contingency table T classifies 592 people according to their eye color and their hair color. From the table T, we want to know whether "Eye color" and "Hair color" are independent.

Eye color\Hair color	Black	Brunette	Red	Blonde	Total
Brown	68	119	26	7	220
Blue	20	84	17	94	215
Hazel	15	54	14	10	93
Green	5	29	14	16	64
Total	108	286	71	127	592

In general, when an n-way contingency table is given, we are interested in n factor interaction. For the sake of simplicity, we explain the case $n = 2$ here. (See [1] for details.) We consider the following $I \times J$ contingency table T_0. There are two random variables X and Y which take values in $X_1, \dots, X_I$ and $Y_1, \dots, Y_J$ respectively.

$X \backslash Y$	Y_1	Y_2	$\cdots$	Y_J	Total
X_1	n_{11}	n_{12}	$\cdots$	n_{1J}	n_{1+}
X_2	n_{21}	n_{22}	$\cdots$	n_{2J}	n_{2+}
$\vdots$	$\vdots$	$\vdots$		$\vdots$	$\vdots$
X_I	n_{I1}	n_{I2}	$\cdots$	n_{IJ}	n_{I+}
Total	n_{+1}	n_{+2}	$\cdots$	n_{+J}	N

Then we suppose the *null hypothesis* H_0 "X and Y are independent," and try to check whether we can reject H_0 or not.

One of the popular methods which test the independence of X and Y is the χ^2 test. In the χ^2 test, we compute the χ^2 statistic

$$\chi^2(T_0) = \sum_{i=1}^{I}\sum_{j=1}^{J} \frac{\left(n_{ij} - \frac{n_{i+}n_{+j}}{N}\right)^2}{\frac{n_{i+}n_{+j}}{N}}.$$

If the hypothesis H_0 is true, then χ^2 statistic has asymptotic χ^2 distributions with degrees of freedom $(I-1)(J-1)$. Thus we compare $\chi^2(T_0)$ with the value $\alpha_{0.05}$ of the right-hand side which is 5% point of χ^2 distribution. If $\chi^2(T_0) > \alpha_{0.05}$, then we conclude that H_0 is rejected, and X and Y are not independent.

However, it is known that, if T_0 is *sparse*, that is, there are many cells with $n_{i+}n_{+j} < 5$, then the χ^2 approximation is bad. Hence χ^2 test is not good for contingency tables such that N/IJ is small.

For such sparse contingency tables, we use *Fisher's exact test.* Let $\mathcal{F}_{T_0}$ denote the set of tables with the same marginal distribution as T_0. For example, in Example 0.1, $\mathcal{F}_T$ is the set of all 4×4 matrix such that the sum of four rows is $(108\ 286\ 71\ 127)$ and the sum of four columns is the transpose of $(220\ 215\ 93\ 64)$. We now assume that H_0 is true and that entries of $\mathcal{F}_{T_0}$ follows the multiple hypergeometric distribution. Here the multiple hypergeometric distribution is defined by

$$P(T) = \frac{\left(\prod_{i=1}^{I} n_{i+}!\right)\left(\prod_{j=1}^{J} n_{+j}!\right)}{N!\prod_{i,j} n_{ij}!}$$

for each $T \in \mathcal{F}_{T_0}$. For Fisher's exact test, we compute $\chi^2(T)$ for *all* $T \in \mathcal{F}_{T_0}$ and P-value

$$P = \sum_{T\in\mathcal{F}_{T_0},\ \chi^2(T)\geq\chi^2(T_0)} P(T)$$

of T_0. If $P < 0.05$, then we conclude that H_0 is rejected, and X and Y are not independent.

Unfortunately, Fisher's exact test also has a problem. In general, it is very difficult to enumerate all elements of $\mathcal{F}_{T_0}$. For example, in Example 0.1, $\mathcal{F}_T$ consists of $1,225,914,276,768,514$ elements. In such a case, it is almost impossible to compute P-value of T_0 exactly.

Here we are in the position to introduce *Markov Chain Monte Carlo method* (called MCMC method for short). For the computation of P-value, we give up above exact calculation and make use of a Markov chain to do the sampling from $\mathcal{F}_{T_0}$.

Note that, if both T and T' belong to $\mathcal{F}_{T_0}$, then $T - T'$ is an integer $I \times J$ matrix such that the sum of all entries of each rows and each columns is zero. Let $\mathcal{M}_{I\times J}$ denote the set of all integer $I \times J$ matrices which satisfies that the sum of all entries of each rows and each columns is zero. Then a *Markov basis* is a finite subset $\{T_1, \ldots, T_\ell\} \subset \mathcal{M}_{I\times J}$ satisfying that, for any $T, T' \in \mathcal{F}_{T_0}$, there exist $T_{i_1}, \ldots, T_{i_A}$ with $\varepsilon_k = \pm 1$ such that

$$T' = T + \sum_{k=1}^{A} \varepsilon_k T_{i_k},$$

$$T + \sum_{k=1}^{a} \varepsilon_k T_{i_k} \in \mathcal{F}_{T_0}$$

for all $1 \leq a \leq A$.

If a Markov basis $\{T_1, \ldots, T_\ell\}$ is given, then we can construct a Markov chain by the following algorithm:

Metropolis–Hastings algorithm

0. Choose $T \in \mathcal{F}_{T_0}$ at random and set $t = T$;
1. Repeat the following:
   1.1. Select T_i from the uniform distribution on $\{T_1, \ldots, T_\ell\}$;
   1.2. Select ε from the uniform distribution on $\{1, -1\}$ (independent of i);
   1.3. If $t + \varepsilon T_i$ is a nonnegative matrix, then set $t = t + \varepsilon T_i$ with probability

$$\min\left\{\frac{P(t + \varepsilon T_i)}{P(t)}, 1\right\}.$$

By Metropolis–Hastings algorithm, we have a set of tables $\{T^{(1)}, T^{(2)}, \ldots\} \subset \mathcal{F}_{T_0}$. By a sampling $\{T^{(i_1)}, T^{(i_2)}, \ldots, T^{(i_s)}\}$ from the sequence, we estimate P-value of T_0 at

$$\frac{1}{s}\sharp\left|\{1 \leq k \leq s\ ;\ \chi^2(T^{(i_k)}) \geq \chi^2(T_0)\}\right|.$$

In Example 0.1, if we have a Markov basis $\mathcal{B} \subset \mathcal{M}_{4\times 4}$ with

$$m_1 = \begin{pmatrix} 1 & 0 & -1 & 0 \\ 0 & 0 & 0 & 0 \\ -1 & 0 & 1 & 0 \\ 0 & 0 & 0 & 0 \end{pmatrix} \in \mathcal{B}, \qquad m_2 = \begin{pmatrix} 0 & 1 & -1 & 0 \\ 0 & 0 & 0 & 0 \\ 0 & 0 & 0 & 0 \\ 0 & -1 & 1 & 0 \end{pmatrix} \in \mathcal{B},$$

the following is a part of a random walk on $\mathcal{F}_T$ by Metropolis–Hastings algorithm:

$$\begin{array}{ccccccc} & \uparrow & & & & \uparrow & \\ \longleftrightarrow & \begin{pmatrix} 68 & 119 & 26 & 7 \\ 20 & 84 & 17 & 94 \\ 15 & 54 & 14 & 10 \\ 5 & 29 & 14 & 16 \end{pmatrix} & \longleftarrow & \pm m_1 & \longrightarrow & \begin{pmatrix} 69 & 119 & 25 & 7 \\ 20 & 84 & 17 & 94 \\ 14 & 54 & 15 & 10 \\ 5 & 29 & 14 & 16 \end{pmatrix} & \longleftrightarrow \\ & \uparrow & & & & \uparrow & \\ & \pm m_2 & & & & \pm m_2 & \\ & \downarrow & & & & \downarrow & \\ \longleftrightarrow & \begin{pmatrix} 68 & 120 & 25 & 7 \\ 20 & 84 & 17 & 94 \\ 15 & 54 & 14 & 10 \\ 5 & 28 & 15 & 16 \end{pmatrix} & \longleftarrow & \pm m_1 & \longrightarrow & \begin{pmatrix} 69 & 120 & 24 & 7 \\ 20 & 84 & 17 & 94 \\ 14 & 54 & 15 & 10 \\ 5 & 28 & 15 & 16 \end{pmatrix} & \longleftrightarrow \\ & \downarrow & & & & \downarrow & \end{array}$$

Let $\mathcal{B} = \{T_1, \dots, T_\ell\}$ be a Markov basis and let $G_{\mathcal{B},T_0}$ denote the graph with the vertex set $\mathcal{F}_{T_0}$ where $m \in \mathcal{F}_{T_0}$ and $m' \in \mathcal{F}_{T_0}$ are joined by an edge if $m - m' \in \mathcal{B} \cup -\mathcal{B}$. The most important point is that, the graph $G_{\mathcal{B},T_0}$ must be connected. Otherwise, there exists an unreachable element of $\mathcal{F}_{T_0}$ in Metropolis–Hastings algorithm. The infinite set $\mathcal{M}_{I\times J}$ is regarded as the set of all integer solutions of some system of linear equations, and hence we can associate $\mathcal{M}_{I\times J}$ with the "toric ideal" $I_{\mathcal{A}_{r_1\cdots r_n}}$.

A configuration in $\mathbb{R}^d$ is a finite set $\mathcal{A} = \{\mathbf{a}_1, \dots, \mathbf{a}_n\} \subset \mathbb{Z}^d$ which is contained in a hyperplane in $\mathbb{R}^d$ without the origin. Let $K[\mathbf{t}, \mathbf{t}^{-1}] = K[t_1, t_1^{-1}, \dots, t_d, t_d^{-1}]$ denote the Laurant polynomial ring in d variables over a field K. We associate a configuration $\mathcal{A} \subset \mathbb{Z}^d$ with the semigroup ring $K[\mathcal{A}] = K[\mathbf{t}^{\mathbf{a}_1}, \dots, \mathbf{t}^{\mathbf{a}_n}] \subset K[\mathbf{t}, \mathbf{t}^{-1}]$, where $\mathbf{t}^{\mathbf{a}} = t_1^{a_1} \cdots t_d^{a_d}$ if $\mathbf{a} = (a_1, \dots, a_d)$. Let $K[\mathbf{x}] = K[x_1, \dots, x_n]$ denote the polynomial ring in n variables over K. The toric ideal $I_\mathcal{A}$ of $\mathcal{A}$ is the kernel of the surjective homomorphism $\pi : K[\mathbf{x}] \longrightarrow K[\mathcal{A}]$ defined by setting $\pi(x_i) = \mathbf{t}^{\mathbf{a}_i}$ for $1 \le i \le n$. A polynomial $f \in K[\mathbf{x}]$ of the form $u - v$, where u and v are monomials, is called a *binomial*. It is known [18] that the toric ideal $I_\mathcal{A}$ is generated by the binomials $u - v$ with $\pi(u) = \pi(v)$.

A configuration $\mathcal{A}$ is called *unimodular* if the initial ideal of $I_\mathcal{A}$ is generated by squarefree monomials with respect to any monomial order. A configuration $\mathcal{A}$ is called *compressed* if the initial ideal of $I_\mathcal{A}$ is generated by squarefree monomials with respect to any reverse lexicographic order. We are interested in the following conditions:

(i) $\mathcal{A}$ is unimodular;
(ii) $\mathcal{A}$ is compressed;
(iii) there exists a monomial order $<$ such that the initial ideal of $I_\mathcal{A}$ with respect to $<$ is generated by squarefree monomials;
(iv) $K[\mathcal{A}]$ is normal.

Then (i) $\Longrightarrow$ (ii) $\Longrightarrow$ (iii) $\Longrightarrow$ (iv) holds and each of the converse of them is false in general. We refer the reader to [12], [14] and [15] for further information.

On the other hand, a binomial f belonging to $I_{\mathcal{A}}$ is called *indispensable* ([20] and [16]) if, for an arbitrary system $\mathcal{F}$ of binomial generators of $I_{\mathcal{A}}$, either f or $-f$ appears in $\mathcal{F}$. If f is indispensable, then $-f$ is indispensable. Hence the set of indispensable binomials is of the form $F \cup -F$, where $F \cap -F = \emptyset$. In abuse of terminology, such a set F will be called *the* set of indispensable binomials of $I_{\mathcal{A}}$.

In the present paper, we study the configuration arising from a $r_1 \times r_2 \times \cdots \times r_n$ contingency table, where $r_1 \geq r_2 \geq \cdots \geq r_n \geq 2$. Let $\mathcal{A}_{r_1 r_2 \cdots r_n}$ be the set of vectors

$$\mathbf{e}^{(1)}_{i_2 i_3 \cdots i_n} \oplus \mathbf{e}^{(2)}_{i_1 i_3 \cdots i_n} \oplus \cdots \oplus \mathbf{e}^{(n)}_{i_1 i_2 \cdots i_{n-1}},$$

where each i_k belongs to $[r_k] = \{1, 2, \ldots, r_k\}$ and $\mathbf{e}^{(k)}_{j_1 j_2 \cdots j_{n-1}}$ is a unit coordinate vector of $\mathbb{Z}^{r_1 \times \cdots r_{k-1} \times r_{k+1} \cdots \times r_n}$. The toric ideal $I_{\mathcal{A}_{r_1 r_2 \cdots r_n}}$ is the kernel of the homomorphism

$$\pi : K[\{x_{i_1 i_2 \cdots i_n} \ ; \ i_k \in [r_k]\}] \longrightarrow K[\{t^{(k)}_{i_1 \cdots i_{k-1} i_{k+1} \cdots i_n} \ ; \ k \in [n], i_k \in [r_k]\}]$$

defined by $\pi(x_{i_1 i_2 \cdots i_n}) = t^{(1)}_{i_2 i_3 \cdots i_n} t^{(2)}_{i_1 i_3 \cdots i_n} \cdots t^{(n)}_{i_1 i_2 \cdots i_{n-1}}$.

Proposition 0.2. *Work with the same notation as above. Then $\mathcal{B} \subset \mathcal{M}_{r_1 \times \cdots \times r_n}$ is a Markov basis for an arbitrary $r_1 \times \cdots \times r_n$ contingency table if and only if the toric ideal $I_{\mathcal{A}_{r_1 \cdots r_n}}$ is generated by the binomials $\mathbf{x}^{\beta^+} - \mathbf{x}^{\beta^-} \in K[\mathbf{x}]$ with $\beta^+ - \beta^- \in \mathcal{B}$.*

Together with known results, there are several original results in the present paper. The present paper is organized as follows. In Section 1, we review basic facts on Gröbner bases of toric ideals. In particular, the notion of both combinatorial pure subrings and the Lawrence lifting will be introduced.

In Section 2, we study indispensable binomials of toric ideals. We give a characterization of indispensable binomials in terms of the reduced Gröbner bases. Given a monomial order $<$ on $K[\mathbf{x}]$, we write $\mathcal{G}_<(I_{\mathcal{A}})$ for the reduced Gröbner basis of $I_{\mathcal{A}}$ with respect to $<$.

Theorem 2.4. *Let f be a binomial belonging to $I_{\mathcal{A}}$. Then the following conditions are equivalent.*

(i) *f is indispensable;*
(ii) *For any lexicographic order $<_{\text{lex}}$ on $K[\mathbf{x}]$, either f or $-f$ belongs to $\mathcal{G}_{<_{\text{lex}}}(I_{\mathcal{A}})$;*
(iii) *For any reverse lexicographic order $<_{\text{revlex}}$ on $K[\mathbf{x}]$, either f or $-f$ belongs to $\mathcal{G}_{<_{\text{revlex}}}(I_{\mathcal{A}})$;*
(iv) *For any monomial order $<$ on $K[\mathbf{x}]$, either f or $-f$ belongs to $\mathcal{G}_<(I_{\mathcal{A}})$.*

In Section 3, we discuss indispensable binomials of toric ideals arising from contingency tables. In particular, we introduce known three classes of contingency tables whose indispensable binomials have been completely determined.

In Section 4, first, we discuss two basic properties of the toric ideal of the configuration $\mathcal{A}_{333}$.

Lemma 4.1. *No reduced Gröbner basis of $I_{\mathcal{A}_{333}}$ coincides with the set of indispensable binomials (=* the *minimal set of binomial generators) of $I_{\mathcal{A}_{333}}$.*

Theorem 4.2. *The configuration $\mathcal{A}_{333}$ is compressed.*

Second, we introduce the results of Boffi–Rossi. In [5], they computed a lexicographic Gröbner basis of the toric ideal $I_{\mathcal{A}_{r_1 33}}$.

In Section 5, we characterize the configurations $\mathcal{A}_{r_1 r_2 \cdots r_n}$ for which there exists a monomial order $<$ such that the reduced Gröbner basis of $I_{\mathcal{A}_{r_1 r_2 \cdots r_n}}$ with respect to $<$ is the set of indispensable binomials of $I_{\mathcal{A}_{r_1 r_2 \cdots r_n}}$.

Theorem 5.1. *Let $n \leq 3$. Then the following conditions are equivalent for $\mathcal{A}_{r_1 r_2 \cdots r_n}$:*

(i) *either $n = 2$ or $r_3 = 2$;*
(ii) *$\mathcal{A}_{r_1 r_2 \cdots r_n}$ is unimodular ;*
(iii) *there exists a monomial order $<$ such that the reduced Gröbner basis of $I_{\mathcal{A}_{r_1 r_2 \cdots r_n}}$ with respect to $<$ is the set of indispensable binomials of $I_{\mathcal{A}_{r_1 r_2 \cdots r_n}}$;*
(iv) *there exists a monomial order $<$ such that the reduced Gröbner basis of $I_{\mathcal{A}_{r_1 r_2 \cdots r_n}}$ with respect to $<$ is a minimal set of binomial generators of $I_{\mathcal{A}_{r_1 r_2 \cdots r_n}}$.*

In Section 6, we study normality of semigroup rings arising from contingency tables. We classify all normal semigroup rings $K[\mathcal{A}_{r_1 r_2 \cdots r_n}]$ except for $K[\mathcal{A}_{553}]$, $K[\mathcal{A}_{543}]$ and $K[\mathcal{A}_{443}]$.

Theorem 6.4. *Work with the same notation as above. Then we have*

$r_1 \times r_2$ *or* $r_1 \times r_2 \times 2 \times \cdots \times 2$	*unimodular*
$r_1 \times 3 \times 3$	*normal*
$5 \times 5 \times 3$ *or* $5 \times 4 \times 3$ *or* $4 \times 4 \times 3$	*UNKNOWN if normal or not*
otherwise, i.e., $n \geq 4$ *and* $r_3 \geq 3$ $n = 3$ *and* $r_3 \geq 4$ $n = 3$, $r_3 = 3$, $r_1 \geq 6$ *and* $r_2 \geq 4$	*not normal*

1. Toric ideals

In the present section, we introduce the basic fact on Gröbner bases of toric ideals. See the survey [9] and the lecture note [18] for details of toric ideals. A configuration in $\mathbb{R}^d$ is a finite set $\mathcal{A} = \{\mathbf{a}_1, \ldots, \mathbf{a}_n\} \subset \mathbb{Z}^d$ which is contained in a hyperplane in $\mathbb{R}^d$ without the origin. Let $K[\mathbf{t}, \mathbf{t}^{-1}] = K[t_1, t_1^{-1}, \ldots, t_d, t_d^{-1}]$ denote the Laurant polynomial ring in d variables over a field K. We associate a configuration $\mathcal{A} \subset \mathbb{Z}^d$ with the semigroup ring $K[\mathcal{A}] = K[\mathbf{t}^{\mathbf{a}_1}, \ldots, \mathbf{t}^{\mathbf{a}_n}] \subset K[\mathbf{t}, \mathbf{t}^{-1}]$, where $\mathbf{t}^{\mathbf{a}} = t_1^{a_1} \cdots t_d^{a_d}$ if $\mathbf{a} = (a_1, \ldots, a_d)$. Let $K[\mathbf{x}] = K[x_1, \ldots, x_n]$ denote the polynomial ring in n variables over K. The toric ideal $I_{\mathcal{A}}$ of $\mathcal{A}$ is the kernel of the surjective homomorphism $\pi : K[\mathbf{x}] \longrightarrow K[\mathcal{A}]$ defined by setting $\pi(x_i) = \mathbf{t}^{\mathbf{a}_i}$ for $1 \leq i \leq n$. A polynomial $f \in K[\mathbf{x}]$ of the form $u - v$, where u and v are monomials, is called a *binomial*.

Proposition 1.1 ([18]). *The toric ideal $I_{\mathcal{A}}$ is generated by the binomials $u - v$ with $\pi(u) = \pi(v)$. Moreover, the reduced Gröbner bases of $I_{\mathcal{A}}$ consists of binomials.*

A binomial $f = u - v$ belonging to $I_{\mathcal{A}}$ is called *primitive* if there is no binomial $g = u' - v' \in I_{\mathcal{A}}$ with $f \neq g$ such that u' divides u and v' divides v. Every primitive binomial is irreducible. The *Graver basis* of $I_{\mathcal{A}}$ is the set of primitive binomials of $I_{\mathcal{A}}$. The Graver basis $Gr_{\mathcal{A}}$ of $I_{\mathcal{A}}$ is finite and is a system of binomial generators of $I_{\mathcal{A}}$. The universal Gröbner basis $\mathcal{U}_{\mathcal{A}}$ of $I_{\mathcal{A}}$ is the union of all reduced Gröbner bases of $I_{\mathcal{A}}$. The support of a monomial u of $K[\mathbf{x}]$ is $\text{supp}(u) = \{x_i : x_i \text{ divides } u\}$ and the support of a binomial $f = u - v$ is $\text{supp}(f) = \text{supp}(u) \cup \text{supp}(v)$. We say that an irreducible binomial f belonging to $I_{\mathcal{A}}$ is a *circuit* of $I_{\mathcal{A}}$ if there is no binomial $g \in I_{\mathcal{A}}$ such that $\text{supp}(g) \subset \text{supp}(f)$ and $\text{supp}(g) \neq \text{supp}(f)$. Every circuit is irreducible and primitive. A binomial $f \in I_{\mathcal{A}}$ is a circuit of $I_{\mathcal{A}}$ if and only if $I_{\mathcal{A}} \cap K[\{x_i : x_i \in \text{supp}(f)\}]$ is generated by f. Let $C_{\mathcal{A}}$ denote the set of circuits of $I_{\mathcal{A}}$.

If $f = u - v$ is a binomial belonging to $I_{\mathcal{A}}$, then we write T_f for the set of those variables t_i such that t_i divides $\pi(u)$ $(= \pi(v))$. Let $K[T_f] = K[\{t_i : t_i \in T_f\}]$ and $\mathcal{A}_f = \{\mathbf{a}_i : \mathbf{t}^{\mathbf{a}_i} \in K[T_f]\}$. The toric ideal $I_{\mathcal{A}_f}$ of $K[\mathcal{A}_f]$ coincides with $I_{\mathcal{A}} \cap K[\{x_i : \pi(x_i) \in K[T_f]\}]$. A binomial $f \in I_{\mathcal{A}}$ is called *fundamental* if $I_{\mathcal{A}_f}$ is generated by f. Let $\mathcal{F}d_{\mathcal{A}}$ denote the set of fundamental binomials of $I_{\mathcal{A}}$.

The *Lawrence lifting* of $\mathcal{A}$ is the configuration

$$\Lambda(\mathcal{A}) = \begin{pmatrix} \mathcal{A} & \mathbf{0} \\ I_n & I_n \end{pmatrix},$$

where I_n is the $n \times n$ identity matrix and $\mathbf{0}$ is the $d \times n$ zero matrix. A configuration $\mathcal{A}$ is called *Lawrence type* if there exists a configuration $\mathcal{B}$ such that $\Lambda(\mathcal{B}) = \mathcal{A}$. The toric ideal $I_{\Lambda(\mathcal{A})}$ is the kernel of homomorphism

$$\pi : K[x_1, \ldots, x_n, y_1, \ldots, y_n] \longrightarrow K[t_1, \ldots, t_d, s_1, \ldots, s_n]$$

defined by setting $\pi(x_i) = \mathbf{t}^{\mathbf{a}_i} s_i$ and $\pi(y_i) = s_i$. It is known that

Proposition 1.2 ([18], Chapter 7). *Work with the same notation as above. Then*

$$I_{\Lambda(\mathcal{A})} = (\mathbf{x}^{\mathbf{a}}\mathbf{y}^{\mathbf{b}} - \mathbf{x}^{\mathbf{b}}\mathbf{y}^{\mathbf{a}} : \mathbf{x}^{\mathbf{a}} - \mathbf{x}^{\mathbf{b}} \in I_{\mathcal{A}})$$
$$Gr_{\Lambda(\mathcal{A})} = \{\mathbf{x}^{\mathbf{a}}\mathbf{y}^{\mathbf{b}} - \mathbf{x}^{\mathbf{b}}\mathbf{y}^{\mathbf{a}} : \mathbf{x}^{\mathbf{a}} - \mathbf{x}^{\mathbf{b}} \in Gr_{\mathcal{A}}\}$$

Moreover, the following sets of binomials coincide:

(i) *the Graver basis of* $I_{\Lambda(A)}$;
(ii) *the universal Gröbner basis of* $I_{\Lambda(A)}$;
(iii) *any reduced Gröbner basis of* $I_{\Lambda(A)}$;
(iv) *any minimal set of binomial generators of* $I_{\Lambda(A)}$.

Hence, in particular, $Gr_{\mathcal{A}}$ *is a finite set.*

Let $\mathcal{A}'$ be a subset of $\mathcal{A}$. The semigroup ring $K[\mathcal{A}']$ is called a *combinatorial pure subring* of $K[\mathcal{A}]$ if $K[\mathcal{A}']$ is of the form $K[\mathcal{A}'] = K[\mathcal{A}] \cap K[t_{i_1}, \ldots, t_{i_s}]$. The following three propositions appear in [11].

Proposition 1.3. *Let* $K[\mathcal{A}']$ *be a combinatorial pure subring of* $K[\mathcal{A}]$ *and let* $\mathcal{G}$ *a Gröbner basis of* $I_{\mathcal{A}}$. *Then* $\mathcal{G} \cap K[\{x_{\mathbf{a}} : \mathbf{a} \in \mathcal{A}'\}]$ *is a Gröbner basis of* $I_{\mathcal{A}'}$.

Proposition 1.4. *Let $K[\mathcal{A}']$ be a combinatorial pure subring of $K[\mathcal{A}]$. If $K[\mathcal{A}]$ is normal, then $K[\mathcal{A}']$ is normal.*

Proposition 1.5. *Let $\Lambda(\mathcal{A})$ be the Lawrence lifting of $\mathcal{A}$. Then, the following conditions are equivalent :*

(i) *$\mathcal{A}$ is unimodular ;*
(ii) *$\Lambda(\mathcal{A})$ is unimodular ;*
(iii) *$K[\Lambda(\mathcal{A})]$ is normal.*

Sketch of Proof. First, (ii) $\Longrightarrow$ (iii) is well-known. Moreover, (i) $\Longleftrightarrow$ (ii) is easy to see. Suppose that $\mathcal{A}$ is not unimodular. Then there exists a circuit $f = \mathbf{x^a} - \mathbf{x^b}$ of $I_\mathcal{A}$ such that either $\mathbf{x^a}$ or $\mathbf{x^b}$ is not squarefree. Then $f' = \mathbf{x^a y^b} - \mathbf{x^b y^a}$ is a circuit of $I_{\Lambda(\mathcal{A})}$ and none of the monomials of f' is squarefree. Let $\mathcal{A}' = \{\mathbf{a}_i \in \mathcal{A}; x_i \in \operatorname{supp}(f)\}$. Then $I_{\Lambda(\mathcal{A}')} = (f')$. Since none of the monomials of f' is squarefree, it follows that $K[\Lambda(\mathcal{A}')]$ is not normal. (See also Lemma 6.1.) Thanks to Proposition 1.4, $K[\Lambda(\mathcal{A})]$ is not normal since $K[\Lambda(\mathcal{A}')]$ is a combinatorial pure subring of $K[\Lambda(\mathcal{A})]$. Thus we have (iii) $\Longrightarrow$ (i) as desired.

On the other hand, the *rth Lawrence lifting* [17] of $\mathcal{A}$ is the configuration arising from the following $(rd+n) \times rn$ matrix:

$$\Lambda^{(r)}(\mathcal{A}) = \begin{pmatrix} \mathcal{A} & \mathbf{0} & \cdots & \mathbf{0} \\ \mathbf{0} & \mathcal{A} & \cdots & \mathbf{0} \\ \vdots & & \ddots & \\ \mathbf{0} & \mathbf{0} & \cdots & \mathcal{A} \\ I_n & I_n & \cdots & I_n \end{pmatrix},$$

where I_n is the $n \times n$ identity matrix and $\mathbf{0}$ is the $d \times n$ zero matrix. Note that $K[\Lambda^{(2)}(\mathcal{A})] \simeq K[\Lambda(\mathcal{A})]$.

Theorem 1.6. *Let $\mathcal{A}'$ be a subset of a configuration $\mathcal{A}$. Then, $K[\Lambda^{(r')}(\mathcal{A}')]$ is a combinatorial pure subring of $K[\Lambda^{(r)}(\mathcal{A})]$ if $2 \leq r' \leq r$.*

By virtue of Proposition 1.4, Proposition 1.5 and Theorem 1.6,

Corollary 1.7. Let $\Lambda^{(r)}(\mathcal{A})$ be the rth Lawrence lifting of $\mathcal{A}$. If $K[\Lambda^{(r)}(\mathcal{A})]$ is normal, then $\mathcal{A}$ is unimodular.

Remark 1.8. The converse of Corollary 1.7 is false for $r \geq 3$ in general. See Proposition 3.1 and Theorem 6.4.

2. Indispensable binomials

A binomial f belonging to $I_\mathcal{A}$ is called *indispensable* if, for any system of binomial generators $\mathcal{F}$ of $I_\mathcal{A}$, either f or $-f$ belongs to $\mathcal{F}$. A binomial $f \in I_\mathcal{A}$ is called *redundant* if f belongs to no minimal system of binomial generators of $I_\mathcal{A}$. In other words, a binomial $f \in I_\mathcal{A}$ is redundant if and only if f is contained in the binomial ideal of $K[\mathbf{x}]$ which is generated by the binomials $g \in I_{\mathcal{A}}$ with $\deg g <$ $\deg f$.

By the results of [18] and [16], we have

Theorem 2.1. *The inclusion relations are as follows:*

$$\begin{array}{ccccc} \mathcal{C}_{\mathcal{A}} & \subset & \mathcal{U}_{\mathcal{A}} & \subset & Gr_{\mathcal{A}} \\ \cup & & \cup & & \cup \\ \mathcal{F}d_{\mathcal{A}} & \subset & \{\textit{indispensable}\} & \subset & \{\textit{not redundant}\} \end{array}$$

Example 2.2.

(a) Let $K[\mathcal{A}] = K[t_1t_2, t_2t_3, t_3t_4, t_4t_5, t_5t_6, t_1t_6, t_2t_6, t_3t_5]$. The binomial $f = x_1x_3x_5 - x_2x_4x_6$ belonging to $I_{\mathcal{A}}$ is a circuit and indispensable. However, f is not fundamental.
(b) Let $K[\mathcal{A}] = K[t_1t_2, t_2t_3, t_1t_3, t_3t_4, t_4t_6, t_3t_5, t_5t_6, t_6t_7, t_7t_8, t_6t_8]$. The binomial $f = x_1x_4x_6x_8x_{10} - x_2x_3x_5x_7x_9$ belonging to $I_{\mathcal{A}}$ is indispensable. However, f is not a circuit.
(c) Let $K[\mathcal{A}] = K[t_1t_2, t_1t_3, t_1t_4, t_2t_3, t_2t_4, t_3t_4]$. The binomial $f = x_1x_6 - x_2x_5$ belonging to $I_{\mathcal{A}}$ is a circuit and not redundant. However, f is not indispensable.
(d) Let $K[\mathcal{A}] = K[t_1t_2, t_2t_3, t_3t_4, t_4t_5, t_5t_6, t_1t_6, t_1t_4]$. Then, the binomial $x_1x_3x_5 - x_2x_4x_6$ belonging to $I_{\mathcal{A}}$ is a circuit and redundant.
(e) Let $K[\mathcal{A}] = K[t_1^2, t_2^2, t_3^2, t_4^2t_5^2, t_1t_2t_5, t_3t_4, t_1t_3t_5, t_2t_4]$. The universal Gröbner basis of $\mathcal{A}$ is $\mathcal{U}_{\mathcal{A}} = \{x_5x_6 - x_7x_8, x_3x_5^2 - x_2x_7^2, x_3x_5x_8 - x_2x_6x_7, x_3x_8^2 - x_2x_6^2, x_1x_3^2x_4 - x_6^2x_7^2, x_1x_3^2x_4x_8 - x_5x_6^3x_7, x_5^2x_6^4 - x_1x_3^2x_4x_8^2, x_6x_7^3x_8 - x_1x_3^2x_4x_5, x_1x_3^2x_4x_5^2 - x_7^4x_8^2, x_1x_2x_3x_4 - x_7^2x_8^2, x_1x_2x_3x_4 - x_5^2x_6^2, x_1x_2^2x_4 - x_5^2x_8^2, x_1x_2^2x_4x_7 - x_5^3x_6x_8, x_1x_2^2x_4x_7^2 - x_5^4x_6^2, x_1x_2^2x_4x_6 - x_5x_7x_8^3, x_7^2x_8^4 - x_1x_2^2x_4x_6^2\}$. The Graver basis of $\mathcal{A}$ is $Gr_{\mathcal{A}} = \mathcal{U}_{\mathcal{A}} \cup \{x_1x_2x_3x_4 - x_5x_6x_7x_8\}$. Moreover, $x_1x_2x_3x_4 - x_5x_6x_7x_8$ is not redundant.

In [21], a binomial $f \in I_{\mathcal{A}}$ is called *1-norm lower degree reducible* if f is generated by two binomials of lower degree. Example 2.3 shows that, in general, there is a redundant binomial which is not 1-norm lower degree reducible.

Example 2.3. Let $d = 6$, $n = 9$, $\mathcal{A} = \{t_1t_2, t_2t_3, t_3t_4, t_4t_5, t_5t_6, t_1t_6, t_1t_3, t_2t_4, t_3t_5\}$. Then the set $\{x_1x_3 - x_7x_8, x_2x_4 - x_8x_9, x_5x_7 - x_6x_9\}$ is a unique minimal set of binomial generators of $I_{\mathcal{A}}$. Hence the binomial $f = x_1x_3x_5 - x_2x_4x_8 \in I_{\mathcal{A}}$ is redundant. However f is not 1-norm reducible.

Given a monomial order $<$ on $K[\mathbf{x}]$, we write $\mathcal{G}_<(I_{\mathcal{A}})$ for the reduced Gröbner basis of $I_{\mathcal{A}}$ with respect to $<$. Since $\mathcal{G}_<(I_{\mathcal{A}})$ is a system of binomial generators of $I_{\mathcal{A}}$, it follows that if a binomial f belonging to $I_{\mathcal{A}}$ is indispensable, then either f or $-f$ belongs to $\mathcal{G}_<(I_{\mathcal{A}})$. More precisely, the indispensable binomial of $I_{\mathcal{A}}$ can be characterized in terms of the reduced Gröbner bases of $I_{\mathcal{A}}$.

Theorem 2.4. *Let f be a binomial belonging to $I_{\mathcal{A}}$. Then the following conditions are equivalent.*

(i) *f is indispensable;*
(ii) *For any lexicographic order $<_{\text{lex}}$ on $K[\mathbf{x}]$, either f or $-f$ belongs to $\mathcal{G}_{<_{\text{lex}}}(I_{\mathcal{A}})$;*
(iii) *For any reverse lexicographic order $<_{\text{revlex}}$ on $K[\mathbf{x}]$, either f or $-f$ belongs to $\mathcal{G}_{<_{\text{revlex}}}(I_{\mathcal{A}})$;*
(iv) *For any monomial order $<$ on $K[\mathbf{x}]$, either f or $-f$ belongs to $\mathcal{G}_{<}(I_{\mathcal{A}})$.*

Proof. We already discussed (i) $\Rightarrow$ (iv). Each of (iv) $\Rightarrow$ (iii) and (iv) $\Rightarrow$ (ii) is obvious. We will show that (ii) $\Rightarrow$ (i) and (iii) $\Rightarrow$ (i).

((ii) $\Rightarrow$ (i)) Given a binomial $f = u_0 - v_0$ belonging to $I_{\mathcal{A}}$ which is *not* indispensable, we choose a system of binomial generators $\mathcal{F}$ of $I_{\mathcal{A}}$ such that neither f nor $-f$ belongs to $\mathcal{F}$. Since $f \in I_{\mathcal{A}}$, for a monomial w of $K[\mathbf{x}]$ with $w \neq 0$ and for a binomial $g \in \mathcal{F}$, one has $wg = u_0 - v$. Let $h = wg - f = v_0 - v \in I_{\mathcal{A}}$.

If $h = 0$, then $f = wg$. Since $f \neq g$, one has $w \neq 1$. Thus f is reducible. Hence f can belong to no reduced Gröbner basis of $I_{\mathcal{A}}$.

Let $h \neq 0$. If h is reducible, then there is an irreducible binomial $h' = u' - v' \in I_{\mathcal{A}}$ with $u' \neq v_0$ such that u' divides v_0. Let $<_{\text{lex}}$ denote a lexicographic order on $K[\mathbf{x}]$ such that $\text{in}_{<_{\text{lex}}}(h') = u'$. Here $\text{in}_{<_{\text{lex}}}(h')$ is the initial monomial of h' with respect to $<_{\text{lex}}$. Hence neither f nor $-f$ belongs to $\mathcal{G}_{<_{\text{lex}}}(I_{\mathcal{A}})$.

Similarly, if wg is reducible, then $f \notin \mathcal{G}_{<'_{\text{lex}}}(I_{\mathcal{A}})$ and $-f \notin \mathcal{G}_{<'_{\text{lex}}}(I_{\mathcal{A}})$ for some lexicographic order $<'_{\text{lex}}$ on $K[\mathbf{x}]$.

Now, suppose that both binomials wg and h are irreducible. In particular $w = 1$. Let $x_i \in \text{supp}(u_0)$ and $x_j \in \text{supp}(v_0)$. Let $<''_{\text{lex}}$ denote a lexicographic order on $K[\mathbf{x}]$ such that each of x_i and x_j is bigger than x_k for all k with $k \neq i$ and $k \neq j$. Since $x_i \notin \text{supp}(v)$ and $x_j \notin \text{supp}(v)$, one has $\text{in}_{<''_{\text{lex}}}(g) = u_0$ and $\text{in}_{<''_{\text{lex}}}(h) = v_0$. Thus $f \notin \mathcal{G}_{<''_{\text{lex}}}(I_{\mathcal{A}})$ and $-f \notin \mathcal{G}_{<''_{\text{lex}}}(I_{\mathcal{A}})$, as desired.

((iii) $\Rightarrow$ (i)) In the above proof of (ii) $\Rightarrow$ (i), each of the lexicographic orders $<_{\text{lex}}$, $<'_{\text{lex}}$ and $<''_{\text{lex}}$ can be chosen as a reverse lexicographic order. Thus if f is not indispensable, then there is a reverse lexicographic order $<_{\text{revlex}}$ on $K[\mathbf{x}]$ such that $f \notin \mathcal{G}_{<_{\text{revlex}}}(I_{\mathcal{A}})$ and $-f \notin \mathcal{G}_{<_{\text{revlex}}}(I_{\mathcal{A}})$. □

Corollary 2.5. If there exists an indispensable binomial $u - v \in I_{\mathcal{A}}$ such that neither u nor v is squarefree, then there exists no regular unimodular triangulation [18, Chapter 8] of the convex hull of $\{\mathbf{a}_1, \ldots, \mathbf{a}_n\}$.

In Section 6, Lemma 6.1 generalize Corollary 2.5. On the other hand, the converse of Corollary 2.5 is false in general. See Example 2.14.

Example 2.6. Let $K[x]$ denote the polynomial ring in one variable over a field K and $I = (x - 1) = (x^2 - x, x^2 - 1)$. Since $\{x - 1\}$ is the only reduced Gröbner basis of I and since $\{x^2 - x, x^2 - 1\}$ is a minimal system of generators of I, Theorem 2.4 cannot be generalized to, e.g., lattice ideals in the obvious way.

In the rest of this section, in order to give examples of indispensable binomials, we introduce the results in [16] on toric ideals arising from finite graphs. Let G be a finite connected graph on the vertex set $[d] = \{1, \ldots, d\}$ with no loop and no multiple edge and $E(G) = \{e_1, \ldots, e_n\}$ the set of edges of G. Let, as before, $K[\mathbf{t}] = K[t_1, \ldots, t_d]$ denote the polynomial ring in d variables over a field K with each $\deg t_i = 1$. We will associate each edge $e = \{i, j\} \in E(G)$ with the squarefree quadratic monomial $\mathbf{t}^e = t_i t_j$ belonging to $K[\mathbf{t}]$. Let $\mathcal{A}_G = \{\mathbf{t}^{e_1}, \ldots, \mathbf{t}^{e_n}\}$ and $K[G] = K[\mathcal{A}_G]$. Let $K[\mathbf{x}] = K[x_1, \ldots, x_n]$ denote the polynomial ring in n variables over K with each $\deg x_i = 1$ and write I_G $(\subset K[\mathbf{x}])$ for the toric ideal of $K[G]$. For notation and terminologies we use in the present section, we follow [13]. In particular if Γ is an even close walk of G, then $f_\Gamma = f_\Gamma^{(+)} - f_\Gamma^{(-)}$ is the binomial [13, p. 512] coming from Γ. One has $f_\Gamma \in I_G$. Moreover, since the Graver basis is a system of binomial generators of I_G, [13, Lemma 3.2] guarantees that

Lemma 2.7. *Let G be a finite connected graph. If $f \in I_G$ is primitive, then we have $f = f_\Gamma$ where Γ is one of the following even closed walk. In particular, I_G is generated by all the binomials f_Γ, where Γ is one of the following even closed walk:*

(i) Γ *is an even cycle* ;
(ii) $\Gamma = (C_1, C_2)$, *where C_1 and C_2 are odd cycles of G having exactly one common vertex* ;
(iii) $\Gamma = (C_1, \Gamma_1, C_2, \Gamma_2)$, *where C_1 and C_2 are odd cycles of G having no common vertex and, where Γ_1 and Γ_2 are walks of G both of which combine a vertex v_1 of C_1 and a vertex v_2 of C_2.*

The following Example 2.8 is essentially discussed in [13].

Example 2.8.

(a) Let G be a finite bipartite graph. For a binomial f belonging to I_G, the following conditions are equivalent: (i) f is indispensable; (ii) f is fundamental; (iii) $f = f_C$ for an even cycle C with no chord.
(b) The toric ideal I_G of a finite bipartite graph G is generated by the fundamental binomials of I_G.

In the study of indispensable binomials of the toric ideal I_G arising from a finite graph G, the most fundamental question is when the binomial f_C coming from an even cycle C is an indispensable binomial of I_G.

Let G be a finite connected graph and $C = (\{v_1, v_2\}, \{v_2, v_3\}, \ldots, \{v_{2q}, v_1\})$ an even cycle of G of length $2q$. A chord $e = \{v_i, v_j\}$ with $1 \leq i < j \leq 2q$ is an *even chord* (resp. *odd chord*) if $j - i$ is odd (resp. even). When $e = \{v_i, v_j\}$ and $e' = \{v_{i'}, v_{j'}\}$ are odd chords of C with $1 \leq i < j \leq 2q$ and $1 \leq i' < j' \leq 2q$, we say that e and e' *cross effectively* in C if either $i < i' < j < j'$ or $i' < i < j' < j$ and if $i' - i$ is odd. (Since each of $j - i$ and $j' - i'$ is even, it follows that each of $j - i'$, $j' - j$ and $j' - i$ is odd.)

Theorem 2.9. *Let G be a finite connected graph and I_G its toric ideal. Given an even cycle C of G of length ≥ 4, the binomial f_C belonging to I_G is indispensable if and only if* (i) *C has no even chord and* (ii) *C has no odd chords e and e' which cross effectively in C.*

A finite connected graph G is called *weakly chordal* if every cycle of G of length 4 has a chord. We will study the toric ideal of the complementary graph of a weakly chordal graph. It follows easily that the complementary graph of a finite connected graph G is weakly chordal if and only if the following condition is satisfied: $(*)$ If e

and e' are edges of G with $e \cap e' = \emptyset$, then there is an edge e'' of G with $e \cap e'' \neq \emptyset$ and with $e' \cap e'' \neq \emptyset$.

Theorem 2.10. *Let G be a finite connected graph whose complementary graph is weakly chordal. Then the toric ideal I_G is generated by the indispensable binomials of I_G if and only if no complete graph of order ≥ 4 is a subgraph of G.*

Remark 2.11. Let G be the graph in [13, Example 2.1]. Then the complementary graph is a cycle of length 5 and hence weakly chordal. Although I_G is generated by indispensable binomials of degree 2, they are not a Gröbner basis with respect to any monomial order since I_G has no quadratic Gröbner basis.

Let G be a finite connected graph on the vertex set $[d] = \{1, \ldots, d\}$ with no loop and no multiple edge and $E(G) = \{e_1, \ldots, e_n\}$ the set of edges of G. If C and C' are cycles of G, then a *bridge* between C and C' is an edge $e = \{i, j\}$ of G, where i (resp. j) is a vertex of C (resp. C') and not a vertex of C' (resp. C). We say that G satisfies the *odd cycle condition* [8] if, for any two odd cycles C and C' of G having no common vertex, there exists a bridge between C and C'. We study toric ideal I_G of a finite connected graph G satisfying the odd cycle condition.

Theorem 2.12. *Let G be a finite connected graph satisfying the odd cycle condition. Then the indispensable binomials f_Γ, where Γ is an even closed walk of G, belonging to the toric ideal I_G can be classified as follows:*

- (α) *Γ is an even cycle C of G such that (i) C has no even chord and that (ii) C has no odd chords e and e' which cross effectively in C;*
- (β) *$\Gamma = (C, C')$ is an even closed walk of G, where C and C' are odd cycles of G having exactly one common vertex i and where either C or C' (say, C') is minimal, such that $C = (\{i_1, i_2\}, \{i_2, i_3\}, \ldots, \{i_{2p-1}, i_1\})$ with $i = i_1$ possesses the properties (i) if $e = \{i_k, i_{k'}\}$ is a chord of C with $1 < k < k' < 2p$, then either $k = 2$ or $k' = 2p-1$ and $k' - k$ is even, and (ii) if $e = \{i_2, i_k\}$ and $e' = \{i_{k'}, i_{2p-1}\}$ are chords of C with $2 < k < 2p$ and $1 < k' < 2p - 1$ such that k is even and k' is odd, then $k' - k = 1$;*
- (γ) *$\Gamma = (C, e, C', e)$ is an even closed walk of G, where C and C' are minimal odd cycles of G having no common vertex and exactly one bridge e between C and C'.*

Theorem 2.12 says that if a finite connected graph G satisfies the odd cycle condition, then for each indispensable binomial $f = u - v$ of I_G, either u or v is squarefree. However, the converse is false in general, even though I_G is generated by the indispensable binomials.

Example 2.13. Let G be the finite graph drawn below. Then G does not satisfy the odd cycle condition. However I_G is generated by the indispensable binomials $f_{\Gamma_1} = x_1x_3 - x_2x_4$ and $f_{\Gamma_2} = x_1x_4x_7x_9 - x_5x_6^2x_8$ where Γ_1 is a cycle of length 4 and $\Gamma_2 = (C, e_6, C', e_6)$ where C and C' are odd cycles of length 3 having no common vertex and exactly one bridge e_6 between C and C'.

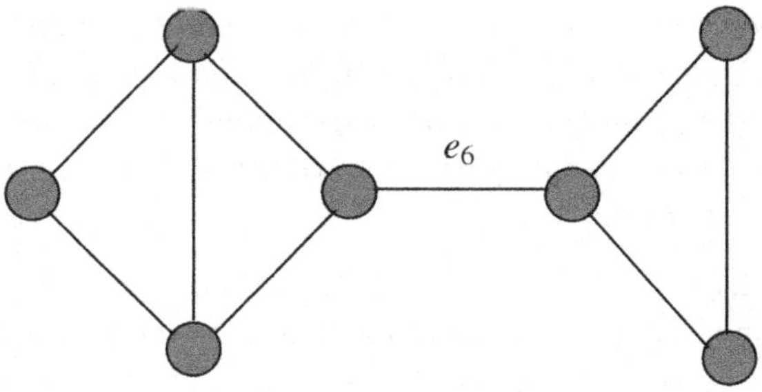

We now give a counter example for the converse of Corollary 2.5.

Example 2.14. Let G be the finite graph drawn below. Then G satisfies the odd cycle condition. Thanks to Theorem 2.12, for each indispensable binomial $f = u - v$ of I_G, either u or v is squarefree. (In this case, I_G is generated by indispensable binomials.) Moreover, in [12], it turns out that there exists no regular unimodular triangulation of the convex hull of the configuration arising from G.

There are 26 indispensable binomials in I_G: 16 even cycles satisfying (α) in Theorem 2.12 (i.e., five even cycles of length 6 with one odd chord, ten even cycles of length 8 with three odd chords, an even cycle of length 10 with five odd chords), 5 even closed walk satisfying (β) in Theorem 2.12 (all of them consist of two odd cycles of length 3 with one common vertex), and 5 even closed walk satisfying (γ) in Theorem 2.12 (all of them consist of two odd cycles of length 3 without common vertex joined by a bridge).

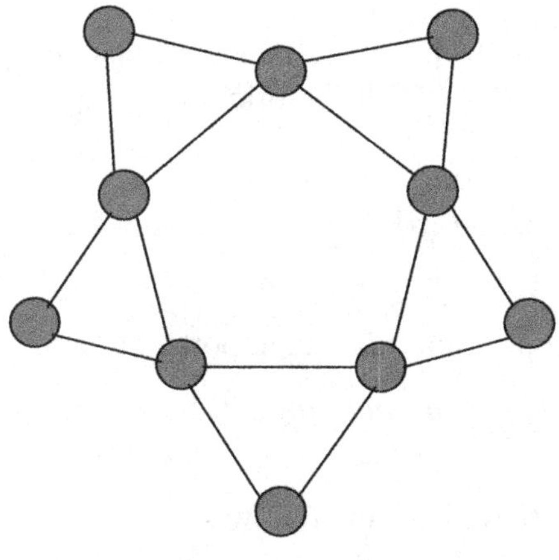

We conclude this section with discussing the existence of indispensable binomials of toric ideals arising from finite graphs. Let, as before, G be a finite connected graph on the vertex set $[d]$ with no loop and no multiple edge and $E(G) = \{e_1, \ldots, e_n\}$ the set of edges of G. Recall that the *degree* of a vertex i of G is a number of edges e of G with $i \in e$.

Proposition 2.15.

(a) *Let G be a complete graph. Then I_G has no indispensable binomial. In particular I_G has no fundamental binomials.*

(b) *If G has a vertex i of degree 1 and if G' is the induced subgraph of G on $[d] \setminus \{i\}$, then I_G has an indispensable binomial (resp. a fundamental binomial) if and only if $I_{G'}$ has an indispensable binomial (resp. a fundamental binomial).*

(c) *Let G be a finite connected graph with no vertex of degree 1. Then $I_G = (0)$ if and only if G is an odd cycle.*

By Proposition 2.15 (b), when we discuss the existence of indispensable binomials (or fundamental binomials), we may assume that G has no vertex of degree 1.

Theorem 2.16. *Let G be a finite connected graph with $I_G \neq (0)$ and suppose that G has no vertex of degree 1. Then the following conditions are equivalent:*

(i) *G is not a complete graph;*
(ii) *I_G possesses a fundamental binomial;*
(iii) *I_G possesses an indispensable binomial.*

3. Contingency tables and indispensable binomials

In the present section, we discuss the configuration $\mathcal{A}_{r_1 r_2 \cdots r_n}$ arising from a $r_1 \times r_2 \times \cdots \times r_n$ contingency table, where $r_1 \geq r_2 \geq \cdots \geq r_n \geq 2$. In particular, we study three classes of $I_{\mathcal{A}_{r_1 r_2 \cdots r_n}}$ whose indispensable binomials have been completely determined. Let $\mathcal{A}_{r_1 r_2 \cdots r_n}$ be the set of vectors

$$\mathbf{e}^{(1)}_{i_2 i_3 \cdots i_n} \oplus \mathbf{e}^{(2)}_{i_1 i_3 \cdots i_n} \oplus \cdots \oplus \mathbf{e}^{(n)}_{i_1 i_2 \cdots i_{n-1}},$$

where each i_k belongs to $[r_k] = \{1, 2, \ldots, r_k\}$ and $\mathbf{e}^{(k)}_{j_1 j_2 \cdots j_{n-1}}$ is a unit coordinate vector of $\mathbb{Z}^{r_1 \times \cdots r_{k-1} \times r_{k+1} \cdots \times r_n}$. The toric ideal $I_{\mathcal{A}_{r_1 r_2 \cdots r_n}}$ is the kernel of the homomorphism

$$\pi : K[\{x_{i_1 i_2 \cdots i_n} \; ; \; i_k \in [r_k]\}] \longrightarrow K[\{t^{(k)}_{i_1 \cdots i_{k-1} i_{k+1} \cdots i_n} \; ; \; k \in [n], i_k \in [r_k]\}]$$

defined by $\pi(x_{i_1 i_2 \cdots i_n}) = t^{(1)}_{i_2 i_3 \cdots i_n} t^{(2)}_{i_1 i_3 \cdots i_n} \cdots t^{(n)}_{i_1 i_2 \cdots i_{n-1}}$.

The following two propositions play an important role in this paper.

Proposition 3.1 ([17]). *The semigroup ring $K[\mathcal{A}_{r_1, \ldots, r_n}]$ is the r_nth Lawrence lifting of $K[\mathcal{A}_{r_1, \ldots, r_{n-1}}]$.*

Thanks to Theorem 1.6, we have the following.

Proposition 3.2. *Suppose that $\mathcal{A}_{r_1, \ldots, r_n}$ and $\mathcal{A}_{s_1, \ldots, s_n}$ satisfy $s_i \leq r_i$ for all $1 \leq i \leq n$. Then $K[\mathcal{A}_{s_1, \ldots, s_n}]$ is a combinatorial pure subring of $K[\mathcal{A}_{r_1, \ldots, r_n}]$.*

By virtue of Proposition 3.1, we can compute the dimension of $K[\mathcal{A}_{r_1, \ldots, r_n}]$. (Corollary 3.3 is a special case of [10, Theorem 2.6].)

Corollary 3.3. The dimension of $K[\mathcal{A}_{r_1, \ldots, r_n}]$ is equal to

$$(-1)^{n-1} + \sum_{k=1}^{n-1} (-1)^{n-k-1} \sum_{i_1 < \cdots < i_k} r_{i_1} \cdots r_{i_k}$$

Proof. If $n = 2$, then it is known that $\dim(\mathcal{A}_{r_1r_2}) = -1+r_1+r_2$. Suppose that the statement holds for $n = m - 1$. Thanks to Proposition 3.1, The semigroup ring $K[\mathcal{A}_{r_1\cdots r_m}]$ is the r_mth Lawrence lifting of $K[\mathcal{A}_{r_1\cdots r_{m-1}}]$. Hence by virtue of the dimension formula in [17, p. 162],

$$\begin{aligned}
&\dim(K[\mathcal{A}_{r_1\cdots r_m}])\\
&= \left((-1)^{m-2} + \sum_{k=1}^{m-2}(-1)^{m-k-2} \sum_{i_1<\cdots<i_k} r_{i_1}\cdots r_{i_k}\right)(r_m - 1) + r_1\cdots r_{m-1}\\
&= (-1)^{m-1} + \sum_{k=1}^{m-1}(-1)^{m-k-1} \sum_{i_1<\cdots<i_k} r_{i_1}\cdots r_{i_k}
\end{aligned}$$

as desired. □

Indispensable binomials have been completely determined for the following three classes of $I_{\mathcal{A}_{r_1r_2\cdots r_n}}$.

Case 1. $n = 2$.

Note that $\mathcal{A}_{r_1r_2} = \mathcal{A}_G$, where G is the complete $r_1 \times r_2$ bipartite graph. Thanks to Example 2.8, we have the following:

Proposition 3.4. *Work with the same notation as above.*

(a) *For a binomial $f \in I_{\mathcal{A}_{r_1r_2}}$, the followings conditions are equivalent: (i) f is indispensable; (ii) f is fundamental; (iii) $f = f_C$ for an even cycle C of G of length 4, i.e., $f = x_{ij}x_{k\ell} - x_{i\ell}x_{kj}$ with $1 \le i, k \le r_1$, $1 \le j, \ell \le r_2$.*
(b) *The toric ideal $I_{\mathcal{A}_{r_1r_2}}$ is generated by indispensable binomials.*

The diagonal monomial order $<_{\text{diag}}$ is a lexicographic order defined by

$$x_{ij} >_{\text{diag}} x_{k\ell} \iff \text{either (i) } i < k \text{ or (ii) } i = k \text{ and } j < \ell.$$

Proposition 3.5 ([18], Theorem 5). *The reduced Gröbner basis of $I_{\mathcal{A}_{r_1r_2}}$ with respect to $<_{\text{diag}}$ consists of indispensable binomials and hence quadratic.*

Since the configuration $\mathcal{A}_{r_1r_2}$ is unimodular, we have $C_{\mathcal{A}_{r_1r_2}} = \mathcal{U}_{\mathcal{A}_{r_1r_2}} = Gr_{\mathcal{A}_{r_1r_2}}$. They consists of the binomials arising from even cycles of G.

Case 2. $n \ge 3$ and $r_1 \times r_2 \times 2 \times \cdots \times 2$.

The configuration $\mathcal{A}_{r_1r_22}$ is the Lawrence lifting of $\mathcal{A}_{r_1r_2}$. Hence the configuration $\mathcal{A}_{r_1r_22\cdots2}$ is obtained by taking the Lawrence lifting several times from $\mathcal{A}_{r_1r_2}$. Thus $\mathcal{A}_{r_1r_22\cdots2}$ is of Lawrence type and unimodular.

Proposition 3.6. *The toric ideal $I_{\mathcal{A}_{r_1r_22\cdots2}}$ is generated by indispensable binomials.*

Since $\mathcal{A}_{r_1r_22\cdots2}$ is of Lawrence type, indispensable binomials of $I_{\mathcal{A}_{r_1r_22\cdots2}}$ is obtained by a primitive binomial of $I_{\mathcal{A}_{r_1r_2}}$, that is, the binomial arising from an even cycle of the complete $r_1 \times r_2$ bipartite graph. For example, from the primitive binomial

$$x_{i_1j_1}x_{i_2j_2}x_{i_3j_3} - x_{i_1j_2}x_{i_2j_3}x_{i_3j_1} \in Gr_{\mathcal{A}_{r_1r_2}},$$

we have the following indispensable binomial of $I_{\mathcal{A}_{r_1 r_2 2}}$

$$x_{i_1 j_1 1} x_{i_2 j_2 1} x_{i_3 j_3 1} \ x_{i_1 j_2 2} x_{i_2 j_3 2} x_{i_3 j_1 2} - x_{i_1 j_2 1} x_{i_2 j_3 1} x_{i_3 j_1 1} \ x_{i_1 j_1 2} x_{i_2 j_2 2} x_{i_3 j_3 2}$$

and the following indispensable binomial of $I_{\mathcal{A}_{r_1 r_2 22}}$

$$x_{i_1 j_1 11} x_{i_2 j_2 11} x_{i_3 j_3 11} x_{i_1 j_2 21} x_{i_2 j_3 21} x_{i_3 j_1 21} \ x_{i_1 j_2 12} x_{i_2 j_3 12} x_{i_3 j_1 12} x_{i_1 j_1 22} x_{i_2 j_2 22} x_{i_3 j_3 22}$$
$$- x_{i_1 j_2 11} x_{i_2 j_3 11} x_{i_3 j_1 11} x_{i_1 j_1 21} x_{i_2 j_2 21} x_{i_3 j_3 21} \ x_{i_1 j_1 12} x_{i_2 j_2 12} x_{i_3 j_3 12} x_{i_1 j_2 22} x_{i_2 j_3 22} x_{i_3 j_1 22}$$

and so on.

Case 3. $r_1 \times 3 \times 3$, $r_1 \times 4 \times 3$, $4 \times 4 \times 4$

Aoki–Takemura computed all indispensable binomials of toric ideals arising from three classes of contingency tables, that is, (i) $\mathcal{A}_{r_1 33}$, (ii) $\mathcal{A}_{r_1 43}$ and (iii) $\mathcal{A}_{444}$.

(i) For $\mathcal{A}_{r_1 33}$, there are 4 types of indispensable binomials. The degree of each binomials is one of 4, 6, 8 and 10.

(1) $x_{123}x_{132}x_{322}x_{333} - x_{122}x_{133}x_{323}x_{332}, \ldots$
(2) $x_{111}x_{122}x_{212}x_{223}x_{313}x_{321} - x_{112}x_{121}x_{213}x_{222}x_{311}x_{323}, \ldots$
(3) $x_{112}x_{123}x_{213}x_{232}x_{322}x_{331}x_{421}x_{433} - x_{113}x_{122}x_{212}x_{233}x_{321}x_{332}x_{423}x_{431}, \ldots$
(4) $x_{113}x_{131}x_{221}x_{233}x_{311}x_{322}x_{423}x_{432}x_{512}x_{533}$
$\quad - x_{111}x_{133}x_{231}x_{223}x_{312}x_{321}x_{422}x_{433}x_{513}x_{532}, \ldots$

(ii) For $\mathcal{A}_{r_1 43}$, there are 21 types of indispensable binomials. The degree of each binomials is one of 4, 6, 8, 9, 10, 12, 14 and 16.

(1) – (4) above and the following:

(5) $x_{111}x_{142}x_{212}x_{221}x_{322}x_{331}x_{432}x_{441} - x_{112}x_{141}x_{211}x_{222}x_{321}x_{332}x_{431}x_{442}$
(6) $x_{111}x_{142}x_{223}x_{241}x_{332}x_{343}x_{412}x_{421}x_{433} - x_{112}x_{141}x_{221}x_{243}x_{333}x_{342}x_{411}x_{423}x_{432}$
(7) $x_{131}x_{143}x_{233}x_{242}x_{311}x_{323}x_{332}x_{413}x_{422}x_{441}$
$\quad - x_{133}x_{141}x_{232}x_{243}x_{313}x_{322}x_{331}x_{411}x_{423}x_{442}$
(8) $x_{131}x_{142}x_{233}x_{241}x_{312}x_{323}x_{413}x_{432}x_{522}x_{543}$
$\quad - x_{132}x_{141}x_{231}x_{243}x_{313}x_{322}x_{412}x_{433}x_{523}x_{542}$
(9) $x_{111}x_{133}x_{213}x_{241}x_{331}x_{342}x_{422}x_{443}x_{523}x_{532}$
$\quad - x_{113}x_{131}x_{211}x_{243}x_{332}x_{341}x_{423}x_{442}x_{522}x_{533}$
(10) $x_{111}x_{122}x_{213}x_{231}x_{321}x_{343}x_{432}x_{441}x_{512}x_{523}x_{533}x_{542}$
$\quad - x_{112}x_{121}x_{211}x_{233}x_{323}x_{341}x_{431}x_{442}x_{513}x_{522}x_{532}x_{543}$
(11) $x_{121}x_{132}x_{223}x_{241}x_{331}x_{343}x_{412}x_{433}x_{513}x_{542}x_{622}x_{643}$
$\quad - x_{122}x_{131}x_{221}x_{243}x_{333}x_{341}x_{413}x_{432}x_{512}x_{543}x_{623}x_{642}$
(12) $x_{111}x_{122}x_{212}x_{231}x_{321}x_{343}x_{433}x_{441}x_{523}x_{542}x_{632}x_{643}$
$\quad - x_{112}x_{121}x_{211}x_{232}x_{323}x_{341}x_{431}x_{443}x_{522}x_{543}x_{633}x_{642}$
(13) $x_{111}x_{122}x_{213}x_{231}x_{321}x_{343}x_{432}x_{441}x_{512}x_{533}x_{623}x_{642}$
$\quad - x_{112}x_{121}x_{211}x_{233}x_{323}x_{341}x_{431}x_{442}x_{513}x_{532}x_{622}x_{633}$
(14) $x_{111}x_{143}x_{212}x_{221}x_{323}x_{331}x_{413}x_{432}x_{522}x_{543}x_{633}^2x_{641}x_{642}$
$\quad - x_{113}x_{141}x_{211}x_{222}x_{321}x_{333}x_{412}x_{433}x_{523}x_{542}x_{631}x_{632}x_{643}^2$
(15) $x_{111}x_{142}x_{213}x_{221}x_{322}x_{331}x_{433}x_{441}x_{523}x_{542}x_{612}x_{632}x_{643}^2$
$\quad - x_{112}x_{141}x_{211}x_{223}x_{321}x_{332}x_{431}x_{443}x_{522}x_{543}x_{613}x_{633}x_{642}^2$

(16) $x_{111}x_{122}x_{213}x_{231}x_{321}x_{343}x_{433}x_{441}x_{512}x_{543}x_{633}x_{642}x_{723}x_{732}$
$- x_{112}x_{121}x_{211}x_{233}x_{323}x_{341}x_{431}x_{443}x_{513}x_{542}x_{632}x_{643}x_{722}x_{733}$

(17) $x_{111}x_{122}x_{213}x_{231}x_{321}x_{343}x_{433}x_{441}x_{512}x_{533}x_{632}x_{643}x_{723}x_{742}$
$- x_{112}x_{121}x_{211}x_{233}x_{323}x_{341}x_{431}x_{443}x_{513}x_{532}x_{633}x_{642}x_{722}x_{743}$

(18) $x_{111}x_{142}x_{221}x_{232}x_{313}x_{331}x_{423}x_{441}x_{512}x_{543}x_{622}x_{643}x_{733}x_{742}$
$- x_{112}x_{141}x_{222}x_{231}x_{311}x_{333}x_{421}x_{443}x_{513}x_{542}x_{623}x_{642}x_{732}x_{743}$

(19) $x_{111}x_{142}x_{221}x_{233}x_{313}x_{331}x_{422}x_{441}x_{512}x_{533}x_{623}x_{642}x_{732}x_{743}$
$- x_{112}x_{141}x_{223}x_{231}x_{311}x_{333}x_{421}x_{442}x_{513}x_{532}x_{622}x_{643}x_{733}x_{742}$

(20) $x_{111}x_{122}x_{213}x_{231}x_{321}x_{343}x_{432}x_{441}x_{512}x_{543}x_{623}x_{632}x_{733}^2x_{742}^2$
$- x_{112}x_{121}x_{211}x_{233}x_{323}x_{341}x_{431}x_{442}x_{513}x_{542}x_{622}x_{633}x_{732}^2x_{743}^2$

(21) $x_{111}x_{122}x_{213}x_{231}x_{321}x_{343}x_{432}x_{441}x_{533}x_{542}x_{633}x_{642}x_{723}x_{732}x_{812}x_{843}$
$- x_{112}x_{121}x_{211}x_{233}x_{323}x_{341}x_{431}x_{442}x_{532}x_{543}x_{632}x_{643}x_{722}x_{733}x_{813}x_{842}$

(iii) For $\mathcal{A}_{444}$, there are 15 types of indispensable binomials. The degree of each binomials is one of 4, 6, 8, 9, 10, 12, 14 and 16.
(1) – (3), (5) – (7) above and the following:

(22) $x_{132}x_{143}x_{234}x_{242}x_{311}x_{324}x_{333}x_{413}x_{421}x_{444}$
$- x_{133}x_{142}x_{232}x_{244}x_{313}x_{321}x_{334}x_{411}x_{424}x_{443}$

(23) $x_{132}x_{144}x_{224}x_{243}x_{313}x_{331}x_{342}x_{411}x_{423}x_{434}$
$- x_{134}x_{142}x_{223}x_{244}x_{311}x_{332}x_{343}x_{413}x_{424}x_{431}$

(24) $x_{113}x_{131}x_{144}x_{224}x_{233}x_{241}x_{314}x_{333}x_{342}x_{423}x_{432}x_{444}$
$- x_{114}x_{133}x_{141}x_{223}x_{231}x_{244}x_{313}x_{332}x_{344}x_{424}x_{433}x_{442}$

(25) $x_{113}x_{122}x_{144}x_{224}x_{232}x_{241}x_{314}x_{333}x_{342}x_{423}x_{431}x_{444}$
$- x_{114}x_{123}x_{142}x_{222}x_{231}x_{244}x_{313}x_{332}x_{344}x_{424}x_{433}x_{441}$

(26) $x_{111}x_{123}x_{142}x_{213}x_{224}x_{231}x_{321}x_{332}x_{344}x_{412}x_{434}x_{443}$
$- x_{112}x_{121}x_{143}x_{211}x_{223}x_{234}x_{324}x_{331}x_{342}x_{413}x_{432}x_{444}$

(27) $x_{111}x_{124}x_{142}x_{213}x_{231}x_{244}x_{321}x_{332}x_{343}x_{412}x_{423}x_{434}$
$- x_{112}x_{121}x_{144}x_{211}x_{234}x_{243}x_{323}x_{331}x_{342}x_{413}x_{424}x_{432}$

(28) $x_{111}x_{122}x_{133}x_{214}x_{232}x_{241}x_{313}x_{321}x_{344}^2x_{424}x_{434}x_{442}x_{443}$
$- x_{113}x_{121}x_{132}x_{211}x_{234}x_{242}x_{314}x_{324}x_{341}x_{343}x_{422}x_{433}x_{444}^2$

(29) $x_{111}^2x_{133}x_{144}x_{223}x_{224}x_{232}x_{242}x_{313}x_{322}x_{341}x_{414}x_{422}x_{431}$
$- x_{113}x_{114}x_{131}x_{141}x_{222}^2x_{233}x_{244}x_{311}x_{323}x_{342}x_{411}x_{424}x_{432}$

(30) $x_{111}x_{122}x_{134}x_{213}x_{232}x_{241}x_{321}x_{333}x_{344}^2x_{414}x_{424}x_{442}x_{443}$
$- x_{114}x_{121}x_{132}x_{211}x_{233}x_{242}x_{324}x_{334}x_{341}x_{343}x_{413}x_{422}x_{444}^2$

Moreover Aoki–Takemura showed that

Proposition 3.7. *Let $\mathcal{A}$ be one of $\mathcal{A}_{r_1 33}$, $\mathcal{A}_{r_1 43}$ and $\mathcal{A}_{444}$. The indispensable binomials of $I_\mathcal{A}$ form a minimal set of binomial generators of $I_\mathcal{A}$. Hence a minimal set of binomial generators is unique for these configurations.*

Conjecture 3.8. The toric ideal of the configuration $\mathcal{A}_{r_1 r_2 \cdots r_n}$ is generated by indispensable binomials.

4. Gröbner bases and indispensable binomials of $I_{\mathcal{A}_{r_1 33}}$

The toric ideal of the configuration $\mathcal{A}_{333}$ is studied in [18], [7] [4] and [2]. Let F_{333} be the set of indispensable binomials of $I_{\mathcal{A}_{333}}$. Then F_{333} consists of (a) 27 binomials of degree 4 and (b) 54 binomials of degree 6 ([18, Theorem 14.13]):

(a) $x_{123}x_{132}x_{322}x_{333} - x_{122}x_{133}x_{323}x_{332}$
(b) $x_{111}x_{122}x_{212}x_{223}x_{313}x_{321} - x_{112}x_{121}x_{213}x_{222}x_{311}x_{323}$

It is known [3] that F_{333} is a *unique* minimal set of generators of $I_{\mathcal{A}_{333}}$.

We now discuss Gröbner bases of $I_{\mathcal{A}_{333}}$. The Graver basis $Gr_{\mathcal{A}_{333}}$ of $I_{\mathcal{A}_{333}}$ is given in [18, Theorem 14.13]. The Graver basis $Gr_{\mathcal{A}_{333}}$ consists of the following binomials together with (a) and (b) above:

(c) $x_{111}x_{123}x_{132}x_{222}x_{231}x_{313}x_{321} - x_{113}x_{122}x_{131}x_{221}x_{232}x_{311}x_{323}$
(d) $x_{112}x_{121}x_{133}x_{222}x_{231}x_{311}x_{323}x_{332}^2 - x_{111}x_{123}x_{132}x_{221}x_{232}x_{312}x_{322}x_{331}x_{333}$
(e) $x_{112}x_{123}x_{131}x_{213}x_{221}x_{232}x_{311}x_{322}x_{333} - x_{113}x_{121}x_{132}x_{211}x_{222}x_{233}x_{312}x_{323}x_{331}$
(f) $x_{111}x_{123}x_{132}x_{213}^2x_{221}x_{231}x_{311}x_{322}x_{333} - x_{113}x_{122}x_{131}x_{211}^2x_{223}x_{233}x_{312}x_{323}x_{331}$
(g) $x_{111}x_{123}^2x_{132}^2x_{213}x_{222}^2x_{231}x_{312}x_{321}x_{333} - x_{113}x_{122}^2x_{131}x_{133}x_{212}x_{221}x_{223}x_{232}x_{311}x_{323}x_{332}$

Hence, for an arbitrary monomial order, the reduced Gröbner basis is a subset of $Gr_{\mathcal{A}_{333}}$. Let $<_{\text{Lex}}$ denote the lexicographic order induced by

$$x_{ijk} <_{\text{Lex}} x_{i'j'k'} \iff \text{the first non-zero component of } (i - i', j - j', k - k') \text{ is negative.}$$

In [4, Theorem 5.1], Boffi and Rossi proved that the reduced Gröbner basis $\mathcal{G}_{<_{\text{Lex}}}$ of $I_{\mathcal{A}_{333}}$ with respect to $<_{\text{Lex}}$ consists of (a) and (b) together with the binomials of degree 7 and of degree 9. This result coincides with that obtained in [7, p. 381]. One can check it by the computation with CoCoA [6]. (Surprisingly, it takes only a few seconds.)

```
Use R ::= Z/(32003)[x[1..9]y[1..9]z[1..9]];
A := [x[1]y[1]z[1], x[1]y[2]z[2], x[1]y[3]z[3],x[2]y[1]z[4], x[2]y[2]z[5],
      x[2]y[3]z[6],x[3]y[1]z[7], x[3]y[2]z[8], x[3]y[3]z[9], x[4]y[4]z[1],
      x[4]y[5]z[2],x[4]y[6]z[3],x[5]y[4]z[4], x[5]y[5]z[5], x[5]y[6]z[6],
      x[6]y[4]z[7], x[6]y[5]z[8], x[6]y[6]z[9],x[7]y[7]z[1], x[7]y[8]z[2],
      x[7]y[9]z[3], x[8]y[7]z[4], x[8]y[8]z[5], x[8]y[9]z[6],
      x[9]y[7]z[7], x[9]y[8]z[8], x[9]y[9]z[9]];
Z := Transposed(Mat([Log(B)|B In A]));
Use S ::= Z/(32003)[t[1..Len(A)]];
J := Toric(Z);
Minimalize(J);
J;
```

On the other hand, since $F_{333} \neq \mathcal{G}_{<_{\mathrm{Lex}}}$, it is natural to ask if F_{333} can be a Gröbner basis with respect to a monomial order. We show that F_{333} cannot be a Gröbner basis with respect to an arbitrary monomial order.

Lemma 4.1. *No reduced Gröbner basis of $I_{\mathcal{A}_{333}}$ coincides with the set F_{333} of indispensable binomials of $I_{\mathcal{A}_{333}}$.*

Proof. Suppose that $F' = F_{333} \cup -F_{333}$ is a Gröbner basis with respect to a monomial order $<$. Let

$$f_{i,j} = x_{i12}x_{i21}x_{j11}x_{j22} - x_{i11}x_{i22}x_{j12}x_{j21} \in F'$$

and

$$g_{i,j} = x_{i22}x_{i33}x_{j23}x_{j32} - x_{i23}x_{i32}x_{j22}x_{j33} \in F',$$

where $1 \leq i, j \leq 3$. Note that $f_{i,j} = -f_{j,i}$ and $g_{i,j} = -g_{j,i}$. If $in_<(f_{i,j}) = x_{i12}x_{i21}x_{j11}x_{j22}$ and $in_<(g_{j,k}) = x_{j22}x_{j33}x_{k23}x_{k32}$, then the S-polynomial is

$$S(f_{i,j}, g_{j,k}) = x_{i12}x_{i21}x_{j11}x_{j23}x_{j32}x_{k22}x_{k33} - x_{j33}x_{k23}x_{k32}x_{i11}x_{i22}x_{j12}x_{j21}.$$

For each binomial h belonging to F', its initial monomial divides none of the two monomials appearing in $S(f_{i,j}, g_{j,k})$. Buchberger's criterion guarantees that F' is not a Gröbner basis. Thus, either $in_<(f_{i,j}) \neq x_{i12}x_{i21}x_{j11}x_{j22}$ or $in_<(g_{j,k}) \neq x_{j22}x_{j33}x_{k23}x_{k32}$.

Without a loss of generality, we may assume that $in_<(f_{1,2})$ is the first monomial $x_{112}x_{121}x_{211}x_{222}$ of $f_{1,2}$. Then $in_<(g_{2,3})$ is the second monomial of $g_{2,3}$, i.e., $in_<(g_{3,2})$ is the first monomial of $g_{3,2}$. Hence $in_<(f_{1,3})$ is the second monomial of $f_{1,3}$, i.e., $in_<(f_{3,1})$ is the first monomial of $f_{3,1}$. Repeating this argument, the initial monomial of $f_{1,2}, g_{3,2}, f_{3,1}, g_{2,1}, f_{2,3}$ and $g_{1,3}$ is their first monomial. Thus the monomial order $<$ satisfies

$$x_{112}x_{121}x_{211}x_{222} > x_{111}x_{122}x_{212}x_{221}$$
$$x_{322}x_{333}x_{223}x_{232} > x_{323}x_{332}x_{222}x_{233}$$
$$x_{312}x_{321}x_{111}x_{122} > x_{311}x_{322}x_{112}x_{121}$$
$$x_{222}x_{233}x_{123}x_{132} > x_{223}x_{232}x_{122}x_{133}$$
$$x_{212}x_{221}x_{311}x_{322} > x_{211}x_{222}x_{312}x_{321}$$
$$x_{122}x_{133}x_{323}x_{332} > x_{123}x_{132}x_{322}x_{333}.$$

However, the product of the left hand side monomials equals to the product of the right hand side monomials. This contradicts the property of a monomial order. Thus F' cannot be a Gröbner basis, as desired. □

On the other hand, since $\mathcal{A}_{333}$ is not unimodular [18, Theorem 14.13] and the initial ideal given in [4] is generated by squarefree monomials, it is natural to ask if $\mathcal{A}_{333}$ is compressed.

Theorem 4.2. *The configuration $\mathcal{A}_{333}$ is compressed.*

Proof. Let $<$ be an arbitrary reverse lexicographic order. Up to the symmetry, there are three binomials f_1, f_2 and f_3 with a non-squarefree monomial belonging to the Graver basis [18, Theorem 14.13] of $\mathcal{A}_{333}$:

Case 1. $f_1 = x_{112}x_{121}x_{133}x_{222}x_{231}x_{311}x_{323}x_{332}^2 - x_{111}x_{123}x_{132}x_{221}x_{232}x_{312}x_{322}x_{331}x_{333}$.
Suppose that $in_<(f_1) = x_{112}x_{121}x_{133}x_{222}x_{231}x_{311}x_{323}x_{332}^2$. The binomials

$$g_1 = x_{112}x_{121}x_{133}x_{311}x_{323}x_{332} - x_{111}x_{123}x_{132}x_{312}x_{321}x_{333},$$

$$g_2 = x_{121}x_{133}x_{222}x_{231}x_{323}x_{332} - x_{123}x_{131}x_{221}x_{232}x_{322}x_{333},$$

$$g_3 = x_{112}x_{121}x_{222}x_{231}x_{311}x_{332} - x_{111}x_{122}x_{221}x_{232}x_{312}x_{331}$$

belong to the Graver basis of $\mathcal{A}_{333}$. Since the first monomial of each of the binomials g_1, g_2 and g_3 divides $in_<(f_1)$ and since each variable in $f_1 - in_<(f_1)$ appears in the second monomial of one of the binomials g_1, g_2 and g_3, it follows that the initial monomial of one of the binomials g_1, g_2 and g_3 divides $in_<(f_1)$.

Case 2. $f_2 = x_{111}x_{123}x_{132}x_{213}^2x_{221}x_{231}x_{311}x_{322}x_{333} - x_{133}x_{122}x_{131}x_{211}^2x_{223}x_{233}x_{313}x_{321}x_{332}$.
Using the symmetry we may assume that

$$in_<(f_2) = x_{111}x_{123}x_{132}x_{213}^2x_{221}x_{231}x_{311}x_{322}x_{333}.$$

The binomials

$$g_1 = x_{111}x_{123}x_{132}x_{213}x_{231}x_{322}x_{333} - x_{113}x_{122}x_{131}x_{211}x_{233}x_{323}x_{332},$$

$$g_2 = x_{123}x_{132}x_{213}x_{221}x_{311}x_{322}x_{333} - x_{122}x_{133}x_{211}x_{223}x_{313}x_{321}x_{332}$$

belong to the Graver basis of $\mathcal{A}_{333}$. By the same argument as in Case 1, the initial monomial of one of the binomials g_1 and g_2 divides $in_<(f_2)$.

Case 3. $f_3 = x_{111}x_{123}^2x_{132}^2x_{213}x_{222}^2x_{231}x_{312}x_{321}x_{333} - x_{113}x_{122}^2x_{131}x_{133}x_{212}x_{221}x_{223}x_{232}x_{311}x_{323}x_{332}$.
Suppose that $in_<(f_3) = x_{111}x_{123}^2x_{132}^2x_{213}x_{222}^2x_{231}x_{312}x_{321}x_{333}$. The binomials

$$g_1 = x_{111}x_{123}x_{132}x_{213}x_{222}x_{231} - x_{113}x_{122}x_{131}x_{211}x_{223}x_{332},$$

$$g_2 = x_{111}x_{123}x_{132}x_{213}x_{222}x_{231}x_{312}x_{321}x_{333} - x_{113}x_{122}x_{131}x_{212}x_{221}x_{233}x_{311}x_{323}x_{332}$$

belong to the Graver basis of $\mathcal{A}_{333}$. By the same argument as in Case 1, the initial monomial of one of the binomials g_1 and g_2 divides $in_<(f_3)$.

Suppose that $in_<(f_3) = x_{113}x_{122}^2x_{131}x_{133}x_{212}x_{221}x_{223}x_{232}x_{311}x_{323}x_{332}$. The binomials

$$g_1 = x_{113}x_{122}x_{131}x_{311}x_{323}x_{332} - x_{111}x_{123}x_{132}x_{313}x_{322}x_{331},$$

$$g_2 = x_{113}x_{122}x_{212}x_{223} - x_{112}x_{123}x_{213}x_{222},$$

$$g_3 = x_{122}x_{131}x_{221}x_{232} - x_{121}x_{132}x_{222}x_{231},$$

$$g_4 = x_{122}x_{133}x_{212}x_{221}x_{311}x_{323}x_{332} - x_{123}x_{132}x_{211}x_{222}x_{312}x_{321}x_{333}$$

belong to the Graver basis of $\mathcal{A}_{333}$. By the same argument as in Case 1, the initial monomial of one of the binomials g_1, g_2, g_3 and g_4 divides $in_<(f_3)$.

Thus the initial ideal of $I_{\mathcal{A}_{333}}$ with respect to $<$ is generated by squarefree monomials, as desired. □

Remark. After submitting the present paper on November 1st 2004, Seth Sullivant showed that $\mathcal{A}_{r_1 33}$ is compressed for $r_1 \geq 3$ in his paper [19, Theorem 5.8] submitted to the Mathematics ArXiv on December 30th 2004.

The rest of the present section, we introduce the results of [5], i.e., the reduced Gröbner basis of the toric ideal $I_{\mathcal{A}_{r_1 33}}$ for an arbitrary $r_1 \geq 3$ is computed explicitly. Let $<_{\text{Lex}}$ denote the lexicographic order induced by

$$x_{ijk} <_{\text{Lex}} x_{i'j'k'} \iff \text{the first non-zero component of } (i - i', j - j', k - k') \text{ is negative}$$

as before. In [5, Section 6], Boffi and Rossi proved that

Proposition 4.3. *Let $\mathcal{G}_{<_{\text{Lex}}}$ be the reduced Gröbner basis of $I_{\mathcal{A}_{r_1 33}}$ with respect to $<_{\text{Lex}}$. Then, for each binomial $f \in \mathcal{G}_{<_{\text{Lex}}}$, there exists $1 \leq i_1 < \cdots < i_5 \leq r_1$ such that $f \in K[\{x_{i_j\, k\, \ell}\}_{1 \leq j \leq 5, 1 \leq k, \ell \leq 3}]$.*

Thanks to Proposition 4.3, $\mathcal{G}_{<_{\text{Lex}}}$ is derived from the reduced Gröbner basis of $I_{\mathcal{A}_{533}}$ with respect to $<_{\text{Lex}}$. Since the initial ideal of $I_{\mathcal{A}_{533}}$ with respect to $<_{\text{Lex}}$ is squarefree, we have the following.

Corollary 4.4. For each $r_1 \geq 3$, the initial ideal of $I_{\mathcal{A}_{r_1 33}}$ with respect to $<_{\text{Lex}}$ is squarefree.

Remark 4.5. The lexicographic Gröbner basis of $I_{\mathcal{A}_{r_1 33}}$ with respect to $<_{\text{Lex}}$ given in [5] is also a Gröbner basis with respect to the reverse lexicographic order $<_{\text{revlex}}$ induced by

$$x_{ijk} <_{\text{revlex}} x_{i'j'k'} \iff \begin{array}{l} \text{(i) } i > i' \\ \text{(ii) } i = i' \text{ and } j > j' \\ \text{(iii) } i = i' \text{ and } j = j' \text{ and } k < k'. \end{array}$$

In fact, the initial monomial of each binomial of the reduced Gröbner basis of $I_{\mathcal{A}_{533}}$ does not change if we consider $<_{\text{revlex}}$ instead of the lexicographic order $<_{\text{Lex}}$. By virtue of Proposition 4.3, this is true for $I_{\mathcal{A}_{r_1 33}}$ with any $r_1 \geq 3$.

5. Gröbner bases consisting of indispensable binomials

By virtue of Theorem 2.4, if the indispensable binomials form a Gröbner basis, then it is reduced. The results in the previous section now enables us to characterize $\mathcal{A}_{r_1 r_2 \cdots r_n}$ whose indispensable binomials form a Gröbner basis.

Theorem 5.1. *Let $n \leq 3$. Then the following conditions are equivalent for $\mathcal{A}_{r_1 r_2 \cdots r_n}$:*

(i) *either $n = 2$ or $r_3 = 2$;*
(ii) *$\mathcal{A}_{r_1 r_2 \cdots r_n}$ is unimodular ;*

(iii) *there exists a monomial order $<$ such that the reduced Gröbner basis of $I_{\mathcal{A}_{r_1r_2\cdots r_n}}$ with respect to $<$ is the set of indispensable binomials of $I_{\mathcal{A}_{r_1r_2\cdots r_n}}$;*
(iv) *there exists a monomial order $<$ such that the reduced Gröbner basis of $I_{\mathcal{A}_{r_1r_2\cdots r_n}}$ with respect to $<$ is a minimal set of binomial generators of $I_{\mathcal{A}_{r_1r_2\cdots r_n}}$.*

Proof. First, (iii) $\Longrightarrow$ (iv) is trivial by the definition of indispensable binomials. If $n = 2$, then $\mathcal{A}_{r_1r_2}$ is unimodular and satisfies (iii) by Proposition 3.5. Let $n = 3$. Since $\mathcal{A}_{r_1r_22}$ is the Lawrence lifting of $\mathcal{A}_{r_1r_2}$, $\mathcal{A}_{r_1r_22}$ is unimodular and the reduced Gröbner basis of $I_{\mathcal{A}_{r_1r_22}}$ with respect to any monomial order is the set of indispensable binomials. Thus (i) $\Longrightarrow$ (ii) and (i) $\Longrightarrow$ (iii) hold. By virtue of [18, Theorem 14.13], one has (ii) $\Longrightarrow$ (i).

Suppose that $\mathcal{A}_{r_1r_2\cdots r_n}$ does not satisfy (i). Then $\mathcal{A}_{333}$ is a combinatorial pure subring [11] of $\mathcal{A}_{r_1r_2\cdots r_n}$. By using Lemma 4.1 together with [11, Proposition 1.1], $\mathcal{A}_{r_1r_2\cdots r_n}$ does not satisfy (iv). Thus one has (iv) $\Longrightarrow$ (i), as desired. □

Note that, if $\mathcal{A}$ does not arise from contingency tables, then each of (ii) $\Longrightarrow$ (iii), (iii) $\Longrightarrow$ (ii), (ii) $\Longrightarrow$ (iv), (iv) $\Longrightarrow$ (iii) is false in general.

Example 5.2. The configuration $\mathcal{A} = \{\mathbf{e}_1 + \mathbf{e}_2, \mathbf{e}_2 + \mathbf{e}_3, \mathbf{e}_3 + \mathbf{e}_4, \mathbf{e}_4 + \mathbf{e}_5, \mathbf{e}_1 + \mathbf{e}_5, \mathbf{e}_1 + \mathbf{e}_6, \mathbf{e}_2 + \mathbf{e}_6, \mathbf{e}_3 + \mathbf{e}_6, \mathbf{e}_4 + \mathbf{e}_6, \mathbf{e}_5 + \mathbf{e}_6\}$ appearing in [13, Example 2.1] is unimodular and its toric ideal $I_{\mathcal{A}}$ is generated by indispensable binomials $x_1x_8 - x_2x_6, x_2x_9 - x_3x_7, x_3x_{10} - x_4x_8, x_4x_6 - x_5x_9, x_1x_{10} - x_5x_7$ which does not form the reduced Gröbner basis with respect to any monomial order. Hence $\mathcal{A}$ satisfies (ii), but does not satisfy (iv) (and (iii)).

Example 5.3. The configuration $\mathcal{A} = \{\mathbf{e}_i + \mathbf{e}_j; 1 \leq i < j \leq 4\}$ is unimodular and its toric ideal $I_{\mathcal{A}}$ is generated by $x_1x_2 - x_5x_6, x_3x_4 - x_5x_6$ which form the reduced Gröbner basis with respect to a monomial order. However, there exists no indispensable binomials of $I_{\mathcal{A}}$. Hence $\mathcal{A}$ satisfies both (ii) and (iv), but does not satisfy (iii).

Example 5.4. The configuration $\mathcal{A} = \{\mathbf{e}_1 + \mathbf{e}_2, \mathbf{e}_2 + \mathbf{e}_3, \mathbf{e}_1 + \mathbf{e}_3, \mathbf{e}_1 + \mathbf{e}_4, \mathbf{e}_4 + \mathbf{e}_5, \mathbf{e}_5 + \mathbf{e}_6, \mathbf{e}_6 + \mathbf{e}_7, \mathbf{e}_5 + \mathbf{e}_7\}$ is not unimodular (in fact, not normal) and its toric ideal $I_{\mathcal{A}}$ is generated by $x_1x_3x_5^2x_7 - x_2x_4^2x_6x_8$ which is the reduced Gröbner basis with respect to any monomial order. Hence $\mathcal{A}$ satisfies (iii) (and (iv)), but does not satisfy (ii).

Conjecture 5.5. The following conditions are equivalent for $\mathcal{A}_{r_1r_2\cdots r_n}$:

(a) either $n = 2$ or $r_n = 2$;
(b) there exists a monomial order $<$ such that the reduced Gröbner basis of $I_{\mathcal{A}_{r_1r_2\cdots r_n}}$ with respect to $<$ is the set of indispensable binomials of $I_{\mathcal{A}_{r_1r_2\cdots r_n}}$;
(c) there exists a monomial order $<$ such that the reduced Gröbner basis of $I_{\mathcal{A}_{r_1r_2\cdots r_n}}$ with respect to $<$ is a minimal set of binomial generators of $I_{\mathcal{A}_{r_1r_2\cdots r_n}}$.

Since $\mathcal{A}_{r_1r_2\cdots r_{n-1}2}$ is the Lawrence lifting of $\mathcal{A}_{r_1r_2\cdots r_{n-1}}$, (a) $\Longrightarrow$ (b) $\Longrightarrow$ (c) holds for Conjecture 5.5.

6. Normal semigroup rings

We now study normal semigroup rings arising from contingency tables. First, we generalize [11, Lemma 3.1].

Lemma 6.1. *Suppose that there exists a binomial $f \in I_{\mathcal{A}}$ such that f is not redundant and none of whose monomials is squarefree. Then $K[\mathcal{A}]$ is not normal.*

Proof. Suppose that $f = x_i^2 u - x_j^2 v$ appears in a minimal set of generators of $I_{\mathcal{A}}$, where u and v are monomials. Since $\pi(x_i^2 u) = \pi(x_j^2 v)$, we have $\pi(x_i u/x_j) = \sqrt{\pi(uv)}$. Suppose that $\pi(x_i u/x_j)$ belongs to $K[\mathcal{A}]$. Then there exists a monomial w such that $\pi(w) = \pi(x_i u/x_j)$. Hence we have $\pi(x_i u) = \pi(x_j w)$. Thus the binomial $g = x_i u - x_j w$ belongs to $I_{\mathcal{A}}$. Since $f = x_i g - x_j h$, where $h = x_i w - x_j v \in I_{\mathcal{A}}$, this contradicts that f appears in a minimal set of generators of $I_{\mathcal{A}}$. Hence $\pi(x_i u/x_j) \notin K[\mathcal{A}]$ and $\pi(x_i u/x_j)^2 = \pi(uv) \in K[\mathcal{A}]$. Thus $K[\mathcal{A}]$ is not normal as required. □

Remark 6.2. Thanks to the proof of Lemma 6.1, we can say that, if there exists a binomial $f \in I_{\mathcal{A}}$ such that f is not *1-norm lower degree reducible* and none of whose monomials is squarefree. Then $K[\mathcal{A}]$ is not normal.

Corollary 6.3. If there exists an indispensable binomial of $I_{\mathcal{A}}$ such that none of whose monomials is squarefree, then $K[\mathcal{A}]$ is not normal.

Theorem 6.4. *Work with the same notation as before. Then we have*

$r_1 \times r_2$ *or* $r_1 \times r_2 \times 2 \times \cdots \times 2$	*unimodular*
$r_1 \times 3 \times 3$	*normal*
$5 \times 5 \times 3$ *or* $5 \times 4 \times 3$ *or* $4 \times 4 \times 3$	*UNKNOWN if normal or not*
otherwise, i.e., $n \geq 4$ *and* $r_3 \geq 3$ $n = 3$ *and* $r_3 \geq 4$ $n = 3$, $r_3 = 3$, $r_1 \geq 6$ *and* $r_2 \geq 4$	*not normal*

Proof. Since $\mathcal{A}_{r_1 r_2 2 \cdots 2}$ is obtained by taking the Lawrence lifting several times from $\mathcal{A}_{r_1 r_2}$, it follows that $\mathcal{A}_{r_1 r_2 2 \cdots 2}$ is unimodular. Since a squarefree initial ideal of $I_{\mathcal{A}_{r_1 33}}$ is given in [5], the semigroup ring $K[\mathcal{A}_{r_1 33}]$ is normal.

Case 1. $n \geq 4$ and $r_3 \geq 3$.

Since $\mathcal{A}_{333}$ is not unimodular, the Lawrence lifting $\mathcal{A}_{3332}$ is not normal by [11, Theorem 3.4]. Similarly, $\mathcal{A}_{3332 \cdots 2}$ is not normal. Moreover, if $n \geq 4$ and $r_3 \geq 3$, then $\mathcal{A}_{3332 \cdots 2}$ is a combinatorial pure subring of $\mathcal{A}_{r_1 r_2 \cdots r_n}$. By virtue of [11, Proposition 1.2], $K[\mathcal{A}_{r_1 r_2 \cdots r_n}]$ is not normal.

Case 2. $n = 3$ and $r_3 \geq 4$.

It is known [3] that there exists an indispensable binomial

$$\begin{aligned} f = {} & x_{111}x_{123}x_{132}x_{214}x_{233}x_{241}x_{312}^2 x_{321}x_{334}x_{344}x_{414}x_{422}x_{443} \\ & - x_{112}x_{121}x_{133}x_{211}x_{234}x_{243}x_{314}^2 x_{322}x_{332}x_{341}x_{412}x_{423}x_{444} \end{aligned}$$

of $I_{\mathcal{A}_{444}}$. None of the monomials of f is squarefree. By using Corollary 6.3, $K[\mathcal{A}_{444}]$ is not normal. Moreover, if $r_3 \geq 4$, then $\mathcal{A}_{444}$ is a combinatorial pure subring of $\mathcal{A}_{r_1 r_2 r_3}$. Hence $K[\mathcal{A}_{r_1 r_2 r_3}]$ is not normal.

Case 3. $n = 3, r_3 = 3, r_1 \geq 6$ and $r_2 \geq 4$.

It is known [3] that there exists an indispensable binomial

$$f = x_{111}x_{143}x_{221}x_{212}x_{331}x_{323}x_{432}x_{413}x_{522}x_{543}x_{641}x_{642}x_{633}^2 \\ - x_{141}x_{113}x_{211}x_{222}x_{321}x_{333}x_{412}x_{433}x_{542}x_{523}x_{631}x_{632}x_{643}^2$$

of $I_{\mathcal{A}_{643}}$. None of the monomials of f is squarefree. By using Corollary 6.3 again, $K[\mathcal{A}_{643}]$ is not normal. Moreover, if $r_3 = 3$, $r_1 \geq 6$ and $r_2 \geq 4$, then $\mathcal{A}_{643}$ is a combinatorial pure subring of $\mathcal{A}_{r_1r_2r_3}$. Hence $K[\mathcal{A}_{r_1r_2r_3}]$ is not normal, as desired. □

Corollary 6.5. Work with the same notation as before. Then the following conditions are equivalent for $K[\mathcal{A}_{r_1r_2\cdots r_n}]$:

(i) The rth Lawrence lifting of $K[\mathcal{A}_{r_1r_2\cdots r_n}]$ is normal for all r ;
(ii) $n = 2$ and either $r_2 = 2$ or $r_1 = r_2 = 3$.

Even though a unique minimal set of generators of $I_{\mathcal{A}_{r_143}}$ is given in [3], it seems to be difficult to know if $K[\mathcal{A}_{543}]$ and $K[\mathcal{A}_{443}]$ are normal.

Question 6.6. Are $K[\mathcal{A}_{553}]$, $K[\mathcal{A}_{543}]$ and $K[\mathcal{A}_{443}]$ normal?

Acknowledgment

The authors are grateful to Serkan Hoşten for his useful suggestions and comments.

References

[1] A. Agresti, *Categorical Data Analysis.* John Wiley & Sons, (2002)

[2] S. Aoki and A. Takemura, Minimal basis for connected Markov chain over $3 \times 3 \times K$ contingency tables with fixed two-dimensional marginals, *Aust. N. Z. J. Stat.* **45** (2003) 229–249

[3] S. Aoki and A. Takemura, The list of indispensable moves of the unique minimal Markov basis for $3 \times 4 \times K$ and $4 \times 4 \times 4$ contingency tables with fixed two-dimensional marginals, *METR Tech. Rep.*, 03-38

[4] G. Boffi and F. Rossi, Gröbner bases related to 3-dimensional transportation problems, *Quaderni matematici Univ. Trieste* II, **482** (2000)

[5] G. Boffi and F. Rossi, *A stability property of certain Gröbner bases.* preprint, February (2003)

[6] A. Capani, G. Niesi and L. Robbiano, *CoCoA, a system for doing Computations in Commutative Algebra.* `http://cocoa.dima.unige.it/`

[7] P. Diaconis and B. Sturmfels, Algebraic algorithms for sampling from conditional distributions, *Ann. Statist.* **26** (1998) 363–397

[8] D. R. Fulkerson, A. J. Hoffman and M. H. McAndrew, Some properties of graphs with multiple edges, *Canad. J. Math.* **17** (1965) 166–177

[9] S. Hoşten, A survey of toric initial ideals in the present book.

[10] S. Hoşten and S. Sullivant, Gröbner bases and polyhedral geometry of reducible and cyclic models, *J. Combin. Theory Ser. A* **100** no. 2 (2002) 277–301

[11] H. Ohsugi, J. Herzog and T. Hibi, Combinatorial pure subrings, *Osaka J. Math.* **37** (2000) 745–757

[12] H. Ohsugi and T. Hibi, A normal (0,1)-polytope none of whose regular triangulations is unimodular, *Discrete Comput. Geom.* **21** (1999) 201–204

[13] H. Ohsugi and T. Hibi, Toric ideals generated by quadratic binomials, *J. Algebra* **218** (1999) 509–527

[14] H. Ohsugi and T. Hibi, Compressed polytopes, initial ideals and complete multipartite graphs, *Illinois J. Math.* **44** (2000) 391–406

[15] H. Ohsugi and T. Hibi, Convex polytopes all of whose reverse lexicographic initial ideals are squarefree, *Proc. Amer. Math. Soc.* **129** no. 9 (2001) 2541–2546

[16] H. Ohsugi and T. Hibi, Indispensable binomials of finite graphs, *J. Algebra Appl.* Vol. 4 no. 4 (2005) 421–434

[17] F. Santos and B. Sturmfels, Higher Lawrence configuraions, *J. Combin. Theory Ser. A* **103** (2003) 151–164

[18] B. Sturmfels, *Gröbner bases and convex polytopes.* Amer. Math. Soc., Providence, RI, (1995)

[19] S. Sullivant. *Compressed polytopes and statistical disclosure limitation.* December 2004, `math.CO/0412535`

[20] A. Takemura and S. Aoki. Some characterizations of minimal Markov basis for sampling from discrete conditional distributions, *Ann. Inst. Statist. Math.* **56** (1) (2003) 1–17

[21] A. Takemura and S. Aoki, Distance reducing Markov bases for sampling from discrete sample space, *Bernoulli* **11** (5) (2005) 793–813

Proc. Int. Conf. – Commutative Algebra and Combinatorics No. 4, 2007, pp. 117–155.

Primary Decompositions

Irena Swanson

Department of Mathematical Sciences, New Mexico State University, Las Cruces, NM 88003, USA;
Current address: Reed College, 3203 SE Woodstock Blvd., Portland, OR 97202, USA
e-mail: iswanson@reed.edu

Abstract. The aim of these lectures is to show that primary ideals, and also primary modules, are an important tool for solving problems, proving theorems, understanding the structure of rings, modules, and ideals, and that there are enough of them to be able to apply the theory.

1. Primary ideals

This section contains the basic definitions. Throughout all rings are commutative with identity, and most of them are Noetherian.

Definition 1.1. An ideal I in a ring R is *primary* if $I \neq R$ and every zerodivisor in R/I is nilpotent.

Examples 1.2. Here are some examples of primary ideals:

(1) Any prime ideal is primary.
(2) For any prime integer p and any positive integer n, $p^n\mathbb{Z}$ is a primary ideal in $\mathbb{Z}$.
(3) More generally, let $\mathfrak{m}$ be a maximal ideal in a Noetherian ring R. Let I be any ideal in R such that $\sqrt{I} = \mathfrak{m}$. Then I is a primary ideal. Namely, if $r \in R$ is a zerodivisor modulo I, then as R/I is Artinian with only one maximal ideal, necessarily the image of r is in this maximal ideal. But then a power of r lies in I.

Lemma 1.3. *Let I be a primary ideal in a ring R. Then $\sqrt{I}$ is a prime ideal.*

Proof. Let $r, s \in R$ such that $rs \in \sqrt{I}$. Then there exists a positive integer n such that $r^n s^n = (rs)^n \in I$. If $s^n \in I$, then $s \in \sqrt{I}$, and we are done. So suppose that $s^n \notin I$. As $r^n s^n \in I$ and as I is primary, r^n is nilpotent on R/I. Thus $r \in \sqrt{I}$. □

Definition 1.4. Let I be a primary ideal in R and $P = \sqrt{I}$. Then I is also called *P-primary.*

The condition that $\sqrt{I}$ be a prime ideal P does not guarantee that I is P-primary, or primary at all. For example, let $R = k[X, Y]$ be a polynomial ring in variables X and Y over a field k. Let I be the ideal (X^2, XY). Then $\sqrt{I} = (X)$ is a prime ideal in R. However, I is not primary as Y is a zerodivisor on R/I, but it is not nilpotent.

In algebraic geometry an algebraic set (a variety) can be decomposed as a union of irreducible algebraic sets. In the standard correspondence between algebraic sets and ideals, algebraic sets correspond to radical ideals and irreducible algebraic sets correspond to prime ideals. Thus the decomposition of algebraic sets into irreducible ones corresponds to writing a radical ideal as an intersection of prime ideals. For example, the zero set of $X^3 - XY^3$ in $\mathbb{R}^2$ can be drawn as

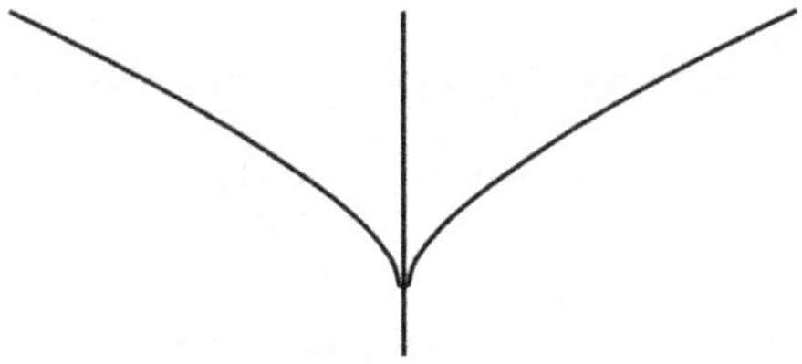

Clearly this algebraic set is the union of the vertical line $X = 0$ and the curve $X^2 - Y^3 = 0$, which correspond to prime ideals (X) and $(X^2 - Y^3)$. Thus $(X^3 - XY^3) = (X) \cap (X^2 - Y^3)$ reflects the primary (even prime) decomposition.

Now here are two ways to draw the zero set of the polynomials in the ideal (X^2, XY):

 and 

In the picture on the right we are emphasizing that the functions X^2, XY vanish at the origin $(0, 0)$ to order 2. Clearly $(X^2, XY) = (X) \cap (X^2, XY, Y^2)$. The vanishing along the line $X = 0$ to order 1 is encoded in the intersectand (X), and the vanishing to second order at the origin is expressed by the intersectand $(X, Y)^2 = (X^2, XY, Y^2)$. Indeed, $(X^2, XY) = (X) \cap (X^2, XY, Y^2)$ is a primary decomposition.

Thus primary decompositions can contain also the information on the vanishing along the components, even the embedded ones. More generally there is the famous Zariski–Nagata theorem:

Theorem 1.5 (Zariski, Nagata). *Let k be an algebraically closed field of characteristic zero and R a polynomial ring over k. Let P be a prime ideal in R. For all positive integers n, let P_n be the set of all elements of R which vanish along the zero set of P to order at least n. Then P_n equals the smallest P-primary ideal containing P^n.*

Remark 1.6. The smallest P-primary ideal containing P^n is called the *nth symbolic power of P*, and it exists in every Noetherian ring. We will prove this existence in Section 3 as a consequence of the existence of primary decompositions.

2. Primary modules

One can, more generally, develop the analogous theory for modules. We proceed to do so not just for the sake of generality, but because we will need the results for modules to establish the desired results for ideals. See for example proof of Proposition 7.5: the result for ideals uses the theory of primary decompositions for modules.

Definition 2.1. Let R be a Noetherian ring and M a finitely generated R-module. A submodule N of M is said to be *primary* if $N \neq M$ and whenever $r \in R, m \in M \backslash N$, and $rm \in N$, then there exists a positive integer n such that $r^n M \subseteq N$.

In other words, N is primary in M if and only if for any $r \in R$, whenever multiplication by r on M/N is not injective, then it is nilpotent as a function.

Observe that N is primary in M if and only if 0 is primary in M/N.

Here are some examples of primary modules:

(1) If I is a primary ideal in a ring R, then I is also a primary submodule of R, when the ring is thought of as a module over itself.
(2) If P is a prime ideal in a ring R, then for any positive integer n, $P \oplus \cdots \oplus P$ (n copies) is a primary submodule of R^n.

Lemma 2.2. *Let $N \subseteq M$ be a primary submodule. Then $\sqrt{N :_R M}$ is a prime ideal.*

Proof. Let $r, s \in R$ such that $rs \in \sqrt{N :_R M}$. Then there exists a positive integer n such that $(rs)^n \in N :_R M$. Thus $r^n s^n M \subseteq N$. If $s^n M \subseteq N$, then $s \in \sqrt{N :_R M}$, and we are done. So suppose that $s^n M \not\subseteq N$. Choose $m \in M$ such that $s^n m \notin N$. Then $r^n(s^n m) \in N$, so that as N is primary in M, multiplication by r^n on M/N is nilpotent. Thus $r \in \sqrt{N :_R M}$. □

Definition 2.3. Let $N \subseteq M$ be a primary submodule. Then N is also called P*-primary*, where P is the prime ideal $P = \sqrt{N :_R M}$.

Primary modules generalize the notion of primary ideals, and here is another connection:

Lemma 2.4. *Let R be a ring, P a prime ideal in R, M an R-module, and N a P-primary submodule of M. Then $N :_R M$ is a P-primary ideal.*

Proof. Set $I = N :_R M$. Then by definition, $\sqrt{I} = P$. Let $r, s \in R$ such that $rs \in I$ and $s \notin P$. As P is a prime ideal, then $r \in P$. Hence for some positive integer n, $r^n \in I$. Choose n to be smallest possible. Then $r^{n-1}M \not\subseteq N$. Choose $m \in M$ such that $r^{n-1}m \notin N$. If $n > 1$, as $sr^{n-1}m \in N$, it follows that $s^l M \subseteq N$ for some positive integer l, whence $s^l \in I \subseteq P$, which is a contradiction. So necessarily $n = 1$, so $r \in I$. □

The converse fails in general:

Example 2.5. Let R be a Noetherian ring, $P \subsetneq Q$ prime ideals, $M = R \oplus R$, $N = P \oplus Q$. Then $N :_R M = P$, but N is not P-primary. Namely, choose $r \in Q \backslash P$, and set $m = (0, 1) \in M \backslash N$. Then $rm \in N$, yet $r^n(1, 0)$ is not in N for any n, so that $r^n M$ is not a submodule of N for any n.

Lemma 2.6. *The intersection of any two P-primary submodules of M is P-primary.*

Proof. Let N, N' be P-primary submodules. Let $r \in R$. Let $m \in M \backslash N \cap N'$ such that $rm \in N \cap N'$. By assumption $m \notin N$ or $m \notin N'$. But both N and N' are primary. If $m \notin N$, then $r^n M \subseteq N$ for all large n, and if $m \notin N'$, then $r^n M \subseteq N'$ for all large n. In any case, a power of r is in P, thus $r \in P$. It follows that $r \in \sqrt{(N : M)} \cap \sqrt{N' : M} = \sqrt{(N : M) \cap (N' : M)} = \sqrt{(N \cap N') : M}$, so that $r^n M \subseteq N \cap N'$ for all large n. Thus $N \cap N'$ is primary to P. □

Lemma 2.7. *Let R be a ring, P a prime ideal, M an R-module, and N a P-primary submodule of M. Then for any $r \in R$,*

$$N :_M r = \begin{cases} N, & \text{if } r \notin P; \\ M, & \text{if } r \in N :_R M; \\ \text{a } P\text{-primary submodule of } M \text{ strictly containing } N, & \text{if } r \in P \setminus (N :_R M). \end{cases}$$

Proof. First assume that r is not in P. Let $m \in (N :_M r) \setminus N$. Then $rm \in N$, and by definition of primary modules, $r^n \in N :_R M \subseteq P$ for some positive integer n. Thus $r \in P$, contradiction. Thus $N :_M r = N$ whenever $r \notin P$.

If $r \in N :_R M$, clearly $N :_M r = M$.

Now assume that $r \in P \setminus (N :_R M)$. Let $x \in R$ and $m \in M \setminus (N :_M r)$ such that $xm \in N :_M r$. Then $rm \notin N$ and $xrm \in N$. As N is P-primary, $x^n M \subseteq N$ for some positive integer n. This means that $x \in P$ and that $x^n M \subseteq N :_M r$, so that $N :_M r$ is P-primary. Furthermore, $r \notin N :_R M$, and by definition of P-primary, $r^n \in N :_R M$ for some n. Let n be minimal such integer. Then $n > 1$, and there exists $m \in M$ such that $r^{n-1} m \notin N$. Thus $N :_M r$ contains N and $r^{n-1} m$, so that it contains N properly. □

Lemma 2.8. *Let R be a ring, P a prime ideal, and M an R-module. Assume that N is a P-primary submodule of M. Then for any $m \in M$,*

$$N :_R m = \begin{cases} R, & \text{if } m \in N; \\ \text{a } P\text{-primary ideal containing } N :_R M, & \text{if } m \notin N. \end{cases}$$

Moreover, if P is finitely generated, there exists $m \in M$ such that $N : m = P$.

Proof. Certainly $N :_R m = M$ if $m \in N$. Now assume that $m \in M \setminus N$. Let $x, y \in R$ such that $xy \in N :_R m$. Assume that $y \notin N :_R m$. Then $ym \notin N$ and $xym \in N$, so that by the definition of primary modules, $x^n M \subseteq N$ for some n. Thus $x^n \in N :_R M \subseteq P$, which proves that $N :_R m$ is a P-primary ideal.

Now assume that P is finitely generated. As N is P-primary in M, $N \neq M$, so by the previous part for any $m \in M \setminus N$, $N :_R m$ is P-primary. As P is finitely generated, there exists a positive integer n such that $P^n \subseteq N :_R m$. Choose n to be the smallest integer with this property. Then $n \geq 1$, and there exists $r \in P^{n-1}$ such that $r \notin N :_R m$. Then $N :_R rm = (N :_R m) :_R r$ is P-primary by Lemma 2.7, and as it contains P, it equals P. □

Lemma 2.9. *Let R be a ring, U a multiplicatively closed subset of R, and $N \subseteq M$ R-modules.*

1. *If N is primary in M and $(N :_R M) \cap U \neq \emptyset$, then $U^{-1}N = U^{-1}M$.*
2. *If N is primary in M and $(N :_R M) \cap U = \emptyset$, then $U^{-1}N$ is primary in $U^{-1}M$.*
3. *There is a one-to-one correspondence between primary submodules N of M such that $(N :_R M)$ is disjoint from U and primary submodules of $U^{-1}M$. The correspondence is given by $N \mapsto U^{-1}N$ for N a submodule in M, and $K \mapsto K \cap M$ for K a submodule of $U^{-1}M$.*

Proof. Part (1) follows trivially. Assume hypotheses of (2). Let $x \in R$, $m \in M$, $u, v \in U$, such that $(x/u)(m/v) \in U^{-1}N$. Assume that $m/v \notin U^{-1}N$. Then $m \notin N$, and there exists $w \in U$ such that $wxm \in N$. Thus $(wx)^n M \subseteq N$ for some integer n, and $wx \in \sqrt{N :_R M}$, which is a prime ideal. But by the assumption on U then $x \in \sqrt{N :_R M}$. Hence x/u is in the radical of $U^{-1}N :_{U^{-1}R} U^{-1}M$, which proves that $U^{-1}N$ is primary to $U^{-1}P$.

Now let K be primary in $U^{-1}M$, and set $N = K \cap M$. Then $N \neq M$. Let $m \in M \backslash N$ and $r \in R$ such that $rm \in N$. Then $m \notin K$ and $rm \in U^{-1}N \subseteq K$. Thus for some positive integer n, $r^n U^{-1}M \subseteq K$. Thus $r^n M \subseteq K \cap M = N$, which proves that N is primary.

Furthermore, it is always true that $N \subseteq U^{-1}N \cap M$ and $U^{-1}(K \cap M) = K$. If $m \in U^{-1}N \cap M$, then $wm \in N$ for some $w \in U$. If $m \notin N$, then $w \in N :_M m$, which is in $\sqrt{N :_R M}$ by Lemma 2.8. But this contradicts the assumption that N is primary to a prime ideal disjoint from U. This establishes the one-to-one correspondence. □

Thus the primary property is preserved under many localizations. It is also preserved under passage to a polynomial extension:

Proposition 2.10. *Let R be a ring, X a variable over R, M an R-module and N a primary submodule. Then $NR[X]$ is a primary $R[X]$-submodule of $MR[X]$. Furthermore, if N is a P-primary submodule of M, then $NR[X]$ is $PR[X]$-primary submodule of $MR[X]$.*

Proof. It is easy to see that $NR[X] \neq MR[X]$. Let $m \in MR[X] \backslash NR[X]$, $r \in R[X]$ such that $rm \in NR[X]$. Write $r = \sum_{i=j_0}^{j_1} r_i X^i$, $m = \sum_{i=k_0}^{k_1} s_i X^i$, for some $r_i \in R$, $s_i \in M$, with $r_{j_1} \neq 0$, $s_{k_1} \notin N$. We proceed by induction on $j_1 - j_0$. As $rm \in NR[X]$, then $r_{j_1} s_{k_1} \in N$. As N is primary in M, there exists a positive integer n such that $r_{j_1}^n \in N :_R M$. In particular, $r_{j_1}^n m \in NR[X]$. Let a be the smallest integer such that $r_{j_1}^a m \in NR[X]$. Then $a \geq 1$ and $r_{j_1}^{a-1} m \notin NR[X]$. Set $m' = r_{j_1}^{a-1} m$. Then $rm' \in NR[X]$, and $r_{j_1} X^{j_1} m' \in NR[X]$. It follows that $(r - r_{j_1} X^{j_1}) m' \in NR[X]$. But $r - r_{j_1} X^{j_1}$ has strictly fewer terms than r, so by induction, $r - r_{j_1} X^{j_1} \in \sqrt{NR[X] :_{R[X]} MR[X]}$. Thus $r \in \sqrt{NR[X] :_{R[X]} MR[X]}$. The last part follows easily. □

It is not true that the passage to a faithfully flat extension preserves the primary property. For example, let R be the ring $\mathbb{R}[X]/(X^2 + 1)$. Let S be $\mathbb{C}[X]/(X^2 + 1)$. Then S is a faithfully flat extension of R. In R, the zero ideal is a prime ideal, hence primary. However, the zero ideal in S is not primary, as $X^2 + 1$ factors over $\mathbb{C}$ into two distinct irreducible factors.

3. Primary decompositions

Definition 3.1. Let R be a ring, M an R-module and N a submodule. A *primary decomposition* of N is an expression of N as a finite intersection of primary submodules of M. In other words, a primary decomposition of N is $N = N_1 \cap \cdots \cap N_s$, where each N_i is primary in M.

Note that N has a primary decomposition $N = N_1 \cap \cdots \cap N_s$ if and only if 0 in M/N has a primary decomposition $0 = N_1/N \cap \cdots \cap N_s/N$.

Primary decompositions exist for Noetherian modules, as we prove below.

Most of the time we will work with ideals, so for emphasis we state explicitly the definition of primary decompositions for ideals:

Definition 3.2. Let I be an ideal in a ring R. A decomposition $I = \bigcap_{i=1}^{s} q_i$ is a *primary decomposition* of I if $q_1, \dots, q_s$ are primary ideals.

One of course wants a constructive method to obtain primary decompositions. The following lemma is a step in that direction:

Lemma 3.3. *Let R be a ring, M a Noetherian R-module, N a submodule of M, and $r \in R$. Then there exists a positive integer n such that $N :_M r^n = N :_M r^{n+1}$, and then for all $m \geq n$, $N :_M r^m = N :_M r^n$ and $N = (N :_M r^m) \cap (N + r^m M)$.*

Proof. For any elements $r, s \in R$, $N :_M r \subseteq N :_M rs$. In particular, $N \subseteq N :_M r \subseteq N :_M r^2 \subseteq \cdots$ is an ascending chain of submodules in M. As M is Noetherian, there exists n such that $N :_M r^n = N :_M r^{n+1}$. Assume that $m > n + 1$. If $s \in N :_M r^m$, then $r^{m-1} rs \in N$, so by induction on m, $r^n rs \in N$, whence $r^n s \in N$. This proves that $N :_M r^n = N :_M r^m$ for all $m \geq n$.

Certainly $N \subseteq (N : r^m) \cap (N + r^m M)$. Let $s \in (N : r^m) \cap (N + r^m M)$. Write $s = a + r^m b$ for some $a \in N$ and $b \in M$. Then $sr^m = ar^m + r^{2m} b \in N$, so that $r^{2m} b \in N$. By the first part then $r^m b \in N$, so that $s = a + r^m b \in N$. Thus $N = (N : r^m) \cap (N + r^m M)$. □

The proof above shows that the ascending chain $N \subseteq N :_M r \subseteq N :_M r^2 \subseteq N :_M r^3 \subseteq N :_M r^4 \cdots$ stabilizes as soon as two consecutive modules in the chain are equal. This makes the computation of the stable value of this chain feasible. This stable value is typically denoted as $N :_M r^\infty$, but no meaning is attached to the notation r^∞. If I is an ideal, the stable value of $I \subseteq I : r \subseteq I : r^2 \subseteq I : r^3 \subseteq \cdots$ is similarly denoted as $I : r^\infty$.

Theorem 3.4. *Let R be a ring, and M a Noetherian R-module. Then any proper submodule N of M has a primary decomposition.*

In particular, every proper ideal I in a Noetherian ring R has a primary decomposition.

Proof. If N is primary, we are done. So assume that N is not primary. Let $r \in R$, $m \in M \setminus N$ such that $rm \in N$ and $r^n M \not\subseteq N$ for any positive integer n. As M is Noetherian, the chain $N \subseteq N :_M r \subseteq N :_M r^2 \subseteq \cdots$ terminates. Choose n such that $N :_M r^n = N :_M r^{n+1} = \cdots$. Let $N' = N :_M r^n$, $N'' = (N + r^n M)$. Then both N' and N'' properly contain N. By Lemma 3.3, $N = N' \cap N''$, so it suffices to find primary decompositions of N' and N''. But this we can do by Noetherian induction. □

We summarize the procedure from the proof above formally for the case of ideals:

Procedure 3.5. Let I be an ideal in a Noetherian ring R, and $r, s \in R$ such that $r \notin \sqrt{I}$, $s \notin I$, $rs \in I$.

1. Find n such that $I : r^\infty = I : r^n$. (See comment after Lemma 3.3.)
2. Set $I' = I + (r^n)$, $I'' = I : r^n$. Then $I = I' \cap I''$.
3. To find a primary decomposition of I, it suffices to find primary decompositions of strictly larger ideals I' and I''.

Note that Procedure 3.5 does not say how one can find the appropriate zero divisors r, s modulo an ideal in order to compute its primary decomposition. Furthermore, the procedure gives no indication on how to determine whether an ideal is primary.

Nevertheless, the procedure can be used successfully to compute some examples:

Example 3.6. Let I be the ideal $I = (X^2, XY)$ in the polynomial ring $k[X, Y]$. Observe that $X \notin I$, $Y \notin \sqrt{I} = (X)$, but that $XY \in I$. So by Procedure 3.5, we need to find $I : Y^\infty$. Clearly $I : Y = (X) = I : Y^\infty$, so, as in the proof, $I = (I : Y^\infty) \cap (I + (Y)) = (X) \cap (X^2, Y)$. But (X) is prime, hence primary, and (X^2, Y) is primary by Example 1.2 (3), as its radical is a maximal ideal. Thus $(X) \cap (X^2, Y) = I$ is a primary decomposition of I.

If we repeat the same procedure with elements X and Y^n in place of X and Y, then we obtain the decomposition $I = (X) \cap (X^2, XY, Y^n)$, which is a primary decomposition by the same reasoning as the one above.

This example shows that primary decompositions are not unique. See Lemma 3.15 for a formalization of this.

If, furthermore, we repeat the procedure above with elements X and $Y(Y-1)$ in place of X and Y, then we obtain the decomposition $I = (X) \cap (X^2, XY, Y(Y-1))$. The component (X) is primary, but the component $J = (X^2, XY, Y(Y-1))$ is not: $Y(Y-1) \in J$, $Y \notin J$ and $Y - 1 \notin \sqrt{J} = (X, Y(Y-1))$. Then $J : (Y-1)^\infty = (X^2, Y) = J : (Y-1)$ is primary, and $J + (Y-1) = (X, Y-1)$ is also primary as it is a prime ideal. Thus by Lemma 3.3 or by Procedure 3.5, $J = (X^2, Y) \cap (X, Y-1)$, whence $I = (X) \cap (X^2, Y) \cap (X, Y-1)$ is also a primary decomposition.

However, observe that the component $(X, Y-1)$ in the last decomposition is redundant as it contains the component (X). We next formalize the notion of irredundancy or minimality:

Definition 3.7. A primary decomposition $N = N_1 \cap \cdots \cap N_s$ of a submodule N of M is *irredundant* or *minimal* if

(1) the prime ideals $\sqrt{N_1 :_R M}, \ldots, \sqrt{N_s :_R M}$ are distinct, and
(2) for all $j = 1, \ldots, s$, $N \neq \bigcap_{i \neq j} N_i$.

Explicitly for ideals, a primary decomposition $I = q_1 \cap \cdots \cap q_s$ of ideal I is an *irredundant* or *minimal* decomposition if $\sqrt{q_1}, \ldots, \sqrt{q_s}$ are all distinct, and for any $i \in \{1, \ldots, s\}$, $\cap_{j \neq i} q_j \neq I$.

Irredundant decompositions exist whenever primary decompositions exist:

Proposition 3.8. *Let R be a ring, M an R-module and N a submodule of M. If N has a primary decomposition, then N has an irredundant primary decomposition.*

Proof. Write $N = N_1 \cap \cdots \cap N_s$, with each N_i primary. By Lemma 2.6, we may combine any two N_i with the same radical to obtain another primary module. Thus once a primary decomposition is given, a primary decomposition satisfying (1) is easy to obtain. If for some j, the condition (2) is not satisfied, then we remove the jth component from the decomposition. By repeating this step, in finitely many steps we obtain a primary decomposition satisfying (1) and (2). □

Proposition 3.9. *Let $N \subseteq M$ be modules over a ring R. Assume that N has an irredundant primary decomposition $N = N_1 \cap \cdots \cap N_s$. Then*

$$\left\{\sqrt{N_i :_R M} \mid i = 1, \ldots, s\right\}$$
$$= \left\{P \in \operatorname{Spec} R \mid P \textit{ is minimal over } N :_R m \textit{ for some } m \in M\right\}.$$

This set is independent of the irredundant primary decomposition and contains all prime ideals in R which are minimal over $N :_R M$.

If R is Noetherian, then

$$\left\{\sqrt{N_i :_R M} \mid i = 1, \ldots, s\right\} = \left\{P \in \operatorname{Spec} R \mid P = N :_R m \textit{ for some } m \in M\right\}.$$

Proof. For $j = 1, \ldots, s$, let $P_j = \sqrt{N_j :_R M}$. By Lemma 2.2, P_j is a prime ideal.

For each $j \in \{1, \ldots, s\}$, choose $m_j \in \cap_{i \neq j} N_i \setminus N_j$. Then $N :_R m_j = \bigcap_i (N_i : m_j) = N_j : m_j$, which is primary to P_j by Lemma 2.8. Thus by the same lemma, if R is Noetherian, there exists $m \in M$ such that $N :_R m = P$. This proves that each P_j is in the set $\{P \in \operatorname{Spec} R \mid P$ is minimal over $N : m$ for some $m \in M\}$, or in case of Noetherian ring, in the set $\{P \in \operatorname{Spec} R \mid P = N : m$ for some $m \in M\}$.

Now let P be a prime ideal minimal over $N : m$ for some $m \in M$. Then P is minimal over $N : m = \bigcap_{i=i}^{s} (N_i : m)$, so necessarily there exists $j \in \{1, \ldots, s\}$ such that P is minimal over $N_j : m$. But then by Lemma 2.8, necessarily $P = P_j$. This proves the displayed equality, and hence that the set is independent of the irredundant primary decomposition.

Now assume that P is minimal over $N :_R M$. As $N :_R M = \bigcap_i (N_i :_R M)$, it follows that P is minimal over some $N_j :_R M$. But then by Lemma 2.2, $P = P_j$. □

The unique set in the theorem above has a name:

Definition 3.10. Let $N \subseteq M$ be modules over a ring R. The set

$$\{P \in \operatorname{Spec} R \mid P \text{ is minimal over } N :_R m \text{ for some } m \in M\}$$

is called *the set of weakly associated primes of M/N*, and is denoted $\widetilde{\operatorname{Ass}}(M/N)$.

If I is an ideal, *the weakly associated primes of the ideal I* are the weakly associated primes of the module R/I.

Note that $\widetilde{\operatorname{Ass}}(M/N)$ does not depend on N and M, but only on the module M/N. Thus to simplify notation, from now on we will usually talk about (weakly) associated primes and primary decompositions of the zero submodule in a module. Sometimes, for clarity, if M is a module over two rings, we will write $\widetilde{\operatorname{Ass}}_R(M)$ to denote the set of weakly associated primes of M as an R-module.

It is possible that the zero submodule of M does not have a primary decomposition, yet one can define associated primes of M.

Remark 3.11. There are several variant definitions of *associated primes* in the literature. The following are from Iroz-Rush [20] and Divaani-Aazar-Tousi [7]:

(1) If P is minimal over $0 :_R m$ for some $m \in M$, then P is, as above, called a *weakly associated prime* of M. Sometimes P is also called a *weak Bourbaki prime* of M.
(2) If P equals $0 :_R m$ for some $m \in M$, then P is called an *associated prime* of M, or a *Bourbaki prime* of M.
(3) If P is a Bourbaki prime of $M/0_P \cap M$, then P is called a *Noether prime* of M.
(4) If $0 :_R m$ is P-primary for some $m \in M$, then P is called a *Zariski-Samuel prime* of M.
(5) If for every $x \in P$ there exists $m \in M$ such that $x \in 0 :_R m \subseteq P$, then P is called a *Krull prime* of M.
(6) If for every finitely generated ideal $I \subseteq P$ there exists $m \in M$ such that $I \subseteq 0 :_R m \subseteq P$, then P is called a *strong Krull prime* of M.
(7) If there exists a multiplicatively closed subset U of R such that $U^{-1}P$ is a maximal ideal in the set of zerodivisors in $U^{-1}R$ on $U^{-1}M$, then P is called a *Nagata prime* of M.
(8) If there exists a submodule N of M such that $P = 0 :_R N$, then P is called a *Divaani-Aazar-Tousi prime* of M.

By convention, the zero module has no associated primes of any type.

All the primes above are the same if R is Noetherian and M is finitely generated, see Exercise 3.12. But these notions need not agree in general. For example, let k be a field and $t, x_1, x_2, \dots, y_1, y_2, \dots$ variables over k. Let

$$R = k[t, x_1, y_1, x_2, y_2, \dots]/(tx_iy_i \mid i)(y_i^i \mid i),$$

$P = (x_i \mid i)R$, and $Q = (y_i \mid i)R$. Then P is minimal over $0 :_R ty_1$, but P is not the radical of $0 :_R m$ for any $m \in M$. Also, Q is the radical of $0 : tx_1y_1$ but Q is not $0 : N$ for any $N \subseteq M$.

For Noetherian rings and modules at least, the following definition of associated primes is more standard:

Definition 3.12. Let $N \subseteq M$ be modules over a ring R. The set

$$\{P \in \operatorname{Spec} R \mid P = N :_R m \text{ for some } m \in M\}$$

is called *the set of associated primes of* M/N, and is denoted $\operatorname{Ass}(M/N)$ or $\operatorname{Ass}_R(M/N)$ to make the underlying ring clear. If I is an ideal, *the associated primes of the ideal* I are the associated primes of the module R/I.

Associated primes and primary decompositions localize in a natural way:

Corollary 3.13. Let M be an R-module, and U any multiplicatively closed subset of R. Then

$$\widetilde{\operatorname{Ass}}(U^{-1}(M)) = \{U^{-1}P \mid P \in \widetilde{\operatorname{Ass}}(M),\ P \cap U = \emptyset\},$$

and if R is Noetherian,

$$\operatorname{Ass}(U^{-1}(M)) = \{U^{-1}P \mid P \in \operatorname{Ass}(M),\ P \cap U = \emptyset\}.$$

If 0 has an irredundant primary decomposition $0 = \bigcap_i N_i$, then

$$U^{-1}0 = \bigcap_{U \cap (N_i :_R M) = \emptyset} U^{-1}N_i$$

is an irredundant primary decomposition if $U^{-1}0 \neq U^{-1}M$.

Proof. Let $P \in \widetilde{\operatorname{Ass}}\, M$ (resp. in $\operatorname{Ass} M$). Then there exists $m \in M$ such that P is minimal over $0 :_R m$ (resp. $P = 0 :_R m$). If $U \cap P = \emptyset$, then $U^{-1}P$ is minimal over $U^{-1}0 :_{U^{-1}R} m$ (resp. $U^{-1}P = U^{-1}0 :_{U^{-1}R} m$). Thus $U^{-1}P \in \widetilde{\operatorname{Ass}}\, U^{-1}M$, (resp. $U^{-1}P \in \operatorname{Ass} U^{-1}M$).

Now let $U^{-1}P \in \widetilde{\operatorname{Ass}}\, U^{-1}M$ (resp. in $\operatorname{Ass} U^{-1}M$). Then $P \cap U = \emptyset$ and there exists $m \in M$ such that $U^{-1}P$ is minimal over $U^{-1}0 :_{U^{-1}R} m$ (resp. $U^{-1}P = U^{-1}0 :_{U^{-1}R} m$). But then P is minimal over $0 :_R m$, and in particular $P \in \widetilde{\operatorname{Ass}}\, M$. If R is Noetherian, then by Proposition 3.9, $P \in \operatorname{Ass}(0 :_R m)$, hence in $\operatorname{Ass} M$.

This proves that the (weakly) associated primes localize as stated.

Now let $0 = \bigcap_i N_i$ be an irredundant primary decomposition. Let $P_i = \sqrt{N_i :_R M}$. By Proposition 3.9, $P_1, \ldots, P_s$ are the weakly associated primes of M. After reindexing, assume that $P_i \cap U = \emptyset$ for $i = 1, \ldots, r$, and that $P_i \cap U \neq \emptyset$ for $i = r+1, \ldots, s$. By Lemma 2.9, $U^{-1}N_i$ is primary to $U^{-1}P_i$ for $i = 1, \ldots, r$, and $U^{-1}N_i$ is $U^{-1}M$ for $i > r$.

Localize the primary decomposition at U: as $U^{-1}0 \neq U^{-1}M, r \geq 1$, and so $U^{-1}0 = \bigcap_{i=0}^{r} U^{-1}N_i$ is a primary decomposition. It is an irredundant primary decomposition because we have proved that the localizations of $P_1, \ldots, P_r$ are (weakly) associated. □

We proved in Proposition 3.9 that $\widetilde{\operatorname{Ass}}(M)$ contains the prime ideals which are minimal over $0 :_R M$. These are special, and form a special subset of $\widetilde{\operatorname{Ass}}(M)$. In general, for any type of associated prime ideals we partition the associated primes as follows:

Definition 3.14. Let R be a ring and M an R-module. Among the associated primes of a module (associated of any type as in Remark 3.11), the ones minimal with respect to inclusion are called the *minimal* prime ideals, the others *embedded* prime ideals.

If $0 = N_1 \cap \cdots \cap N_s$ is an irredundant primary decomposition, then N_i is called an *embedded component* of M if $\sqrt{N_i : M}$ is embedded, and is called a *minimal component* of M if $\sqrt{N_i : M}$ is minimal.

Lemma 3.15. *Let R be a ring and M an R-module in which the zero ideal has a primary decomposition. Each minimal component of M is uniquely determined.*

Explicitly, if P is a minimal prime in $\operatorname{Ass}(M)$, *then the P-primary component of M equals the kernel of the map $M \to M_P$.*

If R is Noetherian and M finitely generated, embedded components are never uniquely determined.

Proof. Let $0 = N_1 \cap \cdots \cap N_s$ be an irredundant primary decomposition. Let $P = \sqrt{N_j :_R M}$. If P is minimal, then $0_P = \bigcap_i (N_i)_P = (N_j)_P$, so that $0_P \cap M = (N_j)_P \cap M = N_j$ by Lemma 2.9, whence the P-primary component is uniquely determined.

Now let P be an embedded prime ideal. Let N_j be a P-primary component. As $\sqrt{N_j :_R M} = P$ and R is Noetherian, there exists a positive integer n such that $P^n M \subseteq N_j$. For any $m \geq n$, set $K_m = P^m M_P \cap M$. By Lemma 2.9, K_m is P-primary. Furthermore, $(K_m)_P = P^m M_P$ is contained in $(N_j)_P$, so that $K_m \subseteq N_j$. It follows that $\bigcap_{i \neq j} N_i \cap K_m \subseteq \bigcap_i N_i = 0$, so that $0 = \bigcap_{i \neq j} N_i \cap K_m$. Thus any K_m can be taken to be a P-primary component.

It remains to show that $K_m \neq K_{m'}$ whenever $m' \neq m$. By Lemma 2.9, $K_m \neq K_{m'}$ if and only if $P^m M_P \neq P^{m'} M_P$. By Nakayama's lemma, $P^m M_P = P^{m'} M_P$ if and only if $P^{\min\{m,m'\}} M_P = 0_P$. But P is not minimal over $0 :_R M$, so PR_P is not minimal over $0_P :_{R_P} M$, so $P^m M_P = 0_P$ is impossible. Thus $K_m \neq K_{m'}$ whenever $m' \neq m$. □

In particular, if P is a prime ideal in a Noetherian ring R, then for every positive integer n, there is a unique P-primary component of P^n. This component is called the nth *symbolic power* of P, and is usually denoted $P^{(n)}$.

Even though the embedded components are highly non-unique, the P-primary component from one decomposition can be used as the P-primary component in any other decomposition:

Theorem 3.16 ("Mix-and-match", Yao [52]). *Let R be a Noetherian ring and M a finitely generated R-module. Let $\{P_1, \ldots, P_s\} = \operatorname{Ass} M$. Assume that for all $i, j = 1, \ldots, s$, N_{ji} is a P_i-primary component of 0 in M, and that we have s primary decompositions of 0 in M:*

$$0 = \bigcap_{i=1}^{s} N_{ji}, \qquad j = 1, \ldots, s.$$

Then $0 = \bigcap_{i=1}^{s} N_{ii}$ is also a primary decomposition.

4. More ways to get associated primes

The only method for computing the associated primes of a module described so far involves the computation of a primary decomposition. This is not an easy task, as we will show in Section 9. Fortunately, there are indirect and efficient methods for computing the associated primes without computing the primary decompositions. We describe some of those methods in this section. Another method can be found in the proof of Theorem 8.4.

For example, the knowledge of zero divisors gives information on associated primes, and vice versa:

Proposition 4.1. *Let R be a ring and M an R-module. Then the set of zero divisors on M is contained in the set $\bigcup_{P \in \widetilde{\operatorname{Ass}}(M)} P$. If the latter set is finite (for example, if the zero submodule of M has a primary decomposition), then the set of zero divisors in M equals $\bigcup_{P \in \widetilde{\operatorname{Ass}}(M)} P$.*

Proof. Let m be a non-zero element of M and $x \in R$ such that $xm = 0$. Then $x \in 0 :_R m$. Thus by the definition of associated primes, the zerodivisor x on M is contained in some weakly associated prime of M.

Now let x be a non-zero element in some weakly associated prime P of M. Then there exists an element $m \in M$ such that P is minimal over $0 :_R m$. By assumption, there are only finitely many prime ideals minimal over $0 :_R m$. Thus there exists $r \in R$ contained in all the prime ideals minimal over $0 :_R m$ but not in P. Then rx is contained in all the prime ideals minimal over $0 :_R m$, so that for some integers n, n', $r^{n'} x^n m = 0$. Choose n, n' such that n is smallest possible. By the choice of r, $n \geq 1$. Then the non-zero element $r^{n'} x^{n-1} m$ is annihilated by x. Thus x is a zerodivisor on M. □

This gives another way to decide whether a given prime ideal P is associated to a module:

Lemma 4.2. *Let R be a ring, P a prime ideal, and M an R-module. If $0_P :_{M_P} P \neq 0_P$, then $PR_P \in \widetilde{\text{Ass}}(M_P)$.*

If $PR_P \in \widetilde{\text{Ass}}(M_P)$ is finitely generated (after localization at P), then $0_P :_{M_P} P \neq 0_P$.

Proof. Without loss of generality we may assume that R is a ring whose only maximal ideal is P.

Assume that $0 :_M P \neq 0$. Let m be non-zero in $0 :_M P$. Then $P \subseteq 0 :_R m$. As m is non-zero, $0 :_R m$ is a proper ideal, so that $P = 0 :_R m$, whence $P \in \text{Ass}(M)$ by Proposition 3.9.

Now assume that $P \in \widetilde{\text{Ass}}(M)$. Then P is minimal over $0 : m$ for some $m \in M$. As P is the maximal ideal and finitely generated, there exists an integer n such that $P^n \subseteq 0 :_R m$. Choose n minimal. Then $n \geq 1$ and $P^{n-1}m$ is a non-zero submodule of $0 :_M P$. □

Associated primes and even primary decompositions can be obtained via contraction:

Lemma 4.3. *Let R be a ring, S an R-algebra, $N \subseteq M$ S-modules, and K an R-submodule of M. Then $\widetilde{\text{Ass}}_R(K/N \cap K) \subseteq \{P \cap R \mid P \in \widetilde{\text{Ass}}_S(M/N)$.*

Furthermore, if N has a primary decomposition $N = N_1 \cap \cdots \cap N_s$ in M, then $N \cap K = (N_1 \cap K) \cap \cdots \cap (N_s \cap K)$, after removing intersectands equal to K, is a primary decomposition of the R-module $N \cap K$ in K.

Proof. Let $P \in \widetilde{\text{Ass}}_R(K/N \cap K)$. By localization we may assume that P is the only maximal ideal in R. There exists $m \in K \setminus N$ such that P is minimal over $N \cap K :_R m$. Then $N :_S m$ is a proper ideal in S which contains P. Any prime ideal Q in S minimal over $N :_S m$ contracts to P and is weakly associated to M/N. This proves the first statement.

Now let N_i be a P-primary submodule of the S-module M. Assume that $N_i \cap K$ is a proper submodule of K. Let $m \in K \setminus N_i$ and $r \in R$ such that $rm \in N_i \cap K$. Then $m \in M \setminus N_i$, and as N_i is primary, there exists a positive integer n such that $r^n M \subseteq N_i$. Thus $r_i^N K \subseteq N_i \cap K$. This proves that $N_i \cap K$ is primary in K. Furthermore, the proof above shows that $\sqrt{N_i \cap K :_R K} \subseteq \sqrt{N_i :_S M} \cap R$, and the other inclusion is trivial. This proves that if N_i is P-primary in M and $K \not\subseteq N_i$, then $N_i \cap K$ is primary in K to the prime ideal $P \cap R$.

Hence $N \cap K = (N_1 \cap K) \cap \cdots \cap (N_s \cap K)$, after removing intersectands K, is a primary decomposition of the R-module $N \cap K$ in K.

One can then also read off the associated primes of the contraction from this:

$$\begin{aligned}\widetilde{\operatorname{Ass}}_R(K/N \cap K) &\subseteq \left\{\sqrt{N_i \cap K :_R K} \mid i = 1, \ldots, s\right\} \\ &\subseteq \left\{\sqrt{N_i :_S M} \cap R \mid i = 1, \ldots, s\right\} \\ &\subseteq \left\{P \cap R \mid P \in \widetilde{\operatorname{Ass}}_S(M/N)\right\}.\end{aligned}$$

□

Associated primes can also be read off from short exact sequences:

Proposition 4.4. *Let R be a ring and*

$$0 \longrightarrow M' \longrightarrow M \longrightarrow M'' \longrightarrow 0$$

a short exact sequences of R-modules. Then

(1) $\widetilde{\operatorname{Ass}}\, M' \subseteq \widetilde{\operatorname{Ass}}\, M$.
(2) $\widetilde{\operatorname{Ass}}\, M \subseteq \widetilde{\operatorname{Ass}}\, M' \cup \widetilde{\operatorname{Ass}}\, M''$.
(3) *If $P \in \widetilde{\operatorname{Ass}}\, M''$ is finitely generated, then $P \in \widetilde{\operatorname{Ass}}\, M \cup \widetilde{\operatorname{Ass}}\, H^1_P(M')$.*

Proof. (1) follows from Lemma 4.3.

Let $P \in \widetilde{\operatorname{Ass}}\, M$. By localization, without loss of generality P is the unique maximal ideal of R. There exists $m \in M$ such that P is minimal over $0 :_R m$. Let m'' be the image of m in M''. If m'' is non-zero, then P is minimal over $0 :_R m''$ so that $P \in \widetilde{\operatorname{Ass}}\, M''$. Now assume that $m'' = 0$. Then $m \in M'$, and P is minimal over $(0_{M'}) :_R m$, so that $P \in \widetilde{\operatorname{Ass}}\, M'$. This proves (2).

Now let $P \in \widetilde{\operatorname{Ass}}\, M''$ be finitely generated. Again without loss of generality by localization P is the only maximal ideal in R. By Lemma 2.8 there exists $m'' \in M''$ such that $P = 0 :_R m''$. Thus m'' is a non-zero element of $H^0_P(M'')$ and $P \in \widetilde{\operatorname{Ass}}\, H^0_P(M'')$. The following is a part of the long exact sequence on local cohomology:

$$H^0_P(M) \longrightarrow H^0_P(M'') \longrightarrow H^1_P(M').$$

If $P \notin \widetilde{\operatorname{Ass}}\, M$, then $H^0_P(M)$ is zero. Thus by (1), $P \in \widetilde{\operatorname{Ass}}\, H^1_P(M')$. □

Associated primes can also be constructed from direct limits:

Proposition 4.5. *Let R be a ring, and $\{M_i\}_{i \in I}$ a direct system of R-modules. Let $P \in \operatorname{Ass}(\varinjlim M_i)$ and assume that PR_P is finitely generated. Then $P \in \bigcup_{i \in I} \widetilde{\operatorname{Ass}}\, M_i$.*

Proof. By localization we may assume that P is the only maximal ideal of R. By Lemma 2.8, there exists $m \in \varinjlim M_i$ such that $P = 0 :_R m$. As P is finitely generated, there exists $i \in I$ such that $m \in M_i$ and $Pm = 0$ in M_i. Hence $P \subseteq (0_{M_i}) :_R m$. But m is non-zero, so that $(0_{M_i}) :_R m$ is a proper ideal, whence it equals P. Thus $P \in \widetilde{\operatorname{Ass}}\, M_i$. □

In special cases, a module has a natural corresponding ideal such that the associated primes of the two are equal:

Proposition 4.6. *Let R be a ring R, M a square matrix with entries in R, and C the cokernel of M. Then the minimal primes over* $(\det M)R$ *are precisely the minimal primes over* $0 :_R C$.

Proof. If M is an $n \times n$ matrix, then $R^n \xrightarrow{M} R^n \to C \to 0$ is exact. For a prime ideal P, $C_P = 0$ if and only if $R^n_P \xrightarrow{M} R^n_P$ is surjective, or equivalently, is an isomorphism. This occurs if and only if $\det M$ is a unit in R_P.

This proves that a prime ideal P contains $\det M$ if and only if $C_P \neq 0$. But C is finitely generated, so then $C_P \neq 0$ if and only if $0 :_R C \subseteq P$. □

5. Witnesses

By definition, every associated prime of an R-module M is minimal over $0 :_R m$ for some $m \in M$. Such m is called a *witness* of P. Witnesses are not unique, but not surprisingly, they carry some structure of the modules.

For example, for an ideal containing a non-zerodivisor, every associated prime has a witness which is a non-zerodivisor:

Proposition 5.1. *Let I be an ideal in a ring R which contains a non-zerodivisor. Assume that* 0 *and I have primary decompositions. Let P be weakly associated to I. Then there exists a non-zerodivisor $r \in R$ such that P is minimal over $I : r$. If R is Noetherian, there exists a non-zerodivisor $r \in R$ such that $P = I : r$.*

Proof. Let $P_1, \dots, P_s$ be the associated primes of 0. Let $I = q \cap J$, where q is primary to P and J is the intersection of the irredundant components of I not primary to P. As I contains a non-zerodivisor, so does J. Thus by Proposition 4.1, $J \not\subseteq P_i$ for all i. As $J \not\subseteq q$, by Prime Avoidance (see Exercise 0.7), there exists $r \in J \setminus (q \cup P_1 \cup \cdots \cup P_s)$. Thus r is a non-zerodivisor, and by Lemma 2.8, $I : r = q : r$ is primary to P. If R is Noetherian, there exists a positive integer n such that $P^n \subseteq I : r$. Choose n minimal with this property. Thus $P^{n-1} \not\subseteq I : r$. Then again by Prime Avoidance, there exists $x \in P^{n-1} \setminus ((I : r) \cup P_1 \cup \cdots \cup P_s)$. Then x is a non-zerodivisor, so rx is a non-zerodivisor, $P \subseteq I : rx$, and $I : rx$ is P-primary. Thus $P = I : rx$. □

The following corollary is used in the exercises to prove that for a finite set of ideals in a Noetherian ring, the set of associated primes of products of their powers is finite, see Exercise 7.38.

Corollary 5.2. Let R be a Noetherian ring, $x \in R$ a non-zerodivisor and $P \in \operatorname{Ass}(R/(x))$. Let $y \in P$ be a non-zerodivisor. Then $P \in \operatorname{Ass}(R/(y))$.

Proof. By Proposition 5.1 there exists a non-zerodivisor r such that $P = (x) : r$. Then $ry \in (x)$, so there exists $s \in R$ such that $ry = sx$. Set $Q = (y) : s$. If $z \in P$, then $zr = ax$ for some $a \in R$. Thus $srz = asx = ary$. As r is a non-zerodivisor, $sz = ay$. Hence $z \in (y) : s = Q$.

Conversely, let $z \in (y) : s$. Write $zs = ay$ for some $a \in R$. Then $zsr = ayr = asx$. But s is also a non-zerodivisor, as $sx = ry$ is a product of non-zerodivisors. Thus $rz = ax$, so that $z \in (x) : r$. This proves that $Q = P$, so that P is associated to (y). □

See Exercise 5.33 for a version of this lemma for non-Noetherian rings.

Naturally, associated primes preserve some structure of the ideals:

Proposition 5.3. *Let G be a totally ordered abelian monoid (e.g., $G = \mathbb{N}^n$ or $G = \mathbb{Z}^n$). Let R be a G-graded ring, and M be an R-submodule of a G-graded R-module. Then every associated prime P of M is homogeneous. Furthermore, if M is G-graded, there exists a homogeneous element $m \in M$ such that $P = 0 : m$.*

Proof. Write $P = 0 : m$ for some $m \in M$. Write $m = \sum_g m_g$, where g varies over a finite subset of G, and each m_g is a homogeneous element of degree g. Let $h = \max\{g \mid m_g \neq 0\}$. Let $p \in P$. Write $p = \sum_g p_g$ (finite sum), with each p_g a homogeneous element of degree g. Let $k = \max\{g \mid p_g \neq 0\}$. Then $p_k m_h$ is the component of $pm = 0$ of degree $k + h$, so that $p_k m_h = 0$.

We claim that there exists a positive integer i such that $p_k^i m = 0$. Suppose that for all $g' > g$, we have proved that $p_k^i m_{g'} = 0$. The term of degree $k + g$ in pm is of the form $\sum_{g' \geq g} p_{k+g-g'} m_{g'}$, so that $p_k^{i+1} m_g = 0$. It follows that for some positive integer i, $p_k^i \in 0 : m = P$, and as P is a prime ideal, then $p_k \in P$. By repeating, we get that every component of p is in P, so that every associated prime of M is homogeneous.

In particular, if M is also G-graded, $P = 0 : m = \cap_g (0 : m_g)$, so that for some g, $P = 0 : m_g$. □

In particular, the only maximal ideal in the polynomial ring $k[X_1, \ldots, X_n]$ which may be associated to homogeneous ideals is $(X_1, \ldots, X_n)$. By Lemma 4.2 one can test whether that maximal ideal is associated.

Another consequence of the proposition above is that all the associated primes of a monomial ideal are monomial, and furthermore one can compute all the associated primes of a monomial ideal: for each of the finitely many primes generated by a subset of the variables, by Lemma 4.2 one can test whether that prime ideal is associated to the monomial ideal.

Example 5.4. Let I be the monomial ideal (X^2, XY) in the polynomial ring $k[X, Y]$. The only prime ideals generated by variables are 0, (X), (Y), (X, Y). Clearly 0 and (Y) are not associated to I as they do not contain it. But by using Lemma 4.2, (X) is associated as $I : (X) = (X, Y)$, so that $(I : (X))_{(X)} \neq I_{(X)}$, and (X, Y) is associated as $(I : (X, Y))_{(X,Y)} = (X)_{(X,Y)} \neq (X^2, XY)_{(X,Y)}$. In fact, in Example 3.6 we saw that I has an irredunant primary decomposition $(X^2, XY) = (X) \cap (X^2, Y)$.

As in the case of this example, the primary components of homogeneous modules have graded structure:

Corollary 5.5. Let R be a Noetherian ring graded by a totally ordered abelian monoid G and let M be a Noetherian G-graded R-module. (M is not necessarily graded.) Then the submodule 0 in M has a primary decomposition in which each primary component is homogeneous.

Proof. By Theorem 3.4, every submodule of M has a primary decomposition. Let $0 = N_1 \cap \cdots \cap N_s$ be an irredundant primary decomposition. For each $i = 1, \ldots, s$, let N_i^* be the submodule of N_i generated by all the homogeneous elements of N_i. Then

$0 = N_1^* \cap \cdots \cap N_s^*$, and it suffices to prove that each N_i^* is primary to $P_i = \sqrt{N_i : M}$. By Proposition 5.3, P_i is homogeneous.

We change notation, let N be a P-primary submodule of M, where P is a homogeneous prime ideal, and let N^* be the submodule of N generated by the homogeneous elements. By the primary assumption, there exists k such that $P^k M \subseteq N$. But P and M are G-graded, hence $P^k M \subseteq N^*$. Let $r \in R$ and $m \in M \subseteq N^*$ such that $rm \in N^*$. By possibly multiplying m by a power of P, without loss of generality we may assume that $Pm \subseteq N^*$, and that furthermore no homogeneous component of m is in $\subseteq N^*$. Under the total grading of G, let r' be the homogeneous component of r of highest degree, and let m' be the homogeneous component of m of highest degree. Then $r'm' \in N^* \subseteq N$. As m' is not in N^*, it is not in N, so that as N is P-primary, $r' \in P$. But then $r'm \in N^*$. By repeating this for $(r - r')m$, we get that $r \in P$, which proves that N^* is P-primary. □

Now we can describe monomial primary ideals and primary decompositions of monomial ideals:

Proposition 5.6. *Let $R = k[X_1, \ldots, X_n]$ be a polynomial ring over a field k and let I be an ideal in R generated by monomials in the X_i. Then*

(1) *I has a unique minimal generating set S consisting of monomials.*
(2) *The radical of I equals*

$$\{X_{i_1} \cdots X_{i_r} \mid \text{for some positive integers } a_1, \ldots, a_r, X_{i_1}^{a_1} \cdots X_{i_r}^{a_r} \in S\}.$$

(3) *For any $i = 1, \ldots, n$, $I : X_i = \sum_{m \in S} (m) : X_i$, where*

$$(X_1^{a_1} \cdots X_n^{a_n}) : X_i = \begin{cases} (X_1^{a_1} \cdots X_n^{a_n}), & \text{if } a_i = 0; \\ ((X_1^{a_1} \cdots X_n^{a_n})/X_i), & \text{if } a_i > 0. \end{cases}$$

(4) *I is primary if and only if for every variable X_i, whenever X_i divides one of the monomial generators of I, then $X_i \in \sqrt{I}$.*
(5) *To find a primary decomposition of I, if I is not primary, find $m \in S$ and a variable X_i such that X_i divides m but $X_i \notin \sqrt{I}$. Let a be any integer greater than or equal to $\max\{e \mid X_i^e$ divides some $m \in S\}$. Then $I = (I + (X_i^a)) \cap (I : X_i^a)$ is a decomposition of I into two strictly larger monomial ideals, and so the primary decomposition of I is obtained from Noetherian induction.*

Proof. The first three parts are elementary. If I is primary then by the definition for every variable X_i, if X_i divides one of the monomial generators of I, then $X_i \in \sqrt{I}$. Now assume that for every variable X_i, if X_i divides one of the monomial generators of I, then $X_i \in \sqrt{I}$. We know that $\sqrt{I}$ is a monomial ideal. If $X_{i_1} \cdots X_{i_r}$ is one of the minimal monomial generators of $\sqrt{I}$ then for some positive integers $a_1, \ldots, a_r$, $X_{i_1}^{a_1} \cdots X_{i_r}^{a_r} \in S$. By assumption then $X_{i_1} \in \sqrt{I}$, so that $r = 1$. This proves that $\sqrt{I}$ is a prime ideal. By Proposition 5.3, all the associated prime ideals of I are monomial prime ideals, and as $\sqrt{I}$ is a prime ideal, each of the associated primes other than $\sqrt{I}$ contains a variable not in $\sqrt{I}$. Let X_i be one of those variables. Then X_i divides some

$m \in S$, so by assumption, $X_i \in \sqrt{I}$, contradicting the choice of X_i. Thus necessarily $\sqrt{I}$ is the only associated prime of I, so that I is primary.

The last part follows from Procedure 3.5 and from part (4). □

Homogeneity can be used in yet another way to compute associated primes:

Proposition 5.7. *Let G be a submonoid of $\mathbb{N}^n$. Let R be a G-graded ring and M a G-graded R-module. Let $P \in \widetilde{\operatorname{Ass}}_{R_0} M_g$. Then there exists $Q \in \widetilde{\operatorname{Ass}}_R M$ such that $Q \cap R_0 = P$.*

Furthermore, if P is minimal over $0 :_{R_0} m$, then Q is minimal over $0 :_R m$.

In particular, if $\widetilde{\operatorname{Ass}}_R(M)$ is a finite set, then $\bigcup_{g\in G} \widetilde{\operatorname{Ass}}_{R_0} M_g$ is a finite set.

Proof. Let $m \in M_g$ such that P is minimal over $0 :_{R_0} m$. By definition m is homogeneous. Set $I = 0 :_R m$. Let $R_+ = \oplus_{g>0} R_g$. Then by the assumption on G, R_+ is a proper ideal in R, and even $I + R_+ \subseteq P + R_+$ are proper ideals in R. Clearly $R/(P + R_+) \cong R_0/P$, so that $P + R_+$ is a prime ideal in R. Let Q be a prime ideal in R minimal over PR and contained in $P + R_+$. Then $Q \cap R_0 = P$. Suppose that Q is not minimal over I. Let Q' be a prime ideal containing I and properly contained in Q. Then by the definition of Q, P is not contained in Q'. But then $I \cap R_0 \subseteq Q' \cap R_0 \subsetneq Q \cap R_0 = P$, contradicting the minimality of P over $I \cap R_0$. Thus Q is minimal over I, so that $Q \subseteq \widetilde{\operatorname{Ass}} M$. □

With the set-up as in the previous proposition, under the assumption that M is Noetherian, there are some stabilization properties of the sets $\operatorname{Ass}_{R_0} M_g$:

Proposition 5.8. *Let R be a Noetherian $\mathbb{N}$-graded ring generated over R_0 by elements of degree 1. Let M be a finitely generated $\mathbb{N}$-graded R-module. Then there exists an integer m such that for all $n \geq m$, $\operatorname{Ass}_{R_0} M_n = \operatorname{Ass}_{R_0} M_m$.*

Proof. As M is Noetherian, by Theorem 3.4, the set $\operatorname{Ass}_R M$ is finite. Thus by the previous proposition, $\bigcup_n \operatorname{Ass}_{R_0} M_n$ is a finite set. It suffices to prove that for each $P \in \operatorname{Spec} R_0$ there exists N such that for all $n \geq N$, $P \in \operatorname{Ass}_{R_0} M_n$ if and only if $P \in \operatorname{Ass}_{R_0} M_N$.

Let $P \in \operatorname{Spec} R_0$. Without loss of generality R_0 is local with maximal ideal P.

Suppose that for some $i \geq 0$, $(0 :_M P) \subseteq (0 :_M R_1^i)$. Let N' be the maximal degree of an element in a minimal homogeneous generating set of $0 :_M P$, and let $N = N' + i$. If $n \geq N$ and $P \in \operatorname{Ass}_{R_0} M_n$, write $P = (0 :_{R_0} b)$ for some $b \in M_n$. Then $b \in 0 :_{M_n} P$, so that as $N - n \geq i$ and by degree count, $b \in R_{N-n}(0 :_M P) = 0$, which is a contradiction, i.e., $P \notin \operatorname{Ass}_{R_0} M_n$ for all $n \geq N$.

Now suppose that for all $i \geq 0$, $(0 :_M P) \not\subseteq (0 :_M R_1^i)$. Fix i such that for all n, $(0 :_M R_1^n) \subseteq (0 :_M R_1^i)$. Let $m \in M$ be a homogeneous element such that $Pm = 0$ and $R_1^i m \neq 0$. Then for all n, $R_1^n m \neq 0$. Let $N = \deg m$. Now for any $n \geq N$, $P = (0 :_{R_0} R_1^{n-\deg m} m)$, so that $P \in \operatorname{Ass}_{R_0} M_n$. □

6. Canonical primary decomposition

We saw in Proposition 5.6 that monomial ideals have primary decompositions in which all primary components are monomial ideals. It is straightforward to show that there is a natural notion of the "largest" possible monomial components:

Proposition 6.1. *Let I be a monomial ideal in a polynomial ring over a field. Let P be associated to I. Then there exists a unique largest monomial P-primary ideal q which appears as a P-primary component in some minimal monomial primary decomposition of I. Furthermore, in any primary decomposition of I, the P-primary component of I may be replaced by q.*

Proof. For $i = 1, 2$, let q_i be a monomial P-primary ideal in a monomial primary decomposition of I, and let I_i be the intersection of the other primary components of I. By Theorem 3.16, $I_1 \cap q_2 = I$. By Proposition 5.6, $q_1 + q_2$ is a P-primary monomial ideal. The intersection of monomial ideals is generated by the least common multiples of monomials taken one from one ideal and another from the second ideal. Thus $I_1 \cap (q_1 + q_2) = I_1 \cap q_1 + I_1 \cap q_2 = I + I = I$. Thus clearly there exists an ideal q as in the statement of the proposition. The last statement follows from Theorem 3.16. □

We have seen that embedded components need not be unique, but there is a notion of canonical primary decomposition, which is unique:

Theorem 6.2 (Ortiz [36]). *Every ideal I in a commutative Noetherian ring admits a unique irredundant primary decomposition*

$$I = q_1 \cap \cdots \cap q_r$$

with the following property: whenever $I = q_1' \cap \cdots \cap q_r'$ is another irredundant primary decomposition of I, with $\sqrt{q_i} = \sqrt{q_i'}$ for $i = 1, \ldots, r$, then for all i,

(1) *nilpotency degree* $\mathrm{nil}(q_i)$ *is less than or equal to* $\mathrm{nil}(q_i')$,
(2) *if* $\mathrm{nil}(q_i) = \mathrm{nil}(q_i')$, *then* $q_i \subseteq q_i'$.

Proof. By Theorem 3.16, it suffices to prove that for each $P \in \mathrm{Ass}(R)$, there exists a P-primary ideal q such that q appears as the P-primary component of I in some primary decomposition of I, the nilpotency degree of q is smallest possible, and if $\mathrm{nil}(q) = \mathrm{nil}(q')$ for some P-primary component q' of I, then $q \subseteq q'$. Let S be the set of all Q-primary components of I with the smallest possible nilpotency degree. Let this nilpotency degree be n. Then S is closed under intersections:

$$Q^n \subseteq \bigcap_{q \in S} q \subseteq Q,$$

and furthermore $\bigcap_{q \in S} q$ is Q-primary. Thus S has a minimal element under inclusion. This element, q_r, satisfies the two conditions of the theorem. □

The construction immediately shows the following:

Corollary 6.3. If q is a canonical component of I, then q equals the q-primary component of $I + (\sqrt{q})^{\mathrm{nil}(q)}$.

7. Associated primes of powers of an ideal

We prove in this section that there are classes of ideals which share the same associated primes. In particular, we prove that for any ideal I in a Noetherian ring R, $\mathrm{Ass}(R/I^n)$

is the same for all large n. It is not true in general that $\operatorname{Ass}(R/I^n) = \operatorname{Ass}(R/I)$ for all n, even for prime ideals:

Example 7.1. Let $R = k[X, Y, Z]$, P the kernel of the map $R \to k[t]$ taking X to t^3, Y to t^4, and Z to t^5. Then $P = (X^3 - YZ, Y^2 - XZ, Z^2 - X^2Y)$ and P is of course a prime ideal. However, P^2 has an embedded component:

$$\begin{aligned}(X^3 - YZ)^2 + (X^2Y - Z^2)(Y^2 - XZ) &= X^6 - 3X^3YZ + X^2Y^3 + XZ^3 \\ &= X(X^5 - 3X^2YZ + XY^3 + Z^3),\end{aligned}$$

so that as X is not in P, $X^5 - 3X^2YZ + XY^3 + Z^3$ is in $P^{(2)}$. Note that under the grading $\deg X = 3, \deg Y = 4, \deg Z = 5$, P is a homogeneous ideal. By degree count, $X^5 - 3X^2YZ + XY^3 + Z^3$ is not in P^2. Thus P^2 must have at least one embedded associated prime ideal. By homogeneity, that embedded prime ideal is necessarily equal to (X, Y, Z), and P^2 has exactly one embedded prime ideal.

This shows that $\operatorname{Ass}(R/I^n)$ is not a constant function as n varies. The function is also not monotone increasing or decreasing. Here is an easily verifiable example:

Example 7.2. Let $R = k[X, Y, U, V]$, $I = (X^4, X^3Y, X^2Y^2U, XY^3, Y^4) \cap (V, X^2)$. Then it is easy to verify that $\operatorname{Ass}(R/I) = \{(X, Y), (X, Y, U), (V, X)\}$ and $\operatorname{Ass}(R/I^2) = \{(X, Y), (V, X), (X, Y, U, V)\}$.

We prove below that $\operatorname{Ass}(R/I^n)$ is eventually constant. We will make use of the extended Rees algebras, a ubiquitous construction in the study of properties of powers of an ideal:

Definition 7.3. Let R be a ring, and I an ideal in R. The *Rees algebra* of I is the subalgebra of $R[t]$, where t is a variable over R, generated over R by all elements of the form at, as a varies over the elements of I. This algebra is denoted $R[It]$.

The *extended Rees algebra* of I is the subalgebra of $R[t, t^{-1}]$ generated over R by t^{-1} and all elements of the form at, as a varies over the elements of I. The extended Rees algebra is denoted $R[It, t^{-1}]$.

It is easy to verify that for all $n \in \mathbb{N}$,

$$R[It] = \bigoplus_{n \geq 0} I^n$$

and that

$$I^n R[It, t^{-1}] \cap R = t^{-n} R[It, t^{-1}] \cap R = I^n, \qquad I^n R[It] \cap R = I^n.$$

Also, if I is generated by $a_1, \ldots, a_s$, then the two kinds of Rees algebras are finitely generated over R:

$$R[It] = R[a_1t, \ldots, a_st], \qquad R[It, t^{-1}] = R[a_1t, \ldots, a_st, t^{-1}].$$

In particular, if R is Noetherian, both Rees algebras are Noetherian, and primary decompositions exist in R, $R[It]$, $R[It, t^{-1}]$.

Similarly, if M is an R-module, then $MR[It]$ and $MR[It,t^{-1}]$ are modules over the Rees algebras $R[It]$ and $R[It,t^{-1}]$, respectively. If M is finitely generated, so are these extended modules. In any case,

$$I^n MR[It,t^{-1}] \cap M = t^{-n} MR[It,t^{-1}] \cap M = I^n M, \quad I^n MR[It] \cap M = I^n M.$$

Thus under the Noetherian assumptions, primary decompositions exist in M, $MR[It]$, $MR[It,t^{-1}]$, and by Lemma 4.3, a primary decomposition of $I^n M$ is a contraction of a primary decomposition of $t^{-n} MR[It,t^{-1}]$.

Proposition 7.4. *Let R be a Noetherian ring, I an ideal and M a finitely generated R-module. Then $\bigcup_n \operatorname{Ass}(I^n M/I^{n+1} M)$ is a finite set and there exists an integer m such that for all $n \geq m$, $\operatorname{Ass}(I^n M/I^{n+1} M) = \operatorname{Ass}(I^m M/I^{m+1} M)$.*

Proof. Let S be the Rees algebra $R[It]$ and N the $\mathbb{N}$-graded S-module $\oplus_n I^n M/I^{n+1} M$. For each non-negative integer m, $N_m = I^m M/I^{m+1} M$. Proposition 5.8 applied to the $\mathbb{N}$-graded ring S finishes the proof. □

A similar proof also shows that $\bigcup_n \operatorname{Ass}(I^n M)$ is a finite set and that there exists an integer m such that for all $n \geq m$, $\operatorname{Ass}(I^n M) = \operatorname{Ass}(I^m M)$. A similar result also holds for the modules $M/I^n M$:

Proposition 7.5. *Let R be a Noetherian ring, I an ideal, and M a finitely generated R-module. Then there exists $m \in \mathbb{N}$ such that for all $n \geq m$, $\operatorname{Ass}(M/I^m M) = \operatorname{Ass}(M/I^n M)$.*

Proof. For each positive integer n,

$$0 \to \frac{I^n M}{I^{n+1} M} \to \frac{M}{I^{n+1} M} \to \frac{M}{I^n M} \to 0$$

is a short exact sequence. By Proposition 4.4, $\operatorname{Ass}(I^n M/I^{n+1} M) \subseteq \operatorname{Ass}(M/I^{n+1} M) \subseteq \operatorname{Ass}(M/I^n M) \cup \operatorname{Ass}(I^n M/I^{n+1} M)$. By Proposition 7.4, there exists an integer k such that for all $n \geq k$, $\operatorname{Ass}(I^n M/I^{n+1} M) = \operatorname{Ass}(I^k M/I^{k+1} M)$. Thus for $n \geq k$,

$$\operatorname{Ass}(I^n M/I^{n+1} M) \subseteq \operatorname{Ass}(M/I^{n+1} M) \subseteq \operatorname{Ass}(M/I^n M) \cup \operatorname{Ass}(I^n/I^{n+1} M).$$

It immediately follows that $\bigcup_n \operatorname{Ass}(M/I^n M)$ is finite. Also, for each prime ideal $P \in \bigcup_n \operatorname{Ass}(M/I^n M)$, if P is associated to infinitely many $M/I^n M$, then it is associated to $M/I^n M$ for all $n \geq k$, and if P is associated to only finitely many $M/I^n M$, let k_P be the largest integer such that P is associated to $M/I^{k_P} M$. Set $m = \max\{k, k_P \mid P\}$. Then by the set-up, for all $n \geq m$, $\operatorname{Ass}(M/I^m M) = \operatorname{Ass}(M/I^n M)$. □

8. Primary decompositions of powers of an ideal

In this section we prove that one can choose primary components of powers of an ideal in such a way that they "do not grow small too fast". This is one of the many results on the linear properties of primary decompositions. The proof presented here depends on a version of the Artin-Rees lemma for Artinian modules, due to Kirby [24]. We present Kirby's results here for completeness.

Proposition 8.1 (Kirby [24]). *Let R be a ring, $X_1, \ldots, X_s$ variables over R, and $M = \sum_{i=-\infty}^{\infty} M_i$ a graded $R[X_1, \ldots, X_s]$-module. (So $X_j M_i \subseteq M_{i+1}$, $RM_i \subseteq M_i$.) Then M is an Artinian $R[X_1, \ldots, X_s]$-module if and only if there exist integers k, l such that*

(1) $M_i = 0$ *for* $i > l$,
(2) $0 :_{M_i} (X_1, \ldots, X_s) = 0$ *for* $i < k$,
(3) M_i *is an Artinian R-module for* $i \in \{k, k+1, \ldots, l\}$.

Proof. First assume that M is Artinian. Certainly then conditions (1) and (3) hold. Also, the submodule $0 :_M (X_1, \ldots, X_s)$ of M is Artinian, so that the descending chain $D_n = \oplus_{i=-\infty}^{-n}(0 :_{M_i} (X_1, \ldots, X_s))$ has to stabilize. As $\bigcap_n D_n = 0$, for all sufficiently large n, $D_n = 0$. This proves (2).

Now assume that M satisfies conditions (1), (2) and (3). If $s = 0$, M is trivially an Artinian R-module. So we may assume that $s > 1$. Set $N_n = \frac{0:_M X_s^{n+1}}{0:_M X_s^n}$. Then N_n is a graded $R[X_1, \ldots, X_s]$-module which is annihilated by X_s, so that N_n is a module over $R[X_1, \ldots, X_{s-1}]$. Clearly N_n satisfies conditions (1), (2) and (3), so that by induction on s, N is Artinian over $R[X_1, \ldots, X_{s-1}]$.

For each positive integer n, define $\varphi_n : N_n \to N_0$ by sending $a + 0 :_M X_s^n$ to $X_s^n a$. Then clearly φ_n is injective and $M \supseteq N_0 \supseteq \varphi_1(N_1) \supseteq \varphi_2(N_2) \supseteq \cdots$.

Now let $A_1 \supseteq A_2 \supseteq A_3 \supseteq \cdots$ be a descending chain of $R[X_1, \ldots, X_s]$ submodules of M. For each i, n, define A_{in} to be the submodule of N_n generated by $A_i \cap (0 :_M X_s^{n+1})$. Then $N_n \supseteq A_{1n} \supseteq A_{2n} \supseteq \cdots$ is a descending chain of $R[X_1, \ldots, X_{s-1}]$-submodules of N_n. But N_n is Artinian, so there exist an Artinian module L_n over $R[X_1, \ldots, X_{s-1}]$ and a positive integer m_n such that for all $k \geq m_n$, $L_n = A_{kn}$. For any $k \geq m_n$, m_{n+1},

$$\varphi_n(L_n) = \varphi_n(A_{kn}) \supseteq \varphi_{n+1}(A_{k,n+1}) = \varphi_{n+1}(L_{n+1}).$$

It follows that $N_0 \supseteq \varphi_1(L_1) \supseteq \varphi_2(L_2) \supseteq \cdots$ is a descending chain of $R[X_1, \ldots, X_{s-1}]$-submodules of N_0. But N_0 is Artinian, so there exist an Artinian module L and a positive integer t such that for all $k \geq t$, $L = \varphi_k(L_k)$.

Now let $T = \max\{m_0, \ldots, m_t\}$. Then by the choices of the m_i, for all $k \geq T$ and $0 \leq n \leq t$, $A_{kn} = A_{k+1,n} = \cdots = L_n$. Now let $k \geq T$ and $n > t$. Then

$$\varphi_n(L_n) \subseteq \varphi_n(A_{kn}) \subseteq \varphi_t(A_{kt}) = \varphi_t(L_t) = \varphi_n(L_n).$$

Thus equality holds throughout, and since φ_n is injective, $L_n = A_{kn}$. This proves that for all $n \geq 0$ and $k \geq T$, $A_{kn} = A_{k+1,n}$.

Finally we prove that this implies that $A_k = A_{k+1}$. Let $\alpha \in A_k$. By condition (1) there exists n such that $X_s^{n+1}\alpha = 0$. Thus α has an image in N_n, hence in A_{kn}. Thus there exists $\beta \in A_{k+1}$ such that $\alpha - \beta \in 0 :_M X_s^n$. If $n = 0$, then $\alpha = \beta \in A_k$. Otherwise assume that $n > 0$. Then $\alpha - \beta \in (0 :_M X_s^n) \cap A_k$, so $\alpha - \beta$ has an image in $A_{k,n-1}$. But then by induction on n, $\alpha - \beta \in A_{k+1}$, whence $\alpha \in A_{k+1}$. This proves that $A_k = A_{k+1}$ for all $k \geq T$. Thus M is Artinian. □

A consequence is the following version of the Artin-Rees lemma for Artinian modules:

Theorem 8.2 (Kirby [24]). *Let R be a ring, I an ideal, M an Artinian R-module, and N a submodule of M. Then there exists an integer k such that for all $n \geq k$,*

$$N + 0 :_M I^n = (N + 0 :_M I^k)) :_M I^{n-k}.$$

Proof. Let S be the set of all finitely generated ideals contained in I. Then as M is Artinian, there exists $J \in S$ such that $0 :_M J \subseteq 0 :_M K$ for all $K \in S$. If $m \notin 0 :_M I$, there exists $a \in I$ such that $am \neq 0$. Then $m \notin 0 :_M (J + (a))$, so that by the choice of J, $m \notin 0 :_M J$. It follows that $0 :_M J = 0 :_M I$. Hence by induction on $n \geq 1$, $0 :_M J^n = 0 :_M I^n$.

Let $J = (a_1, \ldots , a_s)$. Define

$$\overline{N}_n = \begin{cases} \frac{M}{0:_M J^{-n}} & \text{if } n \leq 0; \\ 0 & \text{if } n > 0. \end{cases}$$

Set $\overline{N} = \oplus_{n=-\infty}^{\infty} \overline{N}_n$. Then $\overline{N}$ is a module over $R[X_1, \ldots , X_s]$, if X_i acts by multiplication by a_i. Then $\overline{N}$ satisfies conditions (1), (2) and (3) of Proposition 8.1, so that $\overline{N}$ is an Artinian module over $R[X_1, \ldots , X_s]$. For $n \geq 0$, set

$$K_n = \sum_{i=-\infty}^{-n} \frac{(N + 0 :_M J^n) :_M J^{-n-i}}{0 :_M J^{-i}}.$$

Then K_n is a submodule of $\overline{N}$, and $K_1 \supseteq K_2 \supseteq K_3 \supseteq \cdots$. Thus there exists an integer k such that for all $n \geq k$, $K_k = K_n$. In particular, the graded components of $K_k = K_n$ of degree $i = -n$ yield that $(N + 0 :_M J^k) :_M J^{n-k} = N + 0 :_M J^n$. Finally,

$$\begin{aligned} N + 0 :_M I^n &\subseteq (N + 0 :_M I^k) :_M I^{n-k} \\ &\subseteq (N + 0 :_M J^k) :_M J^{n-k} \\ &= N + 0 :_M J^n \\ &= N + 0 :_M I^n, \end{aligned}$$

so that equality holds throughout. □

Corollary 8.3. Let R be a Noetherian ring, M a finitely generated R-module, E an injective Artinian R-module, I an ideal, and $u \in R$ a non-zerodivisor on M. Then there exists a positive integer k such that for all $n \geq 1$, any R-homomorphism $g : M/u^n M = uM/u^{n+1}M \to E$ with the property $I^n g = 0$ can be extended to an R-homomorphism $\tilde{g} : M/u^{n+1}M \to E$ so that $I^{n+k}\tilde{g} = 0$.

Proof. Let $f : M \to uM/u^{n+1}M \to E$ be the composition. Let $H = \operatorname{Hom}_R(M, E)$. As $\operatorname{Hom}_R(\_, E)$ is exact, H is Artinian and $H = uH$. In particular, there exists $\tilde{f} \in H$ such that $f = u\tilde{f}$.

By Theorem 8.2, there exists an integer k such that for all $n \geq k$, $0 :_H u + 0 :_H I^n = (0 :_H u + 0 :_H I^k) :_H I^{n-k}$. The assumption $I^n g = 0$ implies that

$$\tilde{f} \in (0 :_H u) :_H I^n \subseteq (0 :_H u + 0 :_H I^k) :_H I^n \subseteq 0 :_H u + 0 :_H I^{n+k}.$$

Let $f' \in 0 :_H u$ and $f'' \in 0 :_H I^{k+n}$ such that $\tilde{f} = f' + f''$. Then $f = u\tilde{f} = uf''$. As $u^{n+1} f'' = u^n f(M) = 0$, then f'' defines a map $\tilde{g} : M/u^{n+1}M \to E$ which extends g. Furthermore, $I^{k+n}\tilde{g} = 0$. □

Finally we can prove that primary components of powers of an ideal contain linearly growing powers of their corresponding associated primes:

Theorem 8.4. *Let R be a Noetherian ring, I an ideal and M a finitely generated R-module. Then there exists a positive integer k such that for all positive integers n there exists a primary decomposition*

$$I^n M = N_1 \cap N_2 \cap \cdots \cap N_l$$

such that if $P = \sqrt{I^n M :_R N_i}$, then $P^{kn} M \subseteq N_i$.

This result was first proved for $M = R$ in [46], and the generalization to arbitrary M was proved by Sharp in [41]. The proof in [46] used prime filtrations of quotients of a ring by powers of an ideal. The proof below, due to Sharp, shows another useful method of finding primary decompositions.

Proof of Theorem 8.4. Let $S = R[It, t^{-1}]$, $K = MR[It, t^{-1}]$. Suppose that theorem holds for primary decompositions of $(t^{-1})^n K = t^{-n} K$, i.e., that there exists a positive integer k such that for all positive integers n there exists a primary decomposition

$$t^{-n} K = N_1 \cap N_2 \cap \cdots \cap N_l$$

such that if $P_i = \sqrt{t^{-n} K :_S N_i}$, then $P_i^{kn} K \subseteq N_i$. By discussion on page 136 and by Lemma 4.3, $I^n M = t^{-n} K \cap M = (N_1 \cap M) \cap (N_2 \cap M) \cap \cdots \cap (N_l \cap M)$ is a primary decomposition of $I^n M$ and the associated primes of $I^n M$ are in the set $\{P_i \cap R \mid i\}$. Then $(P_i \cap R)^{kn} M \subseteq P_i^{kn} N \cap M \subseteq N_i \cap M$, so that the theorem also holds for I and M. Thus without loss of generality is suffices to prove the theorem for the case when I is a principal ideal generated by a non-zerodivisor u on M.

For every injective module E, write it as $E = \oplus_P E_P$, where P varies over distinct prime ideals and E_P is the direct sum of injective envelopes of R/P. Let π_P be the natural projection $E \to E_P$.

Now let E_1 be the injective envelope of M/uM and $i_1 : M/uM \to E_1$ the natural inclusion. Then $E_1 = \oplus_P (E_1)_P$, where P varies over a finite set S of prime ideals in R. As N is finitely generated and every element of $(E_1)_P$ is annihilated by a power of P, there exists an integer k such that $P^k \pi_P \circ i_1 = 0$. We may choose k to work for all $P \in S$.

Apply Corollary 8.3 to module M, element u, $I = P$ in S, and $E = (E_1)_P$. By possibly increasing k, we may assume that k works for $(E_1)_P$ as in the Corollary. As S is a finite set, there exists a positive integer k as in the Corollary which works for all

$(E_1)_P$ as P varies over elements of S. But then this same k works for all finite direct sums $(E_1)^n_P$, as P varies over elements of S.

We claim that for all $n \geq 1$ there exist an injective module $E_n = (E_1)^n$ and an inclusion $i_n : M/u^nM \to E_n$ such that for all $P \in S$, $P^{kn}\pi_P \circ i_n = 0$. This already holds for $n = 1$. Suppose that it holds for some positive n. Consider the commutative diagram:

$$\begin{array}{ccccccccc} 0 & \longrightarrow & M/u^nM & \xrightarrow{u} & M/u^{n+1} & \xrightarrow{p} & \longrightarrow & 0 & \\ & & \downarrow i_n & & & & \downarrow i_1 & & \\ 0 & \longrightarrow & E_n & \longrightarrow & E_n \bigoplus E_1 & \longrightarrow & E_1 & \longrightarrow & 0 \end{array}$$

in which the rows are exact. By assumption, for all $P \in S$ and every summand $E_R(R/P)$ of E_n, $P^{kn}\pi_P \circ i_n = 0$. By Corollary 8.3, there exists $j_P : M/u^{n+1}M \to E_P$ extending $\pi_P \circ i_n$ such that $P^{kn+k}j_P = 0$. Set $j : M/u^{n+1} \to E_n$ be the direct sum of all the j_P. Define $i_{n+1} : M/u^{n+1}M \to E_n \oplus E_1 = (E_1)^{n+1} = E_{n+1}$ by $i_{n+1}(m + u^{n+1}M) = j(m) + i_1 \circ p(m)$. By construction, $P^{kn+k}\pi_P \circ i_{n+1} = 0$ for all $P \in S$. This i_{n+1} makes the diagram commute, and as i_1 and i_n are injective, so is i_{n+1}.

Finally, we prove the primary decomposition result. For each $n \geq 1$ and $P \in S$, set $N_{nP} = \ker(\pi_P \circ i_n)$. As i_n is an inclusion and $E_n = \oplus_{P \in S}(E_n)_P$, it follows that $\bigcap_P N_{nP} = 0$ in M/u^nM. Also, by construction, $P^{kn}M \subseteq \ker(\pi_P \circ i_n) = N_{nP}$.

Note the that proof does not indicate how k depends on I. A constructive k was achieved for monomial ideals by Hoa and Trung in [16].

There is a related result that the primary components of I^nM "do not grow too small too slowly", see [47], whose proof relies on a deep result of Izumi and Rees:

Theorem 8.5 (Swanson [47]). *Let R be a Noetherian ring, I and J ideals in R. Assume that for each n there exists k_n such that $I^n : J^\infty \subseteq I^{k_n}$ and that the k_n may be chosen so that* $\lim_n k_n = \infty$. *(In other words, assume that the I-adic topology (topology defined by powers of I) is equivalent to the topology defined by $\{I^n : J^\infty\}_n$.) Then there exists an integer k such that for all n, $I^{kn} : J^\infty \subseteq I^n$.*

In particular, the intersection of those primary components of I^{kn} which do not contain any power of J is contained in I^n. In particular, if I is a prime ideal P, this says that whenever the P-adic topology is equivalent to the symbolic topology, then there exists an integer k such that for all n, $P^{(kn)} \subseteq P^n$. The proof in this generality gives no indication on how k depends on I. But here is a special case, first proved by Ein, Lazarsfeld and Smith in [8] in equicharacteristic 0, and later by Hochster and Huneke in [17] in characteristic p. Below is a basic version of their result (but see the two papers for more general statements):

Theorem 8.6 (Ein, Lazarsfeld, Smith [8]; Hochster, Huneke [17]). *Let R be a regular ring containing a field, and P a prime ideal in R. Then for all n,*

$$P^{(n \cdot \operatorname{ht} P)} \subseteq P^n.$$

9. An algorithm for computing primary decompositions

Given an ideal in a computable ring, can one algorithmically determine its primary decomposition? This answer was already treated to some extent by Grete Hermann in [15]: she proved that there exists a function in n, k, d such that if I is a not necessarily homogeneous ideal in a polynomial ring in n variables over a field, if I is generated by k elements of degree at most d, then there exists a primary decomposition of I such that the generators of its primary components are bounded above by that function. Hermann's function was doubly exponential in n. It is still not known if a doubly exponential bound is indeed necessary.

Since then, several algorithms have been written for primary decompositions and associated primes: Gianni, Trager and Zacharias [11], Eisenbud, Huneke, Vasconcelos [9] Wang [51], Shimoyama, Yokoyama [42]. The best source of information on the algorithms is Decker-Greuel-Pfister's paper [6]. Singular [13] and Macaulay2 [12] have an implementation of the computation of primary decompositions of ideals in polynomial rings.

Below is an outline of the algorithm due to Gianni, Trager and Zacharias [11]. This algorithm is based on the theory of Gröbner bases in polynomial rings $R = A[X_1, \ldots, X_n]$, where A is a principal ideal domain such as $\mathbb{Z}$ or $k[X_0]$ for some field k and variable X_0. The first case of the computation of primary decompositions is of course when $n = 0$. Then the problem reduces to the problem of factorization of integers and polynomials. This is a non-trivial problem. There exist several fairly efficient algorithms for such factorization, due to Berlekamp, Lenstra, Lenstra, Lovasz, Davenport, Trager, etc.

Here is a simplistic approach to the factorization problem in a few special cases. To factor an integer n, one can by brute force check whether any integer between 2 and $\sqrt{n}$ divides n and in this manner obtain a factorization. To factor a polynomial in $k[X]$ if k is a finite field one can again use brute force: there are only finitely many polynomials in $k[X]$ of bounded degree. However, these factorization methods are not efficient.

For more details on factorization, see the excellent book *Modern Computer Algebra* [50] by von zur Gathen and Gerhard. There is a nice tutorial on Berlekamp's algorithm in [25, page 38]. In these notes we ignore the difficulty of factorization and instead assume:

Assumption 9.1. Let A be either $\mathbb{Z}$ or a polynomial ring in one variable over a finitely generated field extension over a finite field or over $\mathbb{Q}$. Thus A is a special type of principal ideal domain. We assume that in all such principal ideal domains, every element is algorithmically factorable into a product of primes.

One can show that for any principal ideal domain A as above, if K is a finitely generated field extension of the field of fractions of A and X is a variable over K, then also every element in $K[X]$ is algorithmically factorable into a product of primes. For notational purposes we call any principal ideal domain A with this property *computable*.

The primary decomposition algorithm relies on induction on the number of variables. In particular, the following lemma will be useful:

Lemma 9.2. *Let B be a Noetherian ring in which primary decompositions and associated primes of ideals are computable. Let $X_1, \dots, X_n$ be variables over B, $C = B[X_1, \dots, X_n]$, and I an ideal in C. Then there exist ideals $I_1, \dots, I_s$ in C such that $I = I_1 \cap \dots \cap I_s$ and each $I_j \cap B$ is primary in B. Furthermore, these ideals I_j are computable.*

Proof. Let $K = I \cap B$. Then K is an ideal in B, so it has a primary decomposition $K = K_1 \cap \dots \cap K_s$. We proceed by induction on s. If $s = 1$, there is nothing to do. So assume $s > 1$. By assumption the radicals $\sqrt{K_j}$ can be computed. By reindexing we may assume that $\sqrt{K_1}$ is minimal over K. By prime avoidance there exists $r \in K_2 \cap \dots \cap K_s \setminus \sqrt{K_1}$, and one can actually compute such an element in B. Now we switch to ideals in C. By Lemma 3.3, there exists m such that $I :_C r^\infty = I :_C r^m$ and $I = (I :_C r^m) \cap (I + (r^m))$. Set $I' = I :_C r^m$, $I'' = I + (r^m)$. Then

$$I' \cap B = (I :_C r^m) \cap B = I \cap B :_B r^m = K_1,$$

which is primary in B. Also, $I'' \cap B = (I + (r^m)) \cap B$ is strictly larger than K, so by Noetherian induction one can decompose I'' as in the statement of the lemma, whence one can decompose I. □

We next review what is computable via *Gröbner bases* in the following context: let B be a Noetherian unique factorization domain, $X_1, \dots, X_n$ variables over B and $C = B[X_1, \dots, X_n]$.

(1) For each $f \in C$, write $f = \sum_\nu f_\nu X^\nu$, a finite sum, with each $f_\nu \in B$.
(2) Impose an order on the monomials in the X_i satisfying

 (i) for all $\nu \neq 0$, $X^\nu > 1$,
 (ii) if $X^\nu > X^\mu$ then for any ρ, $X^{\nu+\rho} > X^{\mu+\rho}$.

 As every ideal in C is finitely generated, necessarily every set of monomials in C has a least element.

 One could take for example the *lexicographic ordering*:

$$X^\nu > X^\mu, \text{ if the first non-zero entry in } \nu - \mu \text{ is positive,}$$

 the *degree lexicographic ordering*:

$$X^\nu > X^\mu, \begin{cases} \text{if } |\nu| > |\mu|; \\ \text{or if } |\nu| = |\mu| \text{ and the first non-zero entry in } \nu - \mu \text{ is positive.} \end{cases}$$

 or the *reverse lexicographic ordering*:

$$X^\nu > X^\mu, \begin{cases} \text{if } |\nu| > |\mu|; \\ \text{or if } |\nu| = |\mu| \text{ and the last non-zero entry in } \nu - \mu \text{ is negative.} \end{cases}$$

(3) Under the given monomial ordering, let the *initial term* of f, in f, be $f_\nu X^\nu$ such that X^ν is maximal among all monomials appearing in f with a non-zero coefficient. If in $f = f_\nu X^\nu$, set

$$\text{lm } f = X^\nu, \text{lc } f = f_\nu.$$

These are called *leading monomial* and *leading coefficient* of f. The initial term, the leading monomial and the leading coefficient of f are computable.

(4) For any ideal I in C, define the *initial ideal* $\text{in}(I)$ of I to be $\{\text{in}\, f \mid f \in I\}$. Similarly, for any set G in C, define the *initial ideal* $\text{in}(G)$ of G to be $\{\text{in}\, f \mid f \in G\}$.
WARNING: if $I = (f_1, \dots, f_s)$, $\text{in}(I)$ is in general strictly larger than $(\text{in}\, f_1, \dots, \text{in}\, f_s)$. See Exercise 9.41.

(5) A set $\{g_1, \dots, g_s\}$ is said to be a *Gröbner basis* of an ideal I if $g_1, \dots, g_s \in I$ and $\text{in}(I) = (\text{in}\, g_1, \dots, \text{in}\, g_s)$.

(6) For any $f, g \in C$, the *S-polynomial* of f and g is

$$S(f, g) = \frac{\text{lcm}(\text{in}\, f, \text{in}\, g)}{\text{in}\, f} f - \frac{\text{lcm}(\text{in}\, f, \text{in}\, g)}{\text{in}\, g} g.$$

The S-polynomial is computable.

(7) Let $G = \{f_1, \dots, f_s\}$ be a subset of C and $f \in C$. If $\text{in}\, f$ is a multiple of some $\text{in}\, f_j$, say $\text{in}\, f = m\, \text{in}\, f_j$, set $f' = f - m f_j$. Then $\text{in}\, f' < \text{in}\, f$. This step is called a *reduction step* of f with respect to G and f' is said to be obtained from f by reduction. Repeat the reduction step on f', then on the polynomial obtained from f' by reduction, etc. The reduction step needs to be repeated only finitely many times before the zero polynomial is obtained in the process, as under the ordering, every set of monomials has a least element. Thus there exists $\tilde{f} \in C$ such that $\tilde{f}$ is obtained from f by applying reduction finitely many times, and either $\tilde{f} = 0$ or $\text{in}\, \tilde{f}$ is not a multiple of any $\text{in}\, f_j$. Then $\tilde{f}$ is called a *reduction* of f with respect to G. A reduction of f with respect to G is computable.
WARNING: $\tilde{f}$ is not uniquely determined. See Exercise 9.41.

(8) For any ideal $I = (f_1, \dots, f_s)$ in C, a Gröbner basis of I is computable. In particular, $\text{in}\, I$ is computable. Namely, the algorithm goes as follows:

   (i) Start with $G = \{f_1, \dots, f_s\}$.
   (ii) For each pair $f, g \in G$, compute $S(f, g)$ and reduce it with respect to G. If the reduction is not zero, add the element to G.
   (iii) Repeat the previous step until all S-polynomials reduce to 0 with respect to G.

The indicated algorithm does terminate as at each step the ideal $\text{in}(G)$ becomes strictly larger, and in a Noetherian ring, we cannot increase this ideal indefinitely. Thus the algorithm terminates in finitely many steps. This proves that a Gröbner basis and the initial ideal of I are both computable.

(9) For any variables $Y_1, \dots, Y_m$ over C, impose on $C[Y_1, \dots, Y_m]$ a monomial order such that for any $f \in C[Y_1, \dots, Y_m] \setminus C$, $\text{in}\, f \notin C$. (Many such orders exist, for example, the lexicographic ordering $Y_1 > Y_2 > \cdots > Y_m > X_1 > \cdots > X_n$.) Let G be a Gröbner basis of an ideal I in $C[Y_1, \dots, Y_m]$. For any $f \in I \cap C$, as $f \in I$, f can be written as $f = \sum_{g \in G} h_g g$ for some $h_g \in C[Y_1, \dots, Y_m]$ such that for all g, $\in (h_g h) \leq \in (f)$. But then by the choice of the order, if $h_g \neq 0$, then $h_g, g \in C$, so that $f \in (G \cap C)$. As $G \cap C \subseteq I \cap C$, this proves that $G \cap C$

is a Gröbner basis of $I \cap C$. Thus in particular for any ideal $I \subseteq C[Y_1, \ldots, Y_m]$, $I \cap C$ is computable.

(10) If I and J are ideals in C, then $I \cap J$ is computable. Namely, if Y is a variable over C, then $I \cap J = (IYC[Y] + J(Y-1)C[Y]) \cap C$, so that by the previous case $I \cap J$ is computable.

(11) If I and J are ideals in C, then $I :_C J$ is computable. Namely, if $J = (f_1, \ldots, f_s)$, then $I :_C J = \cap(I :_C f_j)$. By the previous case it suffices to prove that $I :_C f$ is computable for each $f \in C$. We already know that $I \cap (f)$ is computable. But then by using that $I \cap (f) = f(I : f)$, $I : f$ is computable as division of elements is computable.

(12) For any ideal I and element f in C, $I_f \cap C$ is computable. There are several ways to see this:

   (i) $I_f \cap C = I : f^\infty$. By Lemma 3.3, it suffices to find an integer m such that $I : f^m = I : f^{m+1}$. This integer exists as C is Noetherian.

   (ii) Let Y be a variable over C. Then $I_f \cap C = (IYC[Y] + (fY-1)C[Y]) \cap C$, which is computable.

(13) For any principal prime ideal (p) in B and any ideal I in C, $I_{B\setminus(p)} \cap C$ is computable. Namely, let G be a Gröbner basis of I. For each $g \in G$, write $\operatorname{lc} g = b_g p^{n_g}$ for some non-negative integer n_g and some $b_g \in B\setminus(p)$. Set $b = \prod_g b_g$. Then $I_b \cap C$ is computable and $I_b \cap C \subseteq I_{B\setminus(p)} \cap C$. We want to prove equality in this inclusion. Let $f \in I_{B\setminus(p)} \cap C$. To prove that $f \in I_b \cap C$, it suffices to assume that among all elements in $I_{B\setminus(p)} \cap C$ but not in $f \in I_b \cap C$, f has the least leading monomial. There exists $c \in B\setminus(p)$ such that $cf \in I$. Then $c \operatorname{in} f = \operatorname{in}(cf)$ is a multiple of $\operatorname{in} g = b_g p^{n_g} X^\nu$ for some $g \in G$. As B is a unique factorization domain, $b_g \operatorname{in} f$ is a multiple of $\operatorname{in} g = b_g p^{n_g} X^\nu$. Say $b_g \operatorname{in} f = m \operatorname{in} g$. Then $\operatorname{lm} f = \operatorname{lm}(b_g f) > \operatorname{lm}(b_g f - mg)$. As $b_g f - mg \in I_{B\setminus(p)} \cap C$, by the choice of f, $b_g f - mg \in I_b \cap C$. Hence $b_g f \in I_b \cap C$, whence $f \in I_b \cap C$. Thus $I_{B\setminus(p)} \cap C$ is computable, and equals $I_b \cap C$ as above.

With this background in Gröbner bases, the *Gianni-Trager-Zacharias* algorithm goes as follows. Let A be a computable principal ideal domain. Let $R = A[X_1, \ldots, X_n]$, and I an ideal in R. We want to find a primary decomposition of I and its associated primes.

Case $\mathbf{n = 0}$. The primary decompositions and associated primes are computable by the assumption.

Case $\mathbf{n > 0, I \cap A = 0}$. By item (13) above, $I_{A\setminus\{0\}} \cap R$ is computable and equals $I_a \cap R$ for some computable $a \in A$. By Lemma 3.3 there exists a computable m such that $I : a^\infty = I : a^m$ and $I = (I : a^m) \cap (I + (a^m))$. Then $I : a^m = IR_{A\setminus\{0\}} \cap R$. The ring $R_{A\setminus\{0\}}$ has dimension strictly smaller than R, so a primary decomposition and associated primes of $I : a^m = IR_{A\setminus\{0\}} \cap R$ are computable by induction. But then by Lemma 4.3, a primary decomposition and associated primes of $IR_{A\setminus\{0\}} \cap R$ are computable. Thus it suffices to find a primary decomposition and associated primes of $I + (a^m)$. But the contraction of this ideal to A falls into the next case.

Case $\mathbf{n > 0, I \cap A \neq 0}$. By Lemma 9.2 we may assume that $I \cap A$ is primary. As A is a principal ideal domain, $I \cap A$ is primary to a maximal ideal M. As A is a principal ideal domain, there exists $p \in A$ such that $M = pA$. This breaks into two cases, see below.

Case $\mathbf{n > 0, I \cap A}$ **primary to a maximal ideal** $(\mathbf{p})$, $\dim(\mathbf{R/I}) > \mathbf{0}$. Let G be a Gröbner basis G for I. For each $j = 1, \dots, n$, set

$$G_j = \{g \in G \mid \operatorname{in} g = c_j X_j^m \text{ for some } m \in \mathbb{N}_0, c_j \in A\}.$$

Then G_j is computable. Clearly $X_j \in \sqrt{\operatorname{in}(I)}$ if and only if $X_j \in \sqrt{\operatorname{in}(G_j)}$. These equivalent conditions are computable. If $X_1, \dots, X_n \in \sqrt{\operatorname{in} I}$, then R/I is module-finite over $A/I \cap A$ (generated by finitely many monomials in the X_j), thus integral. But then $\dim(R/I) = \dim(A/I \cap A) = 0$, contradicting the assumption. Thus under the assumption one can compute an integer j such that $X_j \notin \sqrt{\operatorname{in} I}$. Set $A' = A[X_j]$. Then $I \cap A'$ is not zero-dimensional. By item (13) one can compute $b \in A' \backslash pA'$ and an integer m such that $I_{pA'} \cap R = I_b \cap R = I : b^m$. As $I_{pA'}$ is an ideal in the polynomial ring $A'_{pA'}[X_1, \dots, \widehat{X_j}, \dots, X_n]$ over a principal ideal domain $A'_{pA'}$, by induction on dimension a primary decomposition and associated primes of $I_{A' \backslash pA'}$ are computable. Then by Lemma 4.3, a primary decomposition and associated primes of $I_{A' \backslash pA'} \cap R$ are computable. By Lemma 3.3 it remains to compute a primary decomposition and associated primes of $I + (b^m)$. By the assumption that $I \cap A'$ is not zero-dimensional, since $b \in A[X_j] \backslash pA[X_j]$, it follows that $I + (b^m)$ strictly contains I. If it is the whole ring, we are done, otherwise we either repeat this case or the following case on this new ideal $I + (b^m)$. The repetition will happen at most finitely many times by the Noetherian hypothesis.

Case $\mathbf{n > 0, I \cap A}$ **primary to a maximal ideal** $(\mathbf{p})$, $\dim(\mathbf{R/I}) = \mathbf{0}$. Set $R' = A[X_1, \dots, X_{n-1}]$. Then $A/I \cap A \subseteq R'/I \cap R' \subseteq R/I$. As $A/I \cap A$ and R/I are zero-dimensional and R, R' are finitely generated A-algebras, it follows that $\dim(R'/I \cap R') = 0$. By Lemma 9.2 we may assume that $I \cap R'$ is primary, to a maximal ideal M. As R/MR is a principal ideal domain, IR/MR is a principal ideal. From the given generators of I one can compute $g \in I$ such that $gR/MR = IR/MR$. Observe that $g \notin MR$. Then $I \subseteq gR + MR = gR + (\sqrt{I \cap R'})R \subseteq \sqrt{I}$, whence $\sqrt{gR + (I \cap R')R} = \sqrt{I}$. By our factorization assumption, there are computable elements $g_1, \dots, g_s \in R$ such that their images in R/MR are distinct irreducible elements and $g = \prod g_j^{k_j}$ modulo MR. Observe that the ideals $g_j R + MR$ are distinct maximal ideals in R and that $g_j^{l_j} R + I$ is primary to $g_j R + MR$ for all $l_j \in \mathbb{N}$. Furthermore, any prime ideal in R containing I contains M and g, hence is one of the $g_j R + MR$. Thus it remains to find the $(g_j R + MR)$-primary component of I. But this is just $I_{f_j} \cap R$, where $f_j = \prod_{i \neq j} g_i$. Thus a primary decomposition of I is computable.

10. Complexity of associated primes

There are aspects of complexity of associated primes and complexity of primary decompositions other than their computability. This section describes some of them.

For example, the main motivation for Theorem 8.4 came from the theory of tight closure. Here is the set-up: let p be a prime integer and R a ring of characteristic p. For any ideal I and any power $q = p^e$ of p, define $I^{[q]}$ to be the ideal $(i^q \mid i \in I)$. This ideal is called the eth *Frobenius power* of I. Hochster and Huneke developed the theory of tight closure and proved many results in commutative algebra with it. But one basic question about tight closure is still open: does tight closure commute with localization? A partial answer to the question would be provided with a positive answer to the following question:

Question 10.1. For every ideal I in a Noetherian ring of characteristic p, does there exist an integer k such that for all $q = p^e$ there exists a primary decomposition

$$I^{[q]} = q_1 \cap \cdots \cap q_l$$

with $\sqrt{q_i}^{k[q]} \subseteq q_i$ for all i?

In case when I is one-generated, its Frobenius powers are ordinary powers, and then the answer to the question above is yes by Theorem 8.4.

If R is a regular ring, the answer to the question is yes, as then the Frobenius map is flat (see Kunz [27]). A set-back to finding an affirmative answer to the question above in general was provided by the following example of Katzman [21]:

Example 10.2. Let k be a field of positive characteristic p, t, x, y variables over k and $R = k(t)[x, y]/(xy(x-y)(x-ty))$. Then $\bigcup_q \operatorname{Ass}(R/I^{[q]})$ is an infinite set.

This is not giving a negative answer to the question. In fact, Smith and Swanson showed in [45] that Katzman's ideal satisfies the condition in the question. Nevertheless, Katzman's example shows that primary decompositions of Frobenius powers are fairly complex, and that resolving the question above will not be easy.

An example similar to Katzman's, but over an integral domain, was provided by Singh and Swanson [44]:

Example 10.3. Let k be a field of positive characteristic p,

$$R = \frac{K[r, t, u, v, w, x, y, z]}{(u^2x^2 + v^2y^2 + tuxvy + rw^2z^2)},$$

and $I = (x, y, z)R$. Then R is an F-regular unique factorization domain, and the set

$$\bigcup_q \operatorname{Ass} \frac{R}{I^{[q]}} = \bigcup_q \operatorname{Ass} \frac{R}{(I^{[q]})^*}$$

has infinitely many maximal elements. (Superscript $*$ denotes tight closure, and F-regularity is a notion from tight closure. I leave these terms undefined here.)

Thus even in very good rings associated primes of Frobenius powers run wild.

Another aspect of complexity is the degree of generators of primary components of an ideal in a graded ring. Grete Hermann [15] proved that in a polynomial ring in n variables over a field, if an ideal I has k generators and all have degree at most d, then the primary components of I are all generated in degrees which are at most doubly exponential in n. Hermann proved a few other doubly exponential bounds, such as for the ideal membership problem and the degrees of the generators of the syzygy module. Mayr and Meyer [32] proved that the doubly exponential degree bounds are achieved for the ideal membership problem: let $J(n, d)$ be the ideal generated by the following polynomials in the polynomial ring in $10n + 2$ variables over a field:

$$c_{0i}(s - fb_{0i}^d), \quad i = 1, 2, 3, 4;$$

$$s_r - s_{r-1}c_{r-1,1}, \quad r = 1, \ldots, n,$$

$$f_r - s_{r-1}c_{r-1,4}, \quad r = 1, \ldots, n,$$

$$f_{r-1}c_{r-1,1} - s_{r-1}c_{r-1,2}, \quad r = 1, \ldots, n,$$

$$f_{r-1}c_{r-1,4} - s_{r-1}c_{r-1,3}, \quad r = 1, \ldots, n,$$

$$s_{r-1}(c_{r-1,3} - c_{r-1,2}), \quad r = 1, \ldots, n,$$

$$f_{r-1}(c_{r-1,2}b_{r-1,1} - c_{r-1,3}b_{r-1,4}), \quad r = 1, \ldots, n,$$

$$f_{r-1}(c_{r-1,2}b_{r-1,1} - c_{r-1,3}b_{r-1,4}), \quad r = 1, \ldots, n,$$

$$f_{r-1}c_{r-1,2}c_{ri}(b_{r-1,2} - b_{ri}b_{r-1,3}), \quad i = 1, \ldots, 4, \ r = 1, \ldots, n-1,$$

$$f_{n-1}c_{n-1,2}(b_{n-1,2} - b_{n-1,3}).$$

Note that the generators have degrees at most $\max\{d + 2, 4, 5\delta_{n\geq 2}\}$. (Here, δ_P is 1 if P is true, and is 0 otherwise.) The degree 1 element $s_n - f_n$ of S is in $J(n, d)$, and when written as a linear combination of the given generators, the coefficient of $c_{04}(s_0 - f_0b_{01}^d)$ has degree which is doubly exponential in n (see [MM]).

It is not yet known if for every primary decomposition of the Mayr-Meyer ideals, the bound on the degrees of the generators of primary components is indeed doubly exponential. It is known that the number of minimal components is $nd^2 + 20$, and that the number of embedded primes is least $31 + 15d + (d^2 - d)\delta_{n=2} + (n-1)(d^3 - d)\delta_{n>2}$ and possibly doubly exponential in n (see [48], [49]). Bayer, Huneke, and Stillman asked whether the doubly exponential behaviour of the Mayr-Meyer ideals is due to the structure of one component or to the number of components. One could ask the same of the family of ideals $J(n, d) + (c_{ri}, s_r, f_r)^2$, which have much higher height than the $J(n, d)$, yet the same doubly exponential behaviour.

And one more aspect of complexity of primary decompositions is the number of associated or minimal primes over an ideal. Here is an example of permanental ideals: let A be an $n \times n$ matrix. In a Laplace expansion of the determinant of A change all the minus signs to plus. The obtained formula is the *permanent* of A.

Now let m and n be arbitrary positive integers and A an $m \times n$ matrix of indeterminates over a field. It is well-known that the ideal generated by determinants of all the $r \times r$ submatrices of A is a prime ideal. Not much is known for the corresponding results for permanents. Here is a special, and a wild, case, reflecting the computational complexity of permanents versus determinants via primary decompositions and associated primes:

Proposition 10.4 (Laubenbacher-Swanson [28]). *Let k be field of characteristic other than 2. Let $m, n \geq 2$ be integers and A an $m \times n$ matrix of indeterminates over k. Let I be the ideal generated by the permanents of all the 2×2 submatrices of A. Then the number of minimal primes over I is $n\delta_{m\geq 3} + m\delta_{n\geq 3} + \binom{n}{2}\binom{m}{2}$. Also, I has an embedded component if and only if $m, n \geq 3$, and the number of embedded components is exactly one.*

And yet another aspect of the complexity of associated primes has to do with local cohomology: if R is a Noetherian ring, M a finitely generated module, and I an ideal, when do the local cohomology modules $H_I^i(M)$ have finitely many associated primes? If R is a regular ring containing a field, all such local cohomology modules have only finitely many associated primes. This was proved by Huneke and Sharp [19] in case of positive characteristic, and by Lyubeznik [29] otherwise. Lyubeznik [30] also proved that the local cohomology modules have finitely many associated primes if R is an unramified regular local ring of mixed characteristic. Marley [31] proved that if R is a Noetherian local ring and M a finitely generated R-module of dimension at most three, then any local cohomology module $H_I^i(M)$ has only finitely many associated primes. Khashyarmanesh and Salarian [23] and Brodmann and Lashgari Faghani [3] proved that if i is the smallest integer such that $H_I^i(M)$ is not finitely generated, then $H_I^i(M)$ has only finitely many associated primes. Singh [43] constructed the first example for which $H_I^i(M)$ does not have finitely many associated primes. Katzman [22] constructed such an example over a local ring containing a field, and Singh and Swanson [44] constructed an example over a local unique factorization domain.

11. Exercises

Exercises have double numbers: the first number corresponds to the section with that number, the second number simply counts the exercises. Exercises with first number 0 cover some background.

0.1: Let P be a prime ideal, and $q_1, \ldots, q_s$ arbitrary ideals. If P contains $q_1 \cap \cdots \cap q_s$, prove that P contains one of the q_i. If $P = q_1 \cap \cdots \cap q_s$, prove that P equals some q_i.

0.2: Let I and J be ideals in a ring R. Prove that $\sqrt{I \cap J} = \sqrt{I} \cap \sqrt{J}$.

0.3: Let I and J be ideals and x an element in a ring R. Prove that $(I \cap J) : x = (I : x) \cap (J : x)$.

0.4: Let I be an ideal and U a multiplicatively closed subset in a ring R. Prove that $\sqrt{I}U^{-1}R = \sqrt{IU^{-1}R}$.

0.5: Let R be a ring, $N \subseteq M$ R-modules, and U a multiplicatively closed subset of R. Assume that M is finitely generated over R or that N is Noetherian. Prove that $U^{-1}(N :_R M) = U^{-1}N :_{U^{-1}R} U^{-1}M$.

0.6: Let R be a ring, $N \subseteq M$ R-modules, I an ideal in R, and U a multiplicatively closed subset of R. Assume that I is finitely generated or that N is Noetherian. Prove that $U^{-1}(N :_M I) = U^{-1}N :_{U^{-1}M} U^{-1}I$.

0.7: (Prime Avoidance) Let R be a ring, $P_1, \ldots, P_s$ ideals in R, at most two of which are not prime ideals. Let I be an ideal such that $I \not\subseteq P_i$ for $i = 1, \ldots, s$. Then I is not contained in $P_1 \cup \cdots \cup P_s$.

Furthermore, if R is graded by a totally ordered abelian monoid and $P_1, \ldots, P_s$ and I are homogeneous, then there exists a homogeneous element $r \in I$ such that $r \notin P_1 \cup \cdots \cup P_s$.

2.8: Find an example of a primary ideal I such that for some integer n, I^n is not primary.

2.9: Let R be a ring, P a prime ideal, M and M' R-modules, N a P-primary submodule of M, and N' a P-primary submodule of M'. Then $N \oplus N'$ is a P-primary submodule of $M \oplus M'$.

2.10: Let R be a Noetherian ring, P a prime ideal in R, and M an R-module. Assume that P is minimal over $0 :_R m$ for some $m \in M$. Prove that $P = 0 :_R n$ for some $n \in N$.

2.11: Let R be the ring $\mathbb{R}[X, Y]/(X^2 - Y^2 - X^3)$ localized at (X, Y), and S the completion of R in the (X, Y)-adic topology. Then S is a faithfully flat extension of R. Verify that there are prime ideals in R which are not primary in S.

3.12: Let R be a Noetherian ring and M a finitely generated R-module. Prove that the different notions of primes associated to M as in Remark 3.11 are all the same.

3.13: Let R be a Noetherian ring, and M an R-module (not necessarily Noetherian). Prove that $\operatorname{Ass}_R(M) = \widetilde{\operatorname{Ass}}_R(M)$.

3.14: Find a ring R and a module M such that the different notions of primes associated to M (as in Remark 3.11) are all distinct.

3.15: Let R be a ring, M, M' both R-modules.

(i) Prove or disprove: $\operatorname{Ass}(M \oplus M') = \operatorname{Ass}(M) \cup \operatorname{Ass}(M')$.
(ii) Assume that M, M' are contained in an R-module. Prove or disprove: $\operatorname{Ass}(M + M') = \operatorname{Ass}(M) \cup \operatorname{Ass}(M')$.

3.16: Let R be a ring and M an R-module. An R-submodule N of M is said to be *irreducible* if whenever N' and N'' are R-submodules of M such that $N = N' \cap N''$, then either $N = N'$ or $N = N''$.

(i) Prove that every irreducible submodule of a Noetherian module is primary.
(ii) Prove that every submodule of a Noetherian submodule can be written as a finite intersection of irreducible submodules. (This generalizes Theorem 3.4.)

3.17: Let R be a ring, I an ideal and X a variable over R. Let $I = q_1 \cap \cdots \cap q_s$ be a (minimal, irredundant) primary decomposition. Prove that $IR[X] = q_1R[X] \cap \cdots \cap q_sR[X]$ is a (minimal, irredundant) primary decomposition.

3.18: Let R be a ring, I an ideal, J a finitely generated ideal in R, and P a prime ideal. Prove that if $P \in \operatorname{Ass}_R(R/(I : J))$, then $P \in \operatorname{Ass}_R(R/I)$.

3.19: (From Dan Katz) Let R be a Noetherian ring, and I an ideal containing a non-zerodivisor. Let n_0 be a positive integer such that for all $n \geq n_0$, $I^{n+1} : I = I^n$.[1] Prove that for all $n \geq n_0$, $\operatorname{Ass}_R(R/I^n) \subseteq \operatorname{Ass}_R(R/I^{n+1})$.

[1] Such n always exists. A proof can be found in McAdam's book [33], Lemma 1.1 (b), applied to ideals which contain a non-zerodivisor. Namely, let $S = \operatorname{gr}_I(R) = R/I \oplus I/I^2 \oplus I^2/I^3 \oplus \cdots$. Then S is a Noetherian $\mathbb{N}$-graded ring. The ideal $0 :_S I/I^2$ is finitely generated, in degrees at most some integer l. For all $n > l$, $(0 :_S I/I^2) \cap I^n/I^{n+1} = 0$. Conclude that for all $n \geq l$, $(I^{n+1} : I) \cap I^l = I^n$. Now let x be a

(Thus if $\bigcup_n \mathrm{Ass}_R(R/I^n)$ is finite, then $\mathrm{Ass}_R(R/I^n) = \mathrm{Ass}_R(R/I^{n+1})$ for all large n. Cf. Proposition 7.5 and Exercise 7.35.)

3.20: Find an example of an ideal I and a primary decomposition $I = q_1 \cap \cdots \cap q_s$ such that for some positive integer n, $I^n \neq q_1^n \cap \cdots \cap q_s^n$.

3.21: Let R be a Noetherian ring, M a finitely generated R-module, and N an R-module. Prove that

$$\mathrm{Ass}_R \mathrm{Hom}_R(M, N) = \mathrm{Ass}_R(N) \cap \mathrm{Supp}_R(M).$$

3.22: Let R be a ring, P a prime ideal, and $E_R(R/P)$ the injective envelope of R/P. Prove that $\widetilde{\mathrm{Ass}}\, E_R(R/P) = \{P\}$.

3.23: Let R be a Noetherian ring and M a finitely generated R-module.

(i) Prove that there exists a filtration of R-modules

$$0 = M_0 \subseteq M_1 \subseteq \cdots \subseteq M_{n-1} \subseteq M_n = M$$

such that for all $i = 1, \ldots, n$, M_i/M_{i-1} is isomorphic to R/P_i for some prime ideal P_i in R.

(ii) Prove that if P is associated to M, then $P = P_i$ for some i.

3.24: (Yassemi [54]) Let R be a ring and M an R-module. Prove that $\widetilde{\mathrm{Ass}}_R(M)$ is a finite set if and only if there exists a filtration of R-modules

$$0 = M_0 \subseteq M_1 \subseteq \cdots \subseteq M_{n-1} \subseteq M_n = M$$

such that for all $i = 1, \ldots, n$, $\widetilde{\mathrm{Ass}}_R\, M_i/M_{i-1}$ consists of one prime ideal which is in addition weakly associated to M.

3.25: (Yassemi [54]) Let R be a ring, I a proper ideal, and M an R-module. Prove that $\bigcup_n \widetilde{\mathrm{Ass}}_R(M/I^nM) = \bigcup_n \widetilde{\mathrm{Ass}}_R(I^{n-1}M/I^nM)$.

3.26: (Heinzer, Ratliff, Shah [14]) Let $(R, \mathfrak{m})$ be a Noetherian local ring such that $\mathfrak{m} \in \mathrm{Ass}\, R \setminus \mathrm{Min}\, R$. Let S be the set of all $\mathfrak{m}$-primary components of (0) which are not contained properly in any other $\mathfrak{m}$-primary component of (0). Prove that S is an infinite set.

3.27: Let R be a ring, M, N R-modules, and $P \in \mathrm{Ass}_R(M)$. Prove $\mathrm{Ass}_R(N/PN) \subseteq \mathrm{Ass}_R(M \otimes_R N)$. (I saw this exercise in [40].)

3.28: (Ortiz, [36], cf. Theorem 6.2) Let R be a Noetherian ring, and M a finitely generated R-module. For each $P \in \mathrm{Ass}_R M$, and each P-primary component N of 0 in M, set

$$\mathrm{nil}_N(P) = \min\{n \mid P^n \subseteq N :_R M\},$$
$$\mathrm{nil}(P) = \min\{\mathrm{nil}_N(P) \mid N \text{ is a } P\text{-primary component of } 0\}.$$

Prove that there exists a unique irredundant primary decomposition $0 = \bigcap_{i=1}^s N_i$ of 0 in M such that

non-zerodivisor in I. Then for all n, $I^n : I \subseteq I^n : r$. By the Artin-Rees lemma, there exists an integer k such that for all $n \geq k$, $I^n \cap (r) \subseteq rI^{n-k}$. As r is a non-zerodivisor, for all $n \geq k$, $I^n : r \subseteq I^{n-k}$. Hence for $n \geq k + l$, $I^{n+1} : I \subseteq I^{n+1} : r \subseteq I^{n+1-k} \subseteq I^l$, whence $I^{n+1} : I \subseteq (I^{n+1} : I) \cap I^l = I^n$.

(i) For each i, with $P_i = \sqrt{N_i :_R M}$, $\mathrm{nil}_{N_i}(P_i) = \mathrm{nil}(P_i)$.

(ii) If $0 = \bigcap_{i=1}^{s} N_i'$ is any other irredundant primary decomposition, then after relabelling without loss of generality for each $i = 1, \ldots, s$, $P_i = \sqrt{N_i' :_R M}$. If $\mathrm{nil}_{N_i'}(P_i) = \mathrm{nil}(P_i)$, then $N_i \subseteq N_i'$.

4.29: Let R be a Noetherian ring and I an ideal in R. Prove that for any $x \in R$, $\mathrm{Ass}(R/I) \subseteq \mathrm{Ass}(R/(I : x)) \cup \mathrm{Ass}(R/(I + xR))$. (Hint: use Proposition 4.4.)

4.30: Let R be a Noetherian ring and $0 \longrightarrow M' \longrightarrow M \longrightarrow M'' \longrightarrow 0$ a short exact sequence of finitely generated R-modules. Prove that

$$\mathrm{Ass}\, M'' \subseteq \mathrm{Ass}\, M \cup \{P \in \mathrm{Spec}\, R \mid \mathrm{depth}_{R_P} M_P = 1\}.$$

5.31: (Krull's intersection theorem) Let R be a Noetherian ring and I an ideal in R. Prove that $\bigcap_n I^n$ is the intersection of those primary components q of 0 which satisfy the property $q + I = R$. (Hint: you may need to use that $I \bigcap_n I^n = \bigcap_n I^n$.)

5.32: (Yassemi [53]) Let R be a ring, S an R-algebra, and M an S-module. Prove that $\{P \cap R \mid P \in \mathrm{Ass}_S(M)\} \subseteq \mathrm{Ass}_R(M)$.

5.33: Let R be a ring and x a non-zerodivisor in R. Assume that 0 and (x) have primary decompositions. If $P \in \mathrm{Ass}(R/(x))$ and $y \in P$ a non-zerodivisor, then P is associated to some power of (y). If P is finitely generated, then P is associated to (y).

5.34: Let R be a Noetherian $\mathbb{N}^n$-graded ring generated over R_0 by elements of total degree 1. Let M be a finitely generated $\mathbb{N}^n$-graded R-module. Prove that there exists $g_0 \in \mathbb{N}^n$ such that for all $g \in \mathbb{N}^n$, if g is componentwise larger than g_0, then $\mathrm{Ass}_{R_0} M_g = \mathrm{Ass}_{R_0} M_{g_0}$.

7.35: Let R be a Noetherian ring and I an ideal. Then $\bigcup_n \mathrm{Ass}(R/I^n)$ is a finite set. (Hint: this of course follows from Proposition 7.5, which proves more. Try a more elementary approach? Compare with the following two exercises.)

7.36: Let R be a Noetherian ring of prime characteristic p and I an ideal of finite projective dimension such that $\mathrm{Min}\, R/I = \mathrm{Ass}_R(R/I)$. Prove that for every $q = p^e$, $\mathrm{Ass}_R(R/I^{[q]}) = \mathrm{Min}\, R/I$, where $I^{[q]} = (i^q \mid i \in I)$.

7.37: Let R be a Noetherian ring and I an ideal generated by a regular sequence. Prove that for all n, $\mathrm{Ass}_R(R/I^n) = \mathrm{Ass}_R(R/I)$.

7.38: Let R be a ring, and $I_1, \ldots, I_s$ ideals in R. The *multi-Rees algebra* of $I_1, \ldots, I_s$ is the subalgebra of $R[t_1, \ldots, t_s]$, where $t_1, \ldots, t_s$ are variables over R, generated over R by all elements of the form $a_i t_i$, as a_i varies over the elements of I_i. This algebra is also denoted $R[I_1 t_1, \ldots, I_s t_s]$. The *extended multi-Rees algebra* of $I_1, \ldots, I_s$ is the subalgebra of $R[t_i, t_i^{-1} \mid i = 1, \ldots, s]$ generated over R by the t_i^{-1} and all elements of the form $a_i t_i$, as a_i varies over the elements of I_i. The extended Rees algebra is also denoted $R[It, t^{-1}]$. $R[I_1 t_1, \ldots, I_s t_s, t_1^{-1}, \ldots, t_s^{-1}]$

(i) Prove that for all non-negative integers $n_1, \ldots, n_s$,

$$t_1^{-n_1} \cdots t_s^{-n_s} R[I_1 t_1, \ldots, I_s t_s, t_1^{-1}, \ldots, t_s^{-1}] \cap R = I_1^{n_1} \cdots I_s^{n_s}.$$

(ii) Assume that M is an R-module. If S denotes the (extended) Rees algebra of $I_1, \ldots, I_s$, prove that MS is an $\mathbb{N}^s$-graded S-module. In particular, S is a graded.

7.39: Let R be a Noetherian ring, $I_1, \ldots, I_s$ ideals, and M a finitely generated R-module. Let S be the (extended) Rees algebra of $I_1, \ldots, I_s$.

(i) Prove that $\bigcup_e \operatorname{Ass}(MS)_e$ is a finite set, as e varies over s-tuples in $\mathbb{N}^s$.
(ii) Use the shorthand notation $I^e = I_1^{e_1} \cdots I_s^{e_s}$. Prove that $\bigcup_e \operatorname{Ass}(I^e M)$ is a finite set.
(iii) Similarly, prove that $\bigcup_{e,i} \operatorname{Ass}(I^e M / I^e I_i M)$ is a finite set.
(iv) Prove that $\operatorname{Ass}(I^e M)$ is independent of e for e componentwise large.

8.40: Let R be a Noetherian ring, $I_1, \ldots, I_s$ ideals and M a finitely generated R-module. Then there exists a positive integer k such that for all positive integers $n_1, \ldots, n_s$ there exists a primary decomposition

$$I_1^{n_1} M \cap \cdots \cap I_s^{n_s} M = N_1 \cap N_2 \cap \cdots \cap N_l$$

such that for all i, $\sqrt{N_i : M}^{kn} M \subseteq N_i$.

9.41: Let $R = \mathbb{Q}[X, Y]$ endowed with the lexicographic ordering. Let $I = (X^2, XY - Y^2)$.

(i) Prove that $\operatorname{in}(I) = (X^2, XY, Y^3)$.
(ii) Show that $\operatorname{in}(I) \neq (\operatorname{in}(X^2), \operatorname{in}(XY - Y^2)) = (X^2, XY)$.
(iii) Let $G = \{X^2, XY - Y^2\}$. Reduce X^2Y with respect to G in two different ways.

9.42: Let k be a perfect field, $X_1, \ldots, X_n$ variables over k, $R = k[X_1, \ldots, X_n]$. Let I be a zero-dimensional ideal in R. For each $j = 1, \ldots, n$, set $F_j k[X_j] = I \cap k[X_j]$. Let G_j be the square-free part of F_j. Prove that $\sqrt{I} = I + (G_1, \ldots, G_n)$.

9.43: Let k be an infinite field, $X_1, \ldots, X_n$ variables over k, and I a zero-dimensional ideal in $R = k[X_1, \ldots, X_n]$. Assume that I is not primary.

(i) Prove that after a generic linear change of coordinates, i.e., after a mapping $X_j \mapsto \sum_i a_{ij} X_i$, with $a_{ij} \in k$ sufficiently general, I contracted to $k[X_1]$ is not primary.
(ii) Give an example showing that without the generic linear change it can happen that $I \cap k[X_j]$ is primary for all j.

Acknowledgement

I thank the Bhaskaracharya Pratishthana and Harish-Chandra Research Institutes for the hospitality and a great conference, and the NSF for partial support. I thank Bruce Olberding and Kamran Divaani-Aazar for the careful reading of the first version of these notes and for improving the non-Noetherian parts. I thank Winfried Bruns and H. S. Hassanzadeh for their comments and corrections.

References

[1] M. F. Atiyah and I. G. MacDonald, *Introduction to Commutative Algebra*, Addison-Wesley Publishing Company (1969)

[2] M. Brodmann, Asymptotic stability of Ass(M/I^nM), *Proc. Amer. Math. Soc.* **74** (1979) 16–18

[3] M. P. Brodmann and A. Lashgari Faghani, A finiteness result for associated primes of local cohomology modules, *Proc. Amer. Math. Soc.* **128** (2000) 2851–2853

[4] W. Bruns and U. Vetter, Determinantal Rings, *Lecture Notes in Math.*, no. 1327, Springer-Verlag (1988)

[5] J. H. Davenport, Y. Siret and E. Tournier, *Computer algebra. Systems and algorithms for algebraic computation.* Second edition. With a preface by Daniel Lazard. Translated from the French by A. Davenport and J. H. Davenport. With a foreword by Anthony C. Hearn. Academic Press, Ltd., London (1993)

[6] W. Decker, G.-M. Greuel and G. Pfister, *Primary decomposition: algorithms and comparisons*. Algorithmic algebra and number theory (Heidelberg, 1997), 187–220, Springer, Berlin (1999)

[7] K. Divaani-Aazar and M. Tousi, A new definition of associated prime ideals, *Ital. J. Pure Appl. Math.* **8** (2000) 99–113

[8] L. Ein, R. Lazarsfeld and K. E. Smith, Uniform bounds and symbolic powers on smooth varieties, *Invent. Math.* **144** (2001) 241–252

[9] D. Eisenbud, C. Huneke and W. Vasconcelos, Direct methods for primary decomposition, *Invent. Math.* **110** (1992) 207–235

[10] L. Fuchs, W. Heinzer and B. Olberding, Commutative ideal theory without finiteness conditions: primal ideals, *Trans. Amer. Math. Soc.* **357** (2005) 2771–2798

[11] P. Gianni, B. Trager and G. Zacharias, Gröbner bases and primary decompositions of polynomial ideals, *J. Symbolic Comput.* **6** (1988) 149–167

[12] D. Grayson and M. Stillman, Macaulay2. 1996. *A system for computation in algebraic geometry and commutative algebra*, available via anonymous `ftp` from `math.uiuc.edu`

[13] G.-M. Greuel, G. Pfister and H. Schönemann, Singular. 1995. *A system for computation in algebraic geometry and singularity theory.* Available via anonymous `ftp` from `helios.mathematik.uni-kl.de`

[14] W. Heinzer, L. J. Ratliff, Jr. and K. Shah, On the embedded primary components of ideals, *I J. Algebra* **167** (1994) 724–744

[15] G. Hermann, Die Frage der endlich vielen Schritte in der Theorie der Polynomideale, *Math. Ann.* **95** (1926) 736–788

[16] L. T. Hoa and N. V. Trung, On the Castelnuovo-Mumford regularity and the arithmetic degree of monomial ideals, *Math. Z.* **229** (1998) 519–537

[17] M. Hochster and C. Huneke, Comp0arison of symbolic and ordinary powers of ideals, *Invent. Math.* **147** (2002) 349–369

[18] S. Hoşten and R. R. Thomas, The associated primes of initial ideals of lattice ideals, *Math. Res. Lett.* **6** (1999) 83–97

[19] C. Huneke and R. Y. Sharp, Bass numbers of local cohomology modules, *Trans. Amer. Math. Soc.* **339** (1993) 765-779

[20] J. Iroz and D. E. Rush, Associated prime ideals in non-Noetherian rings, *Cand. J. Math.* **36** (1984) 344–360

[21] M. Katzman, Finiteness of $\cup_e \mathrm{Ass}\, F^e(M)$ and its connections to tight closure, *Illinois J. Math.* **40** no. 2 (1996) 330–337

[22] M. Katzman, An example of an infinite set of associated primes of a local cohomology module, *J. Algebra* **252** (2002) 161–166

[23] K. Khashyarmanesh and Sh. Salarian, On the associated primes of local cohomology modules, *Comm. Algebra* **27** (1999) 6191–6198

[24] D. Kirby, Artinian modules and Hilbert polynomials, *Q. J. Math.* Oxford (2) **24** (1973) 47–57

[25] M. Kreuzer and L. Robbiano, *Computational Commutative Algebra* 1, Springer-Verlag (2000)

[26] T. Krick and A. Logar, An algorithm for the computation of the radical of an ideal in the ring of polynomials. AAECC9, *Springer LNCS* **539** (1991) 195–205

[27] E. Kunz, Characterizations of regular local rings of characteristic p, *Amer. J. Math.* **91** (1969) 772–784

[28] R. C. Laubenbacher and I. Swanson, Permanental ideals, *J. Symbolic Comput.* **30** (2000) 195–205

[29] G. Lyubeznik, Finiteness properties of local cohomology modules (an application of D-modules to commutative algebra), *Invent. Math.* **113** (1993) 41–55

[30] G. Lyubeznik, Finiteness properties of local cohomology modules for regular local rings of mixed characteristic: the unramified case, *Comm. Algebra* **28** (2000) 5867–5882

[31] T. Marley, The associated primes of local cohomology modules over rings of small dimension, *Manuscripta Math.* **104** (2001) 519–525

[32] E. Mayr and A. Meyer, The complexity of the word problems for commutative semigroups and polynomial ideals, *Adv. Math.* **46** (1982) 305–329

[33] S. McAdam, Asymptotic Prime Divisors, *Lecture Notes in Math.* 1023, Springer-Verlag (1983)

[34] S. McAdam and P. Eakin, The asymptotic Ass, *J. Algebra* **61** (1979) 71–81

[35] C. Monico, Computing the primary decomposition of zero-dimensional ideals, *J. Symbolic Comput.* **34** (2002) 451–459

[36] V. Ortiz, Sur une certaine decomposition canonique d'un idéal en intersection d'idéaux primaires dans un anneau noetherien commutatif, *C. R. Acad. Sci. Paris* **T. 248, n. 24** (1959) 3385–3387

[37] L. J. Ratliff, Jr., On prime divisors of I^n, n large, *Michigan Math. J.* **23** (1976) 337–352

[38] E. W. Rutman, Gröbner bases and primary decomposition of modules, *J. Symbolic Comput.* **14** (1992) 483–503

[39] E. W. Rutman, Primary decomposition of modules: two variables over a field, *J. Symbolic Comput.* **15** (1993) 267–275

[40] D. Savitt, *Associated Primes and Primary Decomposition.* Available electronically at `www.math.harvard.edu/~dsavitt/250b/primary2.ps`.

[41] R. Y. Sharp, Injective modules and linear growth of primary decompositions, *Proc. Amer. Math. Soc.* **128** (2000) 717–722

[42] T. Shimoyama and K. Yokoyama, Localization and primary decomposition of polynomial ideals, *J. Symbolic Comput.* **22** (1996) 247–277

[43] A. K. Singh, p-torsion elements in local cohomology modules, *Math. Res. Lett.* **7** (2000) 165–176

[44] A. K. Singh and I. Swanson, Associated primes of local cohomology modules and of Frobenius powers, *Internat. Math. Res. Notices* **30** (2004) 1703–1733

[45] K. E. Smith and I. Swanson, Linear bounds on growth of associated primes, *Comm. Algebra* **25** (1997) 3071–3079

[46] I. Swanson, Powers of ideals: primary decompositions, Artin-Rees lemma and regularity, *Math. Ann.* **307** (1997) 299–313

[47] I. Swanson, Linear equivalence of ideal topologies, *Math. Z.* **234** (2000) 755–775

[48] I. Swanson, The minimal components of the Mayr-Meyer ideals, *J. Algebra* **267** (2003) 127–155

[49] I. Swanson, On the embedded primes of the Mayr-Meyer ideals, *J. Algebra* **275** (2004) 143–190

[50] J. von zur Gathen and J. Gerhard, *Modern computer algebra*, Second edition. Cambridge University Press, Cambridge (2003)

[51] D. Wang, An elimination method based on Seidenberg's theory and its applications, Computational algebraic geometry (Nice, 1992), 301–328, *Progr. Math.*, 109, Birkheuser Boston, Boston, MA (1993)

[52] Y. Yao, Primary decomposition: compatibility, independence and linear growth, *Proc. Amer. Math. Soc.* **130** (2002) 1629–1637

[53] S. Yassemi, Weakly associated primes under change of rings. *Comm. Algebra* **26** (1998) 2007–2018

[54] S. Yassemi, Weakly associated prime filtration, *Acta Math. Hungar.* **92** (2001) 179–183

Proc. Int. Conf. – Commutative Algebra and Combinatorics No. 4, 2007, pp. 157–180.

Castelnuovo–Mumford Regularity and Related Invariants

Ngô Việt Trung

Institute of Mathematics, Vien Toan Hoc, 18 Hoang Quoc Viet, 10307 Hanoi, Vietnam
e-mail: nvtrung@math.ac.vn

Abstract. These notes are an introduction to basic aspects of Castelnuovo–Mumford regularity and related topics such as weak regularity, a^*-invariant and partial regularities.

Introduction

Let $R = k[x_1, \dots, x_n]$ be a polynomial ring over a field. Let M be a finitely generated graded R-module. The structure of M is best understood by a minimal graded free resolution of M:

$$0 \longrightarrow F_s \longrightarrow \cdots \longrightarrow F_1 \longrightarrow F_0 \longrightarrow M \longrightarrow 0.$$

Let $b_i(M)$ denote the maximal degree of the generators of F_i. The Castelnuovo–Mumford regularity or regularity of M is usually defined as the number

$$\operatorname{reg}(M) = \max\{b_i(M) - i \mid i = 0, \dots, s\}.$$

The regularity was introduced by Mumford by generalizing geometric idea of Castelnuovo [14]. Originally, it was defined for a coherent sheaf by the vanishing of the sheaf cohomology. This approach has led to the equivalent definition

$$\operatorname{reg}(M) = \max\{a_i(M) + i \mid i \geq 0\},$$

where $a_i(M)$ denotes the largest non-vanishing degree of the ith local cohomology of M with respect to the maximal graded ideal of R.

The regularity is a measure for the complexity of the structure of graded modules [3], [26]. Indeed, several important invariants of graded rings and modules can be estimated by means of the regularity.

The aim of these notes is to introduce the reader to some basic results on the Castelnuovo-Mumford regularity and related invariants, which may serve a later and deeper study. The setting will be very general. We will work with standard graded algebra over an arbitrary Noetherian ring. The regularity is defined by means of the local cohomology modules with respect to the ideal of homogeneous elements of positive degree. This is meant for possible applications, in particular, in the study of blow-up rings of ideals. The reader is assumed to be familiar with basic concepts of Commutative Algebra.

The notes are divided in six sections.

In Section 1 we show why the above two definitions of the regularity are equivalent and discuss their consequences when R is a polynomial ring over a field. We also prove the characterizations of the regularity in terms of the Tor and Ext modules. These definitions and characterizations clearly indicate why the regularity is a measure for the complexity of the underlying module.

In Section 2 we present a characterization of the regularity by means of a sequence of linear forms, which behave like a regular sequence in larger degree. Such a sequence is called a filter-regular sequence. The characterization reduces the computation of the regularity to the computation of the largest non-vanishing degree of simple quotient modules, which have finite length if the base ring is an artinian local ring. The notion of filter-regular sequence will play a crucial role in these notes.

In Section 3 we study the notion of weak regularity which controls the vanishing of the local cohomology modules along a shifted degree. By definition, to compute the regularity one needs to check the vanishing of infinitely many components of the the local cohomology modules. The weak regularity reduces this computation to just only a finite number of components. We shall see that the geometric regularity

$$\text{g-reg}(M) = \max\{a_i(M) + i \mid i \geq 1\},$$

which modifies the notion of regularity in Algebraic Geometry, is a weaker version of the weak regularity. The geometric regularity is easier to handle with and can be used to estimate the regularity.

In Section 4 we introduce the invariant

$$a^*(M) = \max\{a_i(M) \mid i \geq 0\},$$

which is a counterpart of the regularity. In fact, if R is a polynomial ring in n variables, it can be shown that

$$a^*(M) = \max\{b_i(M) - i \mid i = 0, \dots, s\} - n.$$

We may view the $a^*(M)$ as the real regularity and $\operatorname{reg}(M)$ as the shifted regularity of M. The a^*-invariant enjoys many interesting properties. In particular, if R is an algebra over a a local ring, $a^*(M)$ is equal to the largest non-vanishing degree of the local cohomology modules of M with respect to the maximal graded ideal. It is well-known that these local cohomology modules carry more information on the structure of M.

In Section 5 we introduce the partial regularities

$$\operatorname{reg}_t(M) = \max\{a_i(M) + i \mid i \leq t\},$$

$$a_t^*(M) = \max\{a_i(M) \mid i \leq t\}$$

for every integer $t \geq 0$. The partial regularities provide more specific information on the graded structure. In fact, if R is a polynomial ring in n variables, it can be shown that

$$\operatorname{reg}_t(M) = \max\{b_i(M) - i \mid i \geq n - t\},$$

$$a_t^*(M) = \max\{b_i(M) \mid i \geq n - t\} - n.$$

In this case, the partial regularities can be viewed as the regularity and the a^*-invariant of the syzygies of M. We shall see that the partial regularities can be computed also by means of filter-regular sequences.

In Section 6 we show how one can use the technique of Gröbner bases to compute the regularity of ideals in a polynomial ring. We show that if the variables in decreasing order form a filter-regular sequence for the quotient ring, then the regularity does not change when passing to the initial ideal with respect to the reverse lexicographic order. That is for example the case of the generic initial ideal. To compute the regularity of a monomial ideal we introduce simpler invariants which are obtained from the given ideal by certain substitution $x_i = 0, 1$. These invariants have simple combinatorial descriptions and can be computed effectively. The regularity is the maximum of these invariants. As a consequence, if the characteristic of the field is zero, the regularity of the generic initial ideal with respect to the reverse lexicographic order is equal to the maximal degree of the generators. Similar results are proved for the a^*-invariant and the partial regularities.

1. Castelnuovo–Mumford regularity

Let $R = k[x_1, \dots, x_n]$ be a polynomial ring over a field k. Let M be a finitely generated graded R-module.

By Hilbert syzygy theorem, M has a minimal graded free resolution of M of finite length

$$0 \longrightarrow F_s \longrightarrow \cdots \longrightarrow F_1 \longrightarrow F_0 \longrightarrow M \longrightarrow 0,$$

where the maps are all of degree zero. Let $b_i(M)$ denote the maximum degree of the generators of F_i. The *Castelnuovo–Mumford regularity* or *regularity* of M is the number

$$\operatorname{reg}(M) := \max\{b_i(M) - i \mid i = 0, \dots, s\}.$$

Example. We have

- $\operatorname{reg}(R) = 0$.
- $\operatorname{reg}(R/fR) = \deg(f) - 1$ for any homogeneous form $f \in R$.

By the above definition, the graded pieces of M behave similarly past the degree $\operatorname{reg}(M)$. For instance, if we denote by $d(M)$ the maximal degree of the generators of M, then

$$d(M) = b_0(M) \le \operatorname{reg}(M).$$

There are several interesting characterizations of the regularity of which the following are especially useful.

Proposition 1.1.

$$\begin{aligned}\operatorname{reg}(M) &= \max\{t \mid \operatorname{Tor}_i^R(M, k)_{t-i} \neq 0 \text{ for some } i \ge 0\},\\ &= \max\{t \mid \operatorname{Ext}_R^i(M, R)_{-t-i} \neq 0 \text{ for some } i \ge 0\}.\end{aligned}$$

Proof. By the minimality of the given resolution we have $\operatorname{Tor}_i^R(M,k) \cong F_i \otimes k$. Therefore,

$$b_i(M) = \max\{t \mid \operatorname{Tor}_i^R(M,k)_t \neq 0\},$$

which implies the first formula for $\operatorname{reg}(M)$.

To prove the second formula we put $r = \operatorname{reg}(M)$ and $F_i^* = \operatorname{Hom}_R(F_i, R)$. Since F_i has no generators of degree $> r+i$, F_i^* must be zero in degree $< -r-i$. Note that $\operatorname{Ext}_R^i(M,R)$ is the homology of the dual of the resolution of M at F_i^*. Then $\operatorname{Ext}_R^i(M,R)_t = 0$ for $t < -r-i$. Now let i be the largest index such that $r = b_i - i$. Then F_i^* has $R(r+i)$ as a summand, whereas F_{i+1}^* has no summand of the form $R(m)$ with $m \geq r+i$. By the minimality of the resolution, the summand $R(r+i)$ of F_i^* must map to zero in F_{i+1}^*. Moreover, nothing in F_{i-1}^* can map on to the generator of $R(r+i)$ in F_i^*, so it gives a nonzero class in $\operatorname{Ext}_R^i(M,R)$ of degree $-r-i$. Thus,

$$r = \max\{t \mid \operatorname{Ext}_R^i(M,R)_{-t-i} \neq 0 \text{ for some } i \geq 0\},$$

as required. □

The above definition and characterization indicate clearly why the regularity can be used as a measure for the complexity of the structure of M. For more information on this topics we refer to [3], [9], [26].

We shall see that the regularity can be defined for modules over a larger class of graded algebras for which we don't have finite minimal free resolutions.

From now on let $R = \oplus_{n\geq 0} R_n$ be a finitely generated standard graded ring over a noetherian commutative ring R_0, where "standard" means R is generated by elements of R_1.

For any graded R-module M we set

$$\Gamma_{R_+}(M) = \{e \in M \mid R_t e = 0 \text{ for } t \gg 0\}.$$

It is easy to check that $\Gamma_{R_+}(.)$ is a covariant and left exact functor on the category of modules over R. The *local cohomology* functors are the right derived functors of $\Gamma_{R_+}(.)$ For $i \geq 0$ we denote by $H_{R_+}^i(M)$ the ith local cohomology graded module of M.

Remark. $H_{R_+}^0(M) = \Gamma_{R_+}(M) = \cup_{t\geq 0}(0_M : (R_+)^t)$, which is the intersection of all primary components of 0_M whose associated prime ideals do not contain R_+.

Now let M be a finitely generated R-module. It is known that $H_{R_+}^i(M)$ is a graded module with $H_{R_+}^i(M)_t = 0$ for $t \gg 0$. Moreover, if $M \neq 0$, there exists an integer $s \leq \dim M$ such that $H_{R_+}^s(M) \neq 0$ and $H_{R_+}^i(M) = 0$ for $i > s$. For more detail see e.g. [7].

To study the graded structure of $H_{R_+}^i(M)$ we introduce the following notation.

For any graded R-module H with $H_t = 0$ for $t \gg 0$ we set

$$a(H) := \sup\{t \mid H_t \neq 0\}$$

with the convention $a(H) = -\infty$ if $H = 0$. This invariant can be understood as the largest non-vanishing degree of H.

If we set $a_i(M) := a(H^i_{R_+}(M))$, then the *Castelnuovo regularity* of M is defined as the number

$$\operatorname{reg}(M) := \max\{a_i(M) + i \mid i \geq 0\}.$$

From the above properties of local cohomology modules one can see that $\operatorname{reg}(M)$ is a finite number if $M \neq 0$.

Remark. If $M_t = 0$ for $t \gg 0$, we have $H^i_{R_+}(M) = M$ and $H^i_{R_+}(M) = 0$ for $i > 0$. Therefore, $\operatorname{reg}(M) = a_0(M) = \max\{t \mid M_t \neq 0\}$. In this case, $\operatorname{reg}(M)$ is the largest non-vanishing degree of M.

If R is a polynomial ring over a field, the new definition of regularity coincides with the previous one. This is a consequence of the following local duality (see e.g. [7]).

Theorem 1.2 (Local duality). *Let R be a polynomial ring over k in n variables. Then*

$$H^i_{R_+}(M)_t \cong \operatorname{Ext}^{n-i}_R(M, R)_{-t-n}$$

for all i and t.

Now, since $\operatorname{Ext}^j_R(M) = 0$ for $j > n$, we obtain

$$\begin{aligned}\max\{a_i(M) + i \mid i \geq 0\} &= \max\{t \mid \operatorname{Ext}^{n-i}_R(M, R)_{-t-n+i} \neq 0 \text{ for some } i = 0, \dots, n\} \\ &= \max\{t \mid \operatorname{Ext}^j_R(M, R)_{-t-j} \neq 0 \text{ for some } j \geq 0\}.\end{aligned}$$

By Proposition 1.1, this shows that the new definition of regularity coincides with the one in the case R is a polynomial ring over a field.

The new definition of regularity has many advantages. First, it holds for a larger class of graded algebras where not all modules have finite free resolutions. Second, it gives us some flexibility in choosing the base ring.

First, if R is a quotient of a graded algebra S, $H^i_{R_+}(M) = H^i_{S_+}(M)$ for all i. Therefore, the regularity of a finitely graded R-module M does not change if we consider M as a graded S-module.

Second, if there exists a standard graded subalgebra A of R such that R is a finite A-module, every element of R_+ is integral over A. Therefore, R_+ is contained in the radical of A_+R. This implies $H^i_{A_+}(M) = H^i_{R_+}(M)$ for all i. Hence $\operatorname{reg}(M)$ does not change if we consider M as a graded A-module.

Remark. If R is a standard graded algebra over a field, we may represent R as a quotient ring of a polynomial ring over a field. In this case, we can use the definition of regularity by means of a minimal finite resolution again. An alternative way to pass to the case of a polynomial ring is by taking a Noether normalization of R.

Using the definition by means of local cohomology we can easily deduce some basic properties of the regularity. For instance, the regularity behaves well in a short exact sequence.

Lemma 1.3. *Let* $0 \longrightarrow E \longrightarrow M \longrightarrow F \longrightarrow 0$ *be an exact sequence of finitely generated R-modules. Then*

(a) $\operatorname{reg}(E) \leq \max\{\operatorname{reg}(M), \operatorname{reg}(F) + 1\}$,
(b) $\operatorname{reg}(M) \leq \max\{\operatorname{reg}(E), \operatorname{reg}(F)\}$,
(c) $\operatorname{reg}(F) \leq \max\{\operatorname{reg}(M), \operatorname{reg}(E) - 1\}$.

Proof. We only need to consider the derived long exact sequence of local cohomology modules:

$$\cdots \longrightarrow H^{i-1}_{R_+}(F)_t \to H^i_{R_+}(E)_t \to H^i_{R_+}(M)_t \to H^i_{R_+}(F)_t \to H^{i+1}_{R_+}(E)_i \to \cdots.$$

From this it follows that

$$a_i(E) \leq \max\{a_{i-1}(F), a_i(M)\},$$

$$a_i(M) \leq \max\{a_i(E), a_i(F)\},$$

$$a_i(F) \leq \max\{a_i(M), a_{i+1}(E)\}.$$

The assertion now follows from the definition of the regularity. □

For more basic information on the regularity we refer to [7], [8], [9], [15].

2. Filter-regular sequence

In this section we present a characterization of the regularity by means of a sequence of linear forms, which does not involve local cohomology explicitly. This characterization will play a crucial role in the rest of these notes.

Let R be a finitely generated standard graded algebra over a noetherian ring. Let M be a finitely generated graded R-module. We call a homogeneous element $z \in R$ an *M-filter-regular element* if $(0_M : z)_t = 0$ for $t \gg 0$. This means that z behaves like an M-regular element in higher degrees. If $M = R$, we delete the prefix M.

It is easy to see that z is an M-filter-regular element if and only if $z \notin \mathfrak{p}$ for any associated prime ideal $\mathfrak{p} \not\supseteq R_+$ of M. Therefore, M-filter-regular elements of any degree do exist if R_0 is a local ring with infinite residue field.

Using an M-filter-regular element one can easily compute $a_0(M)$.

Lemma 2.1. *Let z be any M-filter-regular linear form. Then*

$$a_0(M) = a(0_M : z) = a(0_M : R_+).$$

Proof. By the above characterization of filter-regular element, $\cup_{i\geq 1}(0_M : z^i)$ is the intersection of all primary components of 0_M whose associated prime ideals do not contain R_+. From this it follows that

$$\cup_{i\geq 1}(0_M : z^i) = \cup_{i\geq 1}(0_M : (R_+)^i).$$

Therefore, $0_M : R_+ \subseteq 0_M : z \subseteq \cup_{i\geq 1}(0_M : (R_+)^i) = H^0_{R_+}(M)$. This implies

$$a(0_M : R_+) \leq a(0_M : z) \leq a_0(M).$$

On the other hand, every element of the largest non-vanishing degree of $H^0_{R_+}(M)$ is contained in $0_M : R_+$. Therefore, $a(0_M : R_+) = a_0(M)$, which together with the above inequalities imply the assertion. □

There is the following relationship between $\operatorname{reg}(M)$ and $\operatorname{reg}(M/zM)$.

Lemma 2.2. *Let z be any M-filter-regular linear form. Then*

$$\operatorname{reg}(M) = \max\{a_0(M), \operatorname{reg}(M/zM)\}.$$

Proof. Since $(0_M : z)_t = 0$ for $t \gg 0$, we have $H^i_{R_+}(0_M : z) = 0$ for $i \geq 1$. Now, from the short exact sequence

$$0 \to (0_M : z) \to M \to M/(0_M : z) \to 0$$

we obtain $H^i_{R_+}(M) \cong H^i_{R_+}(M)$ for $i \geq 1$. From the short exact sequence

$$0 \to M/(0_M : z) \xrightarrow{z} M \to M/zM \to 0$$

we obtain the derived long sequences of local cohomology modules:

$$\cdots \to H^{i-1}_{R_+}(M)_t \to H^{i-1}_{R_+}(M)_t \to H^i_{R_+}(M)_{t-1}$$
$$\to H^i_{R_+}(M)_t \to H^i_{R_+}(M/zM) \to \cdots.$$

As a consequence, the map $H^i_{R_+}(M)_{t-1} \to H^i_{R_+}(M)_t$ is injective for $t > a_{i-1}(M/zM)$. Since $H^i_{R_+}(M)_t = 0$ for $t \gg 0$, this implies $H^i_{R_+}(M)_{t-1} = 0$ for $t > a_{i-1}(M/zM)$. Hence $a_i(M) \leq a_{i-1}(M/zM) - 1$. The above exact sequence of local cohomology modules also implies $a_{i-1}(M/zM) \leq \max\{a_{i-1}(M), a_i(M) + 1\}$. Therefore,

$$a_i(M) + i \leq a_{i-1}(M/zM) + (i-1) \leq \max\{a_{i-1}(M) + (i-1), a_i(M) + i\}$$

for $i > 0$. From this it follows that

$$\max\{a_i(M) \mid i \geq 1\} \leq \operatorname{reg}(M/zM) \leq \operatorname{reg}(M),$$

which immediately implies $\operatorname{reg}(M) = \max\{a_0(M), \operatorname{reg}(M/zM)\}$. □

Since every M-regular element is M-filter-regular, Lemma 2.2 implies the following property.

Corollary 2.3. Assume that z is an M-regular linear form. Then

$$\operatorname{reg}(M) = \operatorname{reg}(M/zM).$$

Proof. The existence of an M-filter-regular form implies that 0_M has no associated prime ideals which contain R_+. This means $H^0_{R_+}(M) = 0$. Hence the assertion follows from Lemma 2.2. □

We call a sequence of homogeneous elements $z_1, \dots, z_s$ an *M-filter-regular sequence* if z_{i+1} is an M/Q_iM-filter-regular element for $i = 0, \dots, s-1$, where $Q_0 = 0$ and $Q_i = (z_1, \dots, z_i)$. If $M = R$, we delete the prefix M.

The above condition means $a((Q_iM : z_{i+1})/Q_iM) < \infty$ for $i = 0, \dots, s-1$. For simplicity we set

$$a(z_1, \dots, z_s; M) = \max\{a((Q_iM : z_{i+1})/Q_iM) |\, i = 0, \dots, s-1\}.$$

Remark. Filter-regular sequences are not permutable in general. For example, let $R = k[x, y, z]/(x) \cap (x^2, y)$. Then z, y is an R-filter-regular sequence, while y, z is not.

We call a homogeneous ideal $Q \subseteq R_+$ an *M-reduction* of R_+ if Q is generated by linear forms and $(M/QM)_t = 0$ for $t \gg 0$.

The above lemmas lead us to the following non-cohomological characterizations of the regularity.

Theorem 2.4. *Let $z_1, \dots, z_s$ be an M-filter-regular sequence of linear forms such that $Q = (z_1, \dots, z_s)$ is an M-reduction of R_+. Then*

$$\operatorname{reg}(M) = \max\{a(z_1, \dots, z_s; M), a(M/QM)\}$$
$$= \max\{a((Q_iM : R_+)/Q_iM) |\, i = 0, \dots, s-1\}.$$

Proof. Applying Lemma 2.2 to the quotient modules $M/Q_{i-1}M$, $i = 1, \dots, r$, we obtain

$$\operatorname{reg}(M) = \max\{a_0(M/Q_0M), \dots, a_0(M/Q_{s-1}M), \operatorname{reg}(M/QM)\}.$$

By Lemma 2.1, we have

$$a_0(M/Q_iM) = a((Q_iM : z_{i+1})/Q_iM) = a((Q_iM : R_+)/Q_iM),$$
$$\operatorname{reg}(M/QM) = a_0(M/QM) = a(QM : R_+/QM).$$

Therefore, the assertion is immediate. □

Example. Let $R = R_0[x_1, \dots, x_n]$ be a polynomial ring over R_0. Then $x_1, \dots, x_n$ is a regular sequence in R. Hence $a(x_1, \dots, x_n) = 0$, which implies $\operatorname{reg}(R) = a(R_0) = 0$.

If R is a graded algebra over a local ring $(R_0, \mathfrak{n})$, we may assume that R_0 has infinite residue field by passing to the algebra $R \otimes R_0[u]_{\mathfrak{n}R_0[u]}$, where u is an indeterminate. In this case, any M-reduction of R_+ can be generated by an M-filter-regular sequence of linear forms. This is an easy consequence of the characterization of M-filter-regular elements by means of the associated primes $\mathfrak{p} \not\supseteq R_+$ of 0_M.

In general, any M-reduction of R_+ can be generated by an M-filter-regular sequence in a flat extension of R.

Lemma 2.5. *Let Q be an M-reduction generated by the linear forms $x_1, \dots, x_s$. For $i = 1, \dots, s$ put $z_i = \sum_{j=1}^{s} u_{ij}x_j$, where $U = \{u_{ij} |\, i, j = 1, \dots, s\}$ is a matrix of indeterminates. Put*

$$R' = R[U, \det(U)^{-1}],\ R' = R \otimes_A A',\ M' = M \otimes_A A'.$$

If we view R' as a standard graded algebra over R' and M' as a graded R'-module, then $z_1, \dots, z_s$ is an M'-filter-regular sequence.

Proof. It suffices that z_1 is M-filter-regular. This will imply that z_i is $(M/(z_1, \dots, z_{i-1})M)$-filter-regular for $i = 2, \dots, s$. We will show that $z_1 \notin P$ for any associated prime ideal $P \not\supseteq R'_+$ of M'. By the definition of R', such a prime ideal P must have the form $\mathfrak{p}R'$ for some associated prime ideal $\mathfrak{p} \not\supseteq R_+$ of M. If $Q \subseteq \mathfrak{p}$, then $(M/\mathfrak{p}M)_n = 0$ for $n \gg 0$ because $M/\mathfrak{p}M$ is a quotient module of M/QM. From this it follows that there is a number t such that $(R_+)^t M \subseteq \mathfrak{p}M$. Since $\operatorname{ann}(M) \subseteq \mathfrak{p}$, this implies $R_+ \subseteq \mathfrak{p}$, a contradiction. So we get $Q \not\subseteq \mathfrak{p}$. Since $Q = (x_1, \dots, x_s)$, this implies $z_1 = u_{11}x_1 + \cdots + u_{1s}x_s \notin \mathfrak{p}R' = P$, as desired. □

Lemma 2.5 allows us to use Theorem 2.4 for the computation of $\operatorname{reg}(M)$. Indeed, since $R' =$ is a flat extension of R, we have $H^i_{R'_+}(M')_n \cong H^i_{R_+}(M)_n \otimes_A A'$ for all n and $i \geq 0$, whence

$$\operatorname{reg}(M) = \operatorname{reg}(M').$$

Let $d(M)$ denote the maximal degree of the homogeneous minimal generators of M. The following consequence of Theorem 2.4 is already known in the case R is a polynomial ring.

Corollary 2.6. $d(M) \leq \operatorname{reg}(M)$.

Proof. It is easily seen that

$$d(M) = d(M/R_+M) = a(M/R_+M).$$

By Lemma 2.5 we may assume that R_+ is generated by an M-filter-regular sequence of linear forms. Applying Theorem 2.4 we obtain

$$a(M/R_+M) \leq \operatorname{reg}(M).$$

Therefore, the assertion is immediate. □

The characterization of the regularity by means of filter-regular sequence has been used successfully in the study of the reduction number and the structure of blow-up algebras [19], [20], [21], [22], [23].

3. Weak regularity

By definition, to estimate the regularity we need to check the vanishing of infinitely many components of the local cohomology modules. We shall see that it suffices to check only a few. For this purpose we introduce the following notions.

Let R be a finitely generated standard graded algebra over a noetherian ring. Let M be a finitely generated graded R-module. For any integer t we say that M is *t-regular* if $H^i_{\mathfrak{m}}(M)_{n-i+1} = 0$ for all $i \geq 0$ and $n \geq t$. It is easy to check that

$$\operatorname{reg}(M) = \min\{t \in \mathbb{Z} \mid M \text{ is } t\text{-regular}\}.$$

We say that M is *weakly t-regular* if $H^i_{\mathfrak{m}}(M)_{t-i+1} = 0$ for $i \geq 0$. In general, weak t-regularity does not imply t-regularity.

Example. Let $M = R = k$ be a field. Then $H^i_{R_+}(M) = 0$ and $H^i_{R_+}(E) = 0$ for $i \geq 1$. Hence M is weakly t-regular but not t-regular for $t \leq -2$.

However, the two notions coincide under some mild restriction.

Theorem 3.1. *Let $t \geq d(M)$. Then M is t-regular if M is weakly t-regular.*

Proof. Using Lemma 2.5 we may assume that there exists an M-filter-regular sequence of linear forms $z_1, \ldots, z_s$ such that $R_+ = (z_1, \ldots, z_s)$.

If $s = 0$, $R_+ = 0$. In this case, $H^0_{R_+}(M) = M$ and $H^i_{R_+}(M) = 0$ for $i > 0$. Therefore, M is t-regular if $M_n = 0$ for $n > t$. The latter condition is clearly satisfied because $t \geq d(M)$ and $R_+ = 0$.

If $s > 0$, we set $z = z_1$. As in the proof of Lemma 2.2 we have the exact sequence

$$H^i_{R_+}(M)_{n-1} \to H^i_{R_+}(M)_n \to H^i_{R_+}(M/zM)_n \to H^{i+1}_{R_+}(M)_{n-1}$$

for $i \geq 0$. Since $H^i_{R_+}(M)_{t-i+1} = 0$ and $H^{i+1}_{R_+}(M)_{t-i} = 0$ we get $H^i_{R_+}(M/zM)_{t-i+1} = 0$ for $i \geq 0$. Hence M/zM is weakly t-regular. Since $d(M/zM) = d(M)$, using induction we may assume that M/zM is t-regular. This implies $\operatorname{reg}(M/zM) \leq t$.

As a consequence, $H^0_{R_+}(M/zM)_n = 0$ and hence $H^0_{R_+}(M/(0_M : z))_{n-1} = H^0_{R_+}(M)_n$ for $n \geq t+1$. On the other hand, since $H^1_{R_+}(0_M : z) = 0$, from the exact sequence

$$0 \to (0_M : z) \to M \to M/(0_M : z) \to 0$$

we can derive that the map $H^0_{R_+}(M) \to H^0_{R_+}(M/(0_M : z))$ is surjective. Therefore, there is a surjective map $H^0_{R_+}(M)_{n-1} \to H^0_{R_+}(M)_n$ for all $n \geq t+1$. Since $H^0_{R_+}(M)_{t+1} = 0$, this implies $H^0_{R_+}(M)_n = 0$ for $n \geq t+1$ or, equivalently, $a_0(M) \leq t$.

Summing up, we get

$$\operatorname{reg}(M) = \max\{a_0(M), \operatorname{reg}(M/zM)\} \leq t.$$

Hence M is t-regular. □

Corollary 3.2. $\operatorname{reg}(M) = \min\{t \geq d(M) \mid M \text{ is weakly } t\text{-regular}\}$.

Proof. Since $\operatorname{reg}(M) \geq d(M)$, we get

$$\begin{aligned}\operatorname{reg}(M) &= \min\{t \geq d(M) \mid M \text{ is } t\text{-regular}\}\\ &= \min\{t \geq d(M) \mid M \text{ is weakly } t\text{-regular}\},\end{aligned}$$

as desired. □

The regularity in algebraic geometry is defined a bit differently than in the algebraic case.

For a moment let R be a polynomial ring over a field and M a finitely generated graded R-module. Let $X = \operatorname{Proj} R$. We denote by $\tilde{M}$ the coherent sheaf associated with M on X and by $H^i(X, \tilde{M}(t))$ the ith sheaf cohomology of the twisted $\mathcal{O}_X$-module $\tilde{M}(t)$, $t \in \mathbb{Z}$. Then $\tilde{M}$ is called *t-regular* if the map $M_n \to H^0(X, \tilde{M}(n))$ is surjective and $H^i(X, \tilde{M}(n-i)) = 0$ for all $n > t$ and $i \geq 1$.

There is the following relationship between the sheaf cohomology of $\tilde{M}$ and the local cohomology of M (see e.g. [BrS]).

Theorem 3.3 (Serre–Grothendieck correspondence). *There are the exact sequence*

$$0 \to H^0_{R_+}(M)_n \to M_t \to H^0(X, \tilde{M}(n)) \to H^1_{R_+}(M)_n \to 0$$

and the isomorphisms $H^i(X, \tilde{M}(n)) \cong H^{i+1}_{R_+}(M)_n$ *for* $i > 0$.

As a consequence, $\tilde{M}$ is t-regular if and only if $H^i_{R_+}(M)_{n-i+1} = 0$ for all $n \geq t$ and $i \geq 1$. This leads to the following definition.

Let R now be a finitely generated standard graded algebra over a Noetherian ring and M a finitely generated graded R-module. We say that M is *geometrically* t*-regular* if $H^i_{R_+}(M)_{n-i+1} = 0$ for all $n \geq t$ and $i \geq 1$. The following result of Mumford shows that we only need to check this condition for $n = t$.

Theorem 3.4. *M is geometrically t-regular if $H^i_{R_+}(M)_{t-i+1} = 0$ for $i \geq 1$.*

Proof. Using Lemma 2.5 we may assume that there exists an M-filter-regular sequence of linear forms $z_1, \ldots, z_s$ such that $R_+ = (z_1, \ldots, z_s)$. As in the proof of Theorem 3.1 the case $s = 0$ is trivial and if $s > 0$, we may assume that $H^i_{R_+}(M/zM)_{n-i+1} = 0$ for $n \geq t$ and $i \geq 1$, where $z = z_1$. This implies that the map $H^i_{R_+}(M)_{n-i} \to H^i_{R_+}(M)_{n-i+1}$ is surjective for $n \geq t$ and $i \geq 1$. Since $H^i_{R_+}(M)_{t-i+1} = 0$, this implies $H^i_{R_+}(M)_{n-i+1} = 0$ for all $n \geq t$ and $i \geq 1$, as desired. □

Theorem 3.4 shows that geometric t-regularity is weaker than weak t-regularity and t-regularity. However, if $H^0_{R_+}(M) = 0$, the three notions coincide. In this case, we may reformulate Theorem 3.4 as follows.

Corollary 3.5. Assume that $H^0_{R_+}(M) = 0$. Then M is t-regular if M is geometrically t-regular.

We call the number

$$\text{g-reg}(M) := \max\{a_i(M) + i \mid i \geq 1\},$$

the *geometric regularity* of M. It is clear that

$$\text{g-reg}(M) = \min\{t \mid M \text{ is geometrically } t\text{-regular}\}.$$

The regularity is related to the geometric regularity by the formula

$$\text{reg}(M) = \max\{a_0(M), \text{g-reg}(M)\}.$$

Comparing with the regularity, the geometric regularity has the advantage that it can be estimated in terms of the geometric regularity of a generic "hyperplane section". For that we need the following observation.

Let R be an algebra over an artinian local ring R_0. For all $t \in \mathbb{Z}$, the graded piece M_t is an R_0-module of finite length, and we can consider the Hilbert function $h_M(t) := \ell(M_t)$. It is well-known that $h_M(t)$ is equal to a polynomial $p_M(t)$ for $t \gg 0$. One calls $p_M(t)$ the Hilbert polynomial of M. The difference $h_M(t) - p_M(t)$ can be expressed in terms of local cohomology modules, due to a result of Serre (see e.g. [7]).

Theorem 3.6 (Serre formula). *Let R be an algebra over an artinian local ring. Then*

$$h_M(t) - p_M(t) = \sum_{i=0}^{\dim M} \ell(H_{R_+}(M)_t).$$

This formula will be used in the proof of the following estimate for the geometric regularity, which is based on an idea of Mumford in [14, pp. 101, proof of Theorem].

Theorem 3.7. *Let R be an algebra over an artinian local ring. Let z be an M-filter-regular linear form. Let $t \ge d(M)$ such that M/zM is geometrically t-regular. Then M is geometrically $(t + p_M(t) - h_{M/L}(t))$-regular, where $L = \Gamma_{R_+}(M)$.*

Proof. We have to show that

$$\text{g-reg}(M) \le t + p_M(t) - h_{M/L}(t).$$

Consider the quotient module M/L. Since $L_n = 0$ for $n \gg 0$, we have $H^i_{R_+}(L) = 0$ and hence $H^i_{R_+}(M) = H^i_{R_+}(M/L)$ for $i \ge 1$. Therefore, g-reg$(M/L) =$ g-reg(M). Moreover, we have $h_M(n) = h_{M/L}(n)$ for $n \gg 0$, which implies $p_M(t) = p_{M/L}(t)$. On the other hand, the relation $d(M/L) \le d(M)$ shows that $t \ge d(M/L)$. Since $((L + zM)/zM)_n = 0$ for $n \gg 0$, $H^i_{R_+}((L + zM)/zM) = 0$ and hence $H^i_{R_+}(M/(L + zM)) \cong H^i_{R_+}(M/zM)$ for $i \ge 1$. Thus, $M/(L + zM)$ is geometrically t-regular like M/zM. So we may replace M by M/L. That means we may assume that $L = 0$ and z is M-regular.

As we have seen in the proof of Lemma 2.2,

$$a_i(M) + i \le a_{i-1}(M/zM) + i - 1$$

for $i \ge 1$. Since M/zM is geometrically t-regular, $a_{i-1}(M/zM) + i - 1 \le t$ for $i \ge 2$. Therefore, $a_i(M) + i \le t$ for $i \ge 2$.

Let s be the smallest integer $\ge t$ such that $H^0_{R_+}(M/zM)_s = 0$. Then M/zM is weakly s-regular. By Theorem 3.1, this implies reg$(M/zM) \le s$. Thus, $a_1(M) + 1 \le a_0(M/zM) \le s$. Since we already have $a_{i-1}(M/zM) + i - 1 \le s$ for $i \ge 2$, we get g-reg$(M) \le s$.

Now, we will show that $s \le t + p_M(t) - h_M(t)$. Since $H^0_{R_+}(M) = L = 0$ and $H^1_{R_+}(M/zM)_n = 0$ for $n \le t$, from the exact sequence

$$0 \to M \xrightarrow{z} M \to M/zM \to 0$$

we obtain the exact sequence

$$0 \to H^0_{R+}(M/zM)_n \to H^1_{R+}(M)_{n-1} \to H^1_{R+}(M)_n \to 0$$

for $n \ge t$. Since $H^0_{R+}(M/zM)_n \ne 0$ for $t \le n < s$, we have $\ell(H^1_{R+}(M)_{n-1}) > \ell(H^1_{R+}(M)_n)$ for $t \le n < s$. Therefore, $s - t \le \ell(H^1_{R+}(M)_t)$. On the other hand, $p_M(t) - h_M(t) = \ell(H^1_{R+}(M)_t)$ by Lemma 3.6. So we obtain $s \le t + p_M(t) - h_M(t)$, as required. □

The above estimate is especially useful in finding bounds for the regularity by means of the degree [12], [13], [16], [17].

4. a^*-invariant

Let R be a finitely generated standard graded algebra over a noetherian ring and M a finitely generated graded R-module. We call the number

$$a^*(M) := \max\{a_i(M) | \; i \geq 0\}$$

the *a^*-invariant* of M. In some sense, it can be considered as the "real" regularity, whereas the Castelnuovo–Mumford regularity is the "shifted" regularity of M. It is obvious that

$$a^*(M) \leq \operatorname{reg}(M) \leq a^*(M) + \dim(M).$$

The a^*-invariant has many interesting properties. First of all, if R is a polynomial ring over a field, $a^*(M)$ can be characterized by means of the shifts in the minimal free resolution of M.

Theorem 4.1. *Let R be a polynomial ring over k in n variables. Then*

$$a^*(M) = \max\{b_i(M) | \; i \geq 0\} - n.$$

Proof. By Theorem 1.2 we have

$$a_i(M) = \max\{t | \; \operatorname{Ext}_R^{n-i}(M, R)_{-t-n} \neq 0\}.$$

Therefore,

$$a^*(M) = \max\{t | \; \operatorname{Ext}_R^{n-i}(M, R)_{-t-n} \neq 0 \text{ for some } i \geq 0\}.$$

Put $m = \max\{b_i | \; i \geq 0\} - n$. For every $i \geq 0$, F_i has no generators of degree $\geq m + n + 1$, so $F_i^* = \operatorname{Hom}_R(F_i, R)$ must be zero in degree $\leq -m - n - 1$. Since $\operatorname{Ext}_R^i(M, R)$ is the homology of the dual of the resolution of M at F_i^*, $\operatorname{Ext}_R^i(M, R)_r = 0$ for $r \geq -m - n - 1$. Now let i be the largest index such that $b_i - n = m$. Then F_i^* has $R(m + n)$ as a summand, whereas F_{i+1}^* has no summand of the form $R(r)$ with $r \geq m + n$. By the minimality of the resolution, the summand $R(m + n)$ of F_i^* must map to zero in F_{i+1}^*. Moreover, nothing in F_{i-1}^* can map on to the generator of $R(m + n)$ in F_i^*, so it gives a nonzero class in $\operatorname{Ext}_R^i(M, R)$ of degree $-m - n$. Thus,

$$\max\{t | \; \operatorname{Ext}_R^i(M, R)_{-t-n} \neq 0 \text{ for some } i \geq 0\} = m,$$

as desired. □

The a^*-invariant also controls the place where the Hilbert function $h_M(t)$ coincides with the Hilbert polynomials $p_M(t)$ of M.

Proposition 4.2. *Let R be graded algebra over an artinian local ring. Then $h_M(n) = p_M(n)$ for $n > a^*(M)$.*

Proof. This is an immediate consequence of Theorem 3.6. □

Similarly as for $\operatorname{reg}(M)$, we can compute $a^*(M)$ by means of a filter-regular sequence of linear forms. For this we need the following reduction.

Lemma 4.3. *Let z be an M-filter-regular linear form. Then*

$$a^*(M) = \max\{a_0(M), a^*(M/zM) - 1\}.$$

Proof. We have seen in the proof of Lemma 2.2 that

$$a_i(M) \le a_{i-1}(M/zM) - 1 \le \max\{a_{i-1}(M) - 1, a_i(M)\}$$

for $i \ge 1$. From this it follows that

$$\max\{a_i(M) \mid i \ge 1\} \le a^*(M/zM) \le a^*(M),$$

which implies the assertion. □

The following consequence of Lemma 4.3 is sometime very useful.

Corollary 4.4. Assume that z is an M-regular linear form. Then

$$a^*(M) = a^*(M/zM) - 1.$$

Proof. The existence of an M-filter-regular form implies $H^0_{R_+}(M) = 0$. Hence the assertion is immediate. □

For any sequence of homogeneous elements $z_1, \dots, z_s$ we set

$$s(z_1, \dots, z_r; M) = \max\{a((Q_iM : z_{i+1})/Q_iM) - i \mid i = 0, \dots, s-1\},$$

where $Q_i = (z_1, \dots, z_i)$ (so $Q_0 = 0$).

Theorem 4.5. *Let $z_1, \dots, z_s$ be an M-filter-regular sequence of linear forms such that $Q = (z_1, \dots, z_s)$ is an M-reduction. Then*

$$\begin{aligned} a^*(M) &= \max\{s(z_1, \dots, z_s; M), a(M/QM) - s\} \\ &= \max\{a((Q_iM : R_+)/Q_iM) - i \mid i = 0, \dots, s-1\}. \end{aligned}$$

Proof. Applying Lemma 4.3 to the quotient modules M/Q_iM, $i = 0, \dots, s-1$, we obtain

$$a^*(M) = \max\{a_0(M/Q_0M), \dots, a_0(M/Q_{s-1}M) - s + 1, \operatorname{reg}(M/QM) - s\}.$$

By Lemma 2.1, we have

$$a_0(M/Q_iM) = a((Q_iM : z_{i+1})/Q_iM) = a((Q_iM : R_+)/Q_iM),$$

$$a^*(M/QM) = a_0(M/QM) = a(QM : R_+/QM),$$

which implies the assertion. □

If R is a graded algebra over a local ring, there are the local cohomology modules $H^i_{\mathfrak{m}}(M)$ with respect to the maximal graded ideal $\mathfrak{m}$ of R, which carry more information on the structure of R than the local cohomology modules $H^i_{R_+}(M)$. For instance, M is a Cohen-Macaulay module if $H^i_{\mathfrak{m}}(M) = 0$ for $i < \dim M$.

The vanishing of the local cohomology modules $H^i_{R_+}(M)$ and $H^i_{\mathfrak{m}}(M)$ can be different. For instance, $H^d_{R_+}(M)$ may vanish, while we always have $H^d_{\mathfrak{m}}(M) \ne 0$, where $d = \dim M$. However, they share the same largest non-vanishing degree in the following sense.

Theorem 4.6. *Let R be a graded algebra over a local ring with maximal graded ideal $\mathfrak{m}$. Then*

$$\max\{a(H^i_{\mathfrak{m}}(M))|\ i \ge 0\} = a^*(M).$$

Proof. For simplicity we set $c(M) = \max\{a(H^i_{\mathfrak{m}}(M))|\ i \ge 0\}$.

First, consider the case $M_n = 0$ for $n \gg 0$. Since $H_{R_+}(M) = M$ and $H^i_{R_+}(M) = 0$ for $i \ge 1$, we have $a^*(M) = a_0(M) = a(M)$. It remains to show that $c(M) = a(M)$.

Set $r = a(M)$ and $m = \min\{n|\ M_n \ne 0\}$. If $r = m$, M concentrates only in degree r. Hence $H^i_{\mathfrak{m}}(M)$ also concentrates in degree r for all $i \ge 0$. Since $H^{\dim M}_{\mathfrak{m}}(M) \ne 0$, we get $c(M) = r$. If $r > m$, we consider the exact sequence

$$0 \longrightarrow M_r \longrightarrow M \longrightarrow M/M_r \longrightarrow 0.$$

By induction on $r - m$, we may assume that $c(M/M_r) = a(M/M_r) < r$. Then $H^i_{R_+}(M/M_r)_n = 0$ for $n > r$ and $i \ge 0$. Therefore, $H^i_{\mathfrak{m}}(M)_n \cong H^i_{\mathfrak{m}}(M_r)_n$ for $n > r$ and $i \ge 0$. Since $c(M_r) = r$, this implies $c(M) = r$.

For the general case we may assume that R_+ is generated by an M-filter-regular sequence of linear forms $z_1, \ldots, z_s$. If $s = 0$, $R_+ = 0$ and hence $M_n = 0$ for $n \gg 0$. If $s > 0$, we set $z = z_1$. We shall first show that

$$c(M) = \max\{c(0_M : z), c(M/zM) - 1\}.$$

For simplicity we set $a = c(0_M : z)$ and $b = c(M/zM) - 1$. Since $H^i_{\mathfrak{m}}(0_M : z)_n = 0$ for $n > a$ and $i \ge 0$, the exact sequence

$$0 \to (0_M : z) \to M \to M/(0_M : z) \to 0$$

implies $H^i_{\mathfrak{m}}(M)_n \cong H^i_{\mathfrak{m}}(M/(0_M : z))_n$ for $n > a$ and $i \ge 0$. Now, from the exact sequence

$$0 \to M/(0_M : z) \xrightarrow{z} M \to M/zM \to 0$$

we obtain the long exact sequence

$$\cdots \to H^{i-1}_{\mathfrak{m}}(M)_{n+1} \to H^{i-1}_{\mathfrak{m}}(M/zM)_{n+1} \to H^i_{\mathfrak{m}}(M)_n \to H^i_{\mathfrak{m}}(M)_{n+1} \to \cdots$$

for $n > a$. For $n > b$, we have $H^{i-1}_{\mathfrak{m}}(M/zM)_{n+1} = 0$. Hence the map $H^i_{\mathfrak{m}}(M)_n \to H^i_{\mathfrak{m}}(M)_{n+1}$ is injective for $n \ge \max\{a, b\}$. Since $H^i_{\mathfrak{m}}(M)_{n+1} = 0$ for $n \gg 0$, this implies $H^i_{\mathfrak{m}}(M)_n = 0$ for $n > \max\{a, b\}$. Therefore, $c(M) \le \max\{a, b\}$.

If $a < b$, we have $H^{i-1}_{\mathfrak{m}}(M)_{b+1} = H^i_{\mathfrak{m}}(M)_{b+1} = 0$ for $i \ge 0$. Hence $H^{i-1}_{\mathfrak{m}}(M/zM)_{b+1} \cong H^i_{\mathfrak{m}}(M)_b$ for $i \ge 1$. Choose i such that $H^{i-1}_{\mathfrak{m}}(M/zM)_{b+1} \ne 0$. Then $H^i_{\mathfrak{m}}(M)_b \ne 0$. So we obtain $c(M) \ge b$ and hence $c(M) = b = \max\{a, b\}$.

If $a \ge b$, we assume to the contrary that $c(M) < \max\{a, b\} = a$. Then $H^i_{\mathfrak{m}}(M)_n = 0$ for $n \ge a$ and $i \ge 0$. Therefore,

$$H^{i-1}_{\mathfrak{m}}(M/zM)_{a+1} = H^i_{\mathfrak{m}}(M/(0_M : z))_a \cong H^{i+1}_{\mathfrak{m}}(0_M : z)_a$$

for $i \geq 1$. Since $a = c(0_M : z)$, there is an index i such that $H^{i+1}_{\mathfrak{m}}(0_M : z)_a \neq 0$. Therefore, $H^{i-1}_{\mathfrak{m}}(M/zM)_{a+1} \neq 0$. This implies $b \geq a+1$, a contradiction.

Now, since $(0_M : z)_n = 0$ for $n \gg 0$, we have $c(0_M : z) = a(0_M : z) = a_0(M)$. By induction on s we may also assume that $c(M/zM) = a^*(M/zM)$. Therefore,

$$c(M) = \max\{a_0(M), a^*(M/zM) - 1\}.$$

By Lemma 4.3, this implies $c(M) = a^*(M)$. □

The a^*-invariant plays an important role in the study of the Cohen-Macaulayness of graded algebras [1], [11], [20]. For more information on the a^*-invariant we refer to [18], [22].

5. Partial regularities

Let R be a finitely generated standard graded algebra over a noetherian ring and M a finitely generated graded R-module. Given an integer $t \geq 0$ we call the numbers

$$\operatorname{reg}_t(M) := \max\{a_i(M) + i \mid i \leq t\},$$

$$a_t^*(M) := \max\{a_i(M) \mid i \leq t\}.$$

the *partial regularities* of M.

The regularity and the a^*-invariant are special cases of the partial regularities because $\operatorname{reg}(M) = \operatorname{reg}_t(M)$, $a^*(M) = a_t^*(M)$ for all $t \geq \max\{i \mid H^i_{R_+}(M) \neq 0\}$.

The meaning of the partial regularities lies on the fact that they yields specific information on the vanishing of the local cohomology modules and, if R is a polynomial ring, on the minimal free resolution.

Theorem 5.1. *Let R be a polynomial ring over k in n variables. Then*

(a) $\operatorname{reg}_t(M) = \max\{b_i(M) - i \mid i \geq n - t\}$,
(b) $a_t^*(M) = \max\{b_i(M) \mid i \geq n - t\} - n$.

Proof. The proof is similar to the proof for the characterization of $\operatorname{reg}(M)$ and $a^*(M)$ in terms of the numbers $b_i(M)$. Hence we omit it. □

Theorem 5.1(a) shows that $\operatorname{reg}_t(M)$ is related to the following notion introduced by Bayer-Charalambous-Popescu [2]:

$$t\text{-}\operatorname{reg}(M) := \max\{b_i(M) - i \mid i \geq t\}.$$

By the above theorem, we have $t\text{-}\operatorname{reg}(M) = \operatorname{reg}_{n-t}(M)$.

Let $\operatorname{Syz}_t(M)$ denote the t-th syzygy module of M (which is defined as the kernel of the map $F_t \to F_{t-1}$ in the minimal free resolution of M as an R-module). As a consequence of Theorem 5.1 we obtain the following relationships between the partial regularity of M and the regularity of its syzygy modules.

Corollary 5.2. Let R be a polynomial ring over a field in n variables. Then

(a) $\operatorname{reg}_t(M) = \operatorname{reg}(\operatorname{Syz}_{n-t}(M)) - n + t$,
(b) $a_t^*(M) = a^*(\operatorname{Syz}_{n-t}(M))$.

Proof. This follows from the fact that

$$0 \to F_r \to \cdots \to F_{n-t+1} \to F_{n-t} \to \operatorname{Syz}_{n-t}(M)$$

is a minimal free resolution of $\operatorname{Syz}_{n-t}(M)$. Indeed, we have

$$b_i(M) = b_{i-n+t}(\operatorname{Syz}_{n-t}(M))$$

for $i \geq n - t$. Putting $j = i - n + t$ we get

$$\operatorname{reg}_t(M) = \max\{b_j(\operatorname{Syz}_{n-t}(M)) - j + n - t))|\ j \geq 0\} = \operatorname{reg}(\operatorname{Syz}_{n-t}(M)) + n - t,$$

$$a_t^*(M) = \max\{b_j(\operatorname{Syz}_{n-t}(M))|\ j \geq 0\} - n = a^*(\operatorname{Syz}_{n-t}(M)).$$

□

One can also compute the partial regularities by means of filter-regular sequences. The main point is the following reduction.

Lemma 5.3. *Let z be an M-filter-regular linear form. For any $t \geq 1$ we have:*

(a) $\operatorname{reg}_t(M) = \max\{a_0(M), \operatorname{reg}_{t-1}(M/zM)\}$,
(b) $a_t(M) = \max\{a_0(M), a_{t-1}(M/zM) - 1\}$.

Proof. As we have seen in the proof of Lemma 2.2,

$$a_i(M) + i \leq a_{i-1}(M/zM) + i - 1 \leq \max\{a_{i-1}(M) + i - 1, a_i(M) + i\}$$

for $i \geq 1$. Taking the maximum of each term for $i \leq t$ we obtain

$$\max\{a_i(M) + i \,|\, i = 1, \ldots, t\} \leq \operatorname{reg}_t(M/zM) \leq \operatorname{reg}_t(M),$$

from which (a) immediately follows.

The first inequalities can be rewritten as

$$a_i(M) \leq a_{i-1}(M/zM) - 1 \leq \max\{a_{i-1}(M) - 1, a_i(M)\},$$

which implies (b) similarly as above. □

Corollary 5.4. Assume that z is an M-regular element. Then

(a) $\operatorname{reg}_t(M) = \operatorname{reg}_{t-1}(M/zM)$,
(b) $a_t(M) = a_{t-1}(M/zM) - 1$.

There are two ways for the computation of $\operatorname{reg}_t(M)$ and $a_t^*(M)$ by using M-filter-regular sequences of length t and $t + 1$.

Theorem 5.5. *Let $z_1, \dots, z_t$ be an M-filter-regular sequence of linear forms. Let $Q_0 = 0$ and $Q_i = (z_1, \dots, z_i)$, $i = 1, \dots, t$. Then*

(a) $\text{reg}_t(M) = \max\{a((Q_i M : R_+)/Q_i M) | \, i = 0, \dots, t\}$.
(b) $a_t^*(M) = \max\{a((Q_i M : R_+)/Q_i M) - i \,|\, i = 0, \dots, t\}$.

Proof. Applying Lemma 5.3 to the quotient module $M/Q_i M$ we get

$$\text{reg}(M) = \max\{a_0(M/Q_i M) | \, i = 0, \dots, t\},$$

$$a^*(M) = \max\{a_0(M/Q_i M) - i \,|\, i = 0, \dots, t\}.$$

Since $a_0(M/Q_i M) = a((Q_i M : R_+)/Q_i M)$ by Lemma 2.1, this implies the assertion. □

Theorem 5.6. *Let $z_1, \dots, z_{t+1}$ be an M-filter-regular sequence of linear forms. Then*

(a) $\text{reg}_t(M) = a(z_1, \dots, z_{t+1}; M)$.
(b) $a_t^*(M) = s(z_1, \dots, z_{t+1}; M)$.

Proof. The assertion can be proved similarly as in the proof of Theorem 5.5 because $a_0(M/Q_i M) = a((Q_i M : z_{i+1})/Q_i M)$ by Lemma 2.1. □

For more information on partial regularities we refer to [2], [24].

6. Regularity of ideals

Let $R = k[x_1, \dots, x_n]$ be a polynomial ring over a field k and I a homogeneous ideal in S. We will show how one can use Gröbner basic technique to compute the regularity of I. The method was developed by Bayer and Stillman [4], [5]. It was later applied in [24] for the computation of the partial regularities.

First, we observe that the computation of the regularities of I can be reduced to the computation of R/I. Indeed, from the minimal free resolution of R/I we can deduce that

$$\text{reg}_t(I) = \text{reg}_t(R/I) + 1,$$

$$a_t^*(I) = a_t^*(R/I).$$

We will use the reverse lexicographic order exclusively. Let $\text{in}(I)$ denote the initial ideal of I. Then there is the following relationship between I and $\text{in}(I)$.

Lemma 6.1. *Let $i = n, \dots, 1$. For every integer $m \geq 0$,*

$$[(I, x_n, \dots, x_{i+1}) : x_i]_m = (I, x_n, \dots, x_{i+1})_m$$

if and only if

$$[(\text{in}(I), x_n, \dots, x_{i+1}) : x_i]_m = (\text{in}(I), x_n, \dots, x_{i+1})_m.$$

Proof. Let $f = g + x_n h_n + \cdots + x_{i+1}h_{i+1}$ be an arbitrary polynomial in $(I, x_n, \ldots, x_{i+1})$, where $g \in I$. If $f \notin (x_n, \ldots, x_{i+1})$, then $\mathrm{in}(f)$ does not contain the variables $x_n, \ldots, x_{i+1}$. This is a property of the reverse lexicographic order. In this case, $\mathrm{in}(f) = \mathrm{in}(g)$. This shows that $\mathrm{in}(I, x_n, \ldots, x_{i+1})$ is contained in $(\mathrm{in}(I), x_n, \ldots, x_{i+1})$. Since the inverse containment is trivial, we obtain

$$\mathrm{in}(I, x_n, \ldots, x_{i+1}) = (\mathrm{in}(I), x_n, \ldots, x_{i+1}).$$

From this it follows that

$$\mathrm{in}((I, x_n, \ldots, x_{i+1}) : x_i) = (\mathrm{in}(I), x_n, \ldots, x_{i+1}) : x_i.$$

Since $\mathrm{in}((I, x_n, \ldots, x_{i+1}) : x_i)_m = \mathrm{in}(I, x_n, \ldots, x_{i+1})_m$ if and only if $[(I, x_n, \ldots, x_{i+1}) : x_i]_m = (I, x_n, \ldots, x_{i+1})_m$, this implies the assertion. □

The assertion of Lemma 6.1 can be reformulated as

$$\begin{aligned} &a(((I, x_n, \ldots, x_{i+1}) : x_i)/(I, x_n, \ldots, x_{i+1})) \\ &\quad = a(((\mathrm{in}(I), x_n, \ldots, x_{i+1}) : x_i)/(\mathrm{in}(I), x_n, \ldots, x_{i+1})). \end{aligned}$$

Since this holds for $i = n, \ldots, 1$, we obtain

$$\begin{aligned} a(x_n, \ldots, x_j; R/I) &= a(x_n, \ldots, x_j; R/\mathrm{in}(I)), \\ s(x_n, \ldots, x_j; R/I) &= s(x_n, \ldots, x_j; R/\mathrm{in}(I)). \end{aligned}$$

Now we can use the characterization of the regularities by means of filter-regular sequences to compare the regularities of R/I and $R/\mathrm{in}(I)$.

Theorem 6.2. *Let $t = 0, \ldots, n-1$. Assume that $x_n, \ldots, x_{n-t-1}$ is a filter-regular sequence in $R/\mathrm{in}(I)$. Then*

(a) $\mathrm{reg}_t(I) = \mathrm{reg}_t(\mathrm{in}(I))$,
(b) $a_t^*(I) = a_t^*(\mathrm{in}(I))$.

Proof. The assumption means

$$a(x_n, \ldots, x_{n-t-1}; R/I) = a(x_n, \ldots, x_{n-t-1}; R/\mathrm{in}(I)) < \infty.$$

Therefore, $x_n, \ldots, x_{n-t-1}$ is also filter-regular in (R/I). Now we can apply Theorem 5.6 to the sequence $x_n, \ldots, x_{n-t-1}$ in R/I and $R/\mathrm{in}(I)$ and obtain

$$\begin{aligned} \mathrm{reg}_t(R/I) &= a(x_n, \ldots, x_{n-t-1}; R/I) \\ &= a(x_n, \ldots, x_{n-t-1}; R/\mathrm{in}(I)) = \mathrm{reg}_t(R/\mathrm{in}(I)), \\ a_t(R/I) &= s(x_n, \ldots, x_{n-t-1}; R/I) \\ &= s(x_n, \ldots, x_{n-t-1}; R/\mathrm{in}(I)) = a_t(R/\mathrm{in}(I)), \end{aligned}$$

which imply the assertion. □

Corollary 6.3. Let k be an infinite field. Let $\operatorname{Gin}(I)$ denote the generic initial ideal of I with respect to the reverse lexicographic order. Then

(a) $\operatorname{reg}_t(I) = \operatorname{reg}_t(\operatorname{Gin}(I))$,
(b) $a_t^*(I) = a_t^*(\operatorname{Gin}(I))$.

Proof. Since k is infinite, we may use a generic choice of the variables to assume that $x_n, \dots, x_1$ is a filter-regular sequence in R/I. Since $\operatorname{Gin}(I)$ is the initial ideal of I for generic choices of the variables, the assertion follows from Theorem 6.2. □

In the following, we will present a practical method for the computation of the regularities of I which is based on the substitutions $x_i = 0, 1$ of $\operatorname{in}(I)$ for some variables x_i.

Let J be an arbitrary monomial ideal. For $i = 0, \dots, n-1$ set $R_i := k[x_1, \dots, x_{n-i}]$. Let J_i be the ideal of R_i obtained from J by the substitution $x_n = \cdots = x_{n-i+1} = 0$. Let $\tilde{J}_i$ denote the ideal of R_i obtained from J_i by the substitution $x_{n-i} = 1$.

The ideals J_i and $\tilde{J}_i$ can be easily computed from the generators of J. In fact, if $J = (f_1, \dots, f_s)$, where $f_1, \dots, f_s$ are monomials in S, then J_i is generated by the monomials f_j not divided by any of the variables $x_{n-i+1}, \dots, x_n$ and $\tilde{J}_i$ by those monomials obtained from the latter by setting $x_{n-i} = 1$.

Lemma 6.4. $a(\tilde{J}_i/J_i) = a(((J, x_n, \dots, x_{n-i+1}) : x_{n-i})/(J, x_n, \dots, x_{n-i+1}))$.

Proof. By the definition of the ideals J_i and $\tilde{J}_i$ we have

$$\tilde{J}_i/J_i \cong \cup_{j\geq 1}((J, x_n, \dots, x_{n-i+1}) : x_{n-i}^j)/(J, x_n, \dots, x_{n-i+1}).$$

Therefore, it suffices to show that

$$\begin{aligned} &a(\cup_{j\geq 1}((J, x_n, \dots, x_{n-i+1}) : x_{n-i}^j)/(J, x_n, \dots, x_{n-i+1})) \\ &\quad = a(((J, x_n, \dots, x_{n-i+1}) : x_{n-i})/(J, x_n, \dots, x_{n-i+1})). \end{aligned}$$

But this follows from the fact that every element of the largest non-vanishing degree of

$$\cup_{j\geq 1}((J, x_n, \dots, x_{n-i+1}) : x_{n-i}^j)/(J, x_n, \dots, x_{n-i+1})$$

is also contained in $((J, x_n, \dots, x_{n-i+1}) : x_{n-i})/(J, x_n, \dots, x_{n-i+1})$. □

For $i = 0, \dots, n$ we set

$$c_i(I) := a(\tilde{J}_i/J_i),$$

where $J = \operatorname{in}(I)$. Note that the ideal J_i can be obtained by first evaluating the ideal I at the substitution $x_n = \cdots = x_{n-i+1} = 0$ and then computing the initial ideal of the evaluated ideal.

There is the following characterization of the regularities of R/I by means of the invariants $c_i(I)$.

Theorem 6.5. *Assume that* $c_i(I) < \infty$ *for* $i = 0, \dots, t$. *Then*

(a) $\operatorname{reg}_t(I) = \max\{c_i(I) | \; i = 0, \dots, t\} + 1$,
(b) $a_t^*(I) = \max\{c_i(I) - i \; | \; i = 0, \dots, t\}$.

Proof. By Lemma 6.4, the assumption implies that $x_n, \dots, x_{n-t-1}$ is filter-regular in $R/\operatorname{in}(I)$. Now we can apply Theorem 6.2 and obtain

$$\operatorname{reg}_t(I) = \operatorname{reg}_t(\operatorname{in}(I)) = \operatorname{reg}_t(R/\operatorname{in}(I)) + 1,$$

$$a_t^*(I) = a_t^*(\operatorname{in}(I)) = a_t^*(R/\operatorname{in}(I)).$$

Therefore, the assertion follows from Theorem 5.6 and Lemma 6.4. □

Example. Let $R = k[x_1, x_2, x_3, x_4]$ and

$$I = (x_1x_2 - x_3x_4, x_1x_3^2 - x_2^3, x_1^2x_3 - x_2^2x_4, x_1^3 - x_2x_4^2).$$

Then $\dim R/I = 2$ and $\operatorname{in}(I) = (x_1x_2, x_2^3, x_1^2x_3, x_1^3)$. We have

$$J_0 = (x_1x_2, x_2^3, x_1^2x_3, x_1^3), \;\; \tilde{J}_0 = J_0,$$

$$J_1 = (x_1x_2, x_2^3, x_1^2x_3, x_1^3), \;\; \tilde{J}_1 = (x_1x_2, x_2^3, x_1^2),$$

$$J_2 = (x_1x_2, x_2^3, x_1^3), \;\; \tilde{J}_2 = (x_1).$$

From this it follows that $c_0(I) = -\infty$, $c_1(I) = 2$, $c_2(I) = 2$, which implies $\operatorname{reg}(R/I) = \operatorname{reg}_2(R/I) = 2$.

We would like to mention that the numbers $c_i(I)$ can be described combinatorially in terms of the lattice vectors of the generators of $\operatorname{in}(I)$. This description together with the formulae of Theorem 6.5 give an effective method for the computation of the regularities of R/I. For detail we refer to [6], [25].

Finally, we will use the above method to characterize the regularities of I in terms of the minimal generators of $\operatorname{Gin}(I)$.

Let $\mathcal{B}$ be the Borel subgroup of $\operatorname{GL}(n, k)$ consisting of the upper triangular invertible matrices. Let B act on R by $\pi(f) = f(\pi \cdot (x_1, \dots, x_n)^T)$ for all $\pi \in B$, $f \in R$. A monomial ideal I is called *Borel-fixed* if $g(I) = I$ for all $\pi \in \mathcal{B}$. Our interest in this notion arises from the fact that $\operatorname{Gin}(I)$ is a Borel-fixed ideal (see [10]).

In the case of characteristic zero, Borel-fixed ideals can be characterized as follows (see [4]).

Proposition 6.6. *Assume that* $\operatorname{char}(k) = 0$. *Let* J *be a monomial ideal. Then* J *is Borel-fixed if and only if whenever* $x_1^{p_1} \cdots x_n^{p_n} \in J$, *then*

$$x_1^{p_1} \cdots x_i^{p_i+q} \cdots x_j^{p_j-q} \cdots x_n^{p_n} \in J$$

for each $1 \le i < j \le n$ *and* $0 \le q \le p_j$.

For $A = (p_1, \dots, p_n)$ let x^A denote the monomial $x_1^{p_1} \cdots x_n^{p_n}$ and

$$m(x^A) = \max\{i | \; p_i \neq 0\}.$$

Moreover, we denote by $\operatorname{Min}(J)$ the set of the minimal monomial generators of J.

Lemma 6.7. *Assume that* $\operatorname{char}(k) = 0$. *Let J be a Borel-fixed monomial ideal. Then*

$$a(\tilde{J}_i/J_i) = \max\{\deg(x^A)|\ x^A \in \operatorname{Min}(J),\ m(x^A) = n-i\} - 1.$$

Proof. Let $r = \max\{\deg(x^A)|\ x^A \in \operatorname{Min}(A),\ m(x^A) = n-i\} - 1$. Let x^A be an arbitrary element of $\operatorname{Min}(J)$ with $m(x^A) = n-i$. Write $x^A = x^B x_{n-i}$. Then $x^B \in \tilde{J}_i \setminus J_i$. From this it follows that $r \le a(\tilde{J}_i/J_i)$. It remains to show that $r \ge a(\tilde{J}_i/J_i)$.

Assume to the contrary that there is a monomial $x^C \in \tilde{J}_i \setminus J_i$ with $\deg x^C > r$. Then there exists an integer $m \ge 1$ such that $x^C x_{n-i}^m \in J$ and $m(x^C x_{n-i}^m) = n-i$. Since $\deg x^C x_{n-i}^m \ge r+2$, $x^C x_{n-i}^m$ is not a minimal generator of J. Therefore we can find a monomial $x^D \in J$ such that $x^C x_{n-i} = x^D x_h$ for some $h \le n-i$. Without restriction we may assume that m is the smallest degree such that $x^C x_{n-i}^m \in J$. Since $x^C x_{n-i}^{m-1} \notin J$, $h \ne i$. Thus, x^D is divisible by x_{n-i}^m and we may write $x^D = x^E x_{n-i}^m$. It follows that $x^C = x^E x_h$. By Lemma 6.6, this implies $x^C \in J$, a contradiction. □

Theorem 6.8. *Assume that* $\operatorname{char}(k) = 0$. *Then*

(a) $\operatorname{reg}_t(I) = \max\{\deg x^A|\ x^A \in \operatorname{Min}(\operatorname{Gin}(I)),\ m(x^A) \ge n-t\}$,
(b) $a_t^*(I) = \max\{\deg(x^A) + m(x^A)|\ x^A \in \operatorname{Min}(\operatorname{Gin}(I)),\ m(x^A) \ge n-t\} - n - 1$.

Proof. Set $J = \operatorname{Gin}(I)$. By Lemma 6.7,

$$c_i(I) = \max\{\deg(x^A)|\ x^A \in \operatorname{Min}(\operatorname{Gin}(I)),\ m(x^A) = n-i\} - 1 < \infty.$$

It is easy to check that

$$\max\{c_i(I)|\ i = 0, \dots, t\} = \max\{\deg x^A|\ x^A \in \operatorname{Min}(\operatorname{Gin}(I)),\ m(x^A) \ge n-t\} - 1,$$

$$\max\{c_i(I) - i|\ i = 0, \dots, t\}$$
$$= \max\{\deg(x^A) + m(x^A)|\ x^A \in \operatorname{Min}(\operatorname{Gin}(I)),\ m(x^A) \ge n-t\} - n - 1.$$

Therefore, the assertion follows from Theorem 6.5. □

Corollary 6.9. Assume that $\operatorname{char}(k) = 0$. Then

(a) $\operatorname{reg}(I) = \max\{\deg(x^A)|\ x^A \in \operatorname{Min}(\operatorname{Gin}(I))\}$,
(b) $a^*(I) = \max\{\deg(x^A) + m(x^A)|\ x^A \in \operatorname{Min}(\operatorname{Gin}(I))\} - n - 1$.

Proof. Since $\operatorname{reg}(I) = \operatorname{reg}_n(I)$ and $a^*(I) = a_n^*(I)$, the assertion follows from Theorem 6.8. □

Acknowledgement

The author is grateful to the organizers of the International Conference on Commutative Algebra and Combinatorics at Harish-Chandra Research Institute, Allahabad, December 2003, for generous support and hospitality. Moreover, he is indebted to Irena Swanson for her reading of the first draft and for the suggestions which have brought this version in life.

References

[1] I. M. Aberbach, C. Huneke and N. V. Trung, Reduction numbers, Briançon-Skoda theorems and the depth of Rees rings, *Compositio Math.* **97** (1995) 403–434

[2] D. Bayer, H. Charalambous and S. Popescu, Extremal Betti numbers and applications to monomial ideals, *J. Algebra* **221** (1999) 497–512

[3] D. Bayer and D. Mumford, What can be computed in algebraic geometry?. In: D. Eisenbud and L. Robbiano (eds.), *Computational algebraic geometry and commutative algebra*, Proceedings, Cortona 1991, Cambridge University Press, (1993) 1–48

[4] D. Bayer and M. Stillman, A criterion for detecting m-regularity, *Invent. Math.* **87** (1987) 1–11

[5] D. Bayer and M. Stillman, A theorem on refining division orders by the reverse lexicographic order, *Duke Math. J.* **55** (1987) 321–328

[6] I. Bermejo and P. Gimenez, On the Castelnuovo–Mumford regularity of projective curves, *Proc. Amer. Math. Soc.* **128** (2000) 1293–1299

[7] M. Brodmann and R. Y. Sharp, *Local cohomology: an algebraic introduction with geometric applications*. Cambridge University Press, (1998)

[8] D. Eisenbud, *Commutative algebra with a viewpoint toward algebraic geometry*. Springer, (1994)

[9] D. Eisenbud and S. Goto, Linear free resolutions and minimal multiplicities, *J. Algebra* **88** (1984) 89–133

[10] A. Galligo, Théorème de division et stabilité en géometrie analytique locale, *Ann. Inst. Fourier* **29** (1979) 107–184

[11] E. Hyry, The diagonal subring and the Cohen-Macaulay property of a multigraded ring, *Trans. Amer. Math. Soc.* **351** (1999) 2213–2232

[12] S. Kleiman, Théorie des intersections et théorème de Riemann-Roch. In: SGA 6, *Lecture Notes in Math.* **225**, Springer, (1971)

[13] C. H. Linh, Upper bound for the Castelnuovo–Mumford regularity of associated graded modules, *Comm. Algebra* **33** (2005) no. 6, 1817–1831

[14] D. Mumford, *Lectures on curves on an algebraic surface*. Princeton University Press, New Jersey, (1966)

[15] A. Ooishi, Castelnuovo's regularity of graded rings and modules, *Hiroshima Math. J.* **12** (1982) 627–644

[16] M. E. Rossi, N. V. Trung and G. Valla. Castelnuovo–Mumford regularity and extended degree, *Trans. Amer. Math. Soc.* **355** (2003) 1773–1786

[17] M. E. Rossi, N. V. Trung and G. Valla, Castelnuovo–Mumford regularity and finiteness of Hilbert functions. In: *Commutative algebr*, Lect. Notes in Pure and Appl. Math. **244**, CRC Press, (2005)

[18] R. Sharp, Bass numbers in the graded case, a-invariant formula, and an analogue of Falting's annihilator theorem, *J. Algebra* **222** (1999) 246–270

[19] N. V. Trung. Reduction exponent and degree bounds for the defining equations of a graded ring, *Proc. Amer. Math. Soc.* **102** (1987) 229–236

[20] N. V. Trung, Reduction number, a-invariant and Rees algebras of ideals having small analytic deviation. In: *Comm. algebra* (Trieste, 1992), 245–262, World Sci. Publishing, NJ, (1994)

[21] N. V. Trung, The Castelnuovo regularity of the Rees algebra and the associated graded ring, *Trans. Amer. Math. Soc.* **350** (1998) 2813–2832

[22] N. V. Trung, The largest non-vanishing degree of graded local cohomology modules, *J. Algebra* **215** (1999) 481–499

[23] N. V. Trung, Castelnuovo–Mumford regularity and analytic deviation of ideals, *J. London Math. Soc.* **62** (2) (2000) 41–55

[24] N. V. Trung, Gröbner bases, local cohomology and reduction number, *Proc. Amer. Math. Soc.* **129** (1) (2001) 9–18

[25] N. V. Trung, Evaluations of the initial ideal and Castelnuovo–Mumford regularity, *Proc. Amer. Math. Soc.* **130** (5) (2002) 1265–1274

[26] W. Vasconcelos, Cohomological degrees of graded modules. In: Six lectures on commutative algebra, *Progr. Math.* **166**, Birkhäuser, Boston, (1998) 345–392

Made in United States
North Haven, CT
05 November 2026

21764893R00174